Origin of Anti-Tumor Immunity Failure in Mammals

Origin of Anti-Tumor Immunity Failure in Mammals

Ivan Bubanovic

Medica Centre
Nis, Serbia and Montenegro

Translated by
Marija Rusimovic

Springer Science+Business Media, LLC

ISBN 978-1-4757-7970-7   ISBN 978-0-306-48630-2 (eBook)
DOI 10.1007/978-0-306-48630-2

© 2004 Springer Science+Business Media New York

Originally published by Kluwer Academic/Plenum Publishers, New York in 2004.
Softcover reprint of the hardcover 1st edition 2004

http://www.wkap.nl/

10 9 8 7 6 5 4 3 2 1

A C.I.P. record for this book is available from the Library of Congress

"… sciences long ago recognized that observations are not superior to hypotheses in generating scientific progress nor are hypotheses superior to observations. Both are necessary."

David F. Horrobin
(1939-2003)

PREFACE

The history of science has shown the majority of hypotheses to be wrong. Sharp scientific criticism and strictly controlled experimental studies reject most of the hypotheses, leaving behind only a small number of assumptions and ideas. Nevertheless, each logical assumption should have its rightful place on the scientific "battlefield" supposed to assess its validity and determine its final fate. Even when a hypothesis is wrong, it still finds its place in the entire efforts of the humankind towards attaining the scientific truth. Namely, the wrong hypotheses serve largely to illuminate the way towards the correct ones or, at least, to show which way not to follow. Correct or not, ideas and hypotheses are necessary for the progress of science. They epitomize the efforts of human thought to elucidate nature without experimental verification and in the circumstances of scant data availability. Finally, hypotheses and ideas represent a symbiotic creation of our knowledge and imagination, the two most impressive appearances in the evolution of humans.

It is thoroughly unknown whether Charles Darwin was indeed familiar with the phenomenon of similarity between microcosm and macrocosm claimed by many philosophers and physicists to be convincing and fascinating. Whatever his knowledge might have been in this regard, the possibility that Darwin's approach to defining the phenomena in biology is applicable to both planetary and cellular levels is fairly real. Namely, many physiological and pathological phenomena like those found in immunology and oncology, such as various forms of immune tolerance and immunomodulation, clonal selection of immune and tumor cells, modification of subset relationships between immune cells in tissues and organs, rest on the basic postulates of Darwinism, i.e. randomness of change, negative and positive selection, set-up of ecological niches, and the like.

The body of any animal can be viewed as a society or "ecosystem" whose individual members are cells, reproducing by cell division and organized into collaborative assemblies or tissues. In this "ecosystem", the cells are born, live and die under various forms of selection pressure like territorial limitation, population size, source of nutrients provided, infectious agents, etc. The body is a highly organized society of cells whose main task is the maintenance of homeostasis of the whole organism. The failure of control mechanisms which make the cell the unit of society, marking the beginning of its "asocial" behaviour, is most frequently a malignant alteration. This process is not abrupt, nor is it based on the single event. It is, rather, a long-term process characterized mainly by mutation, competition and natural selection operating within the population of cells. The basic mechanisms controlling the cell sociability represent the first defence line against the altered cells, while the second line of defence is supposed to be made up of the immune system cells. Speaking in Darwinian terms, within the "ecosystem" of organism, cells of the immune system operate as "predators" of the altered and mutated cells or cells infected by the intracellular parasites.

The biological phenomena whose mechanisms are, at present, explored and largely understood, certainly had their own evolution. Searching for the origin and details of the evolution of "advanced solutions" as well as selection pressures that might justify their emergence and existence, we often fail to see that many such phenomena are, in fact, co-evolutionary by-products of "evolutionary innovations". In other words, the evolutionary emergence of "advanced solutions" is sometimes, if not always, accompanied by certain by-products and by the co-evolution of compensatory mechanisms acting as a counterbalance to these.

An example of the evolution of "advanced solutions" is the evolution of adoptive immunity, and co-evolution of auto-immunity and alloimmunity. Alongside with the diversification of the mechanisms of adoptive immunity, auto-immunity and alloimmunity gain attribute of the evolutionary by-products and become sources of selection pressure. To that effect, alloimmunity could be a source of very strong selection pressure in mammals, simply because it is directly connected with the reproductive efficacy. At the same time, new forms of selection pressure that are connected with adoptive immunity gave rise to new mechanisms controlling killer machinery of the immune system. Finally, the last in a line of by-products in the processes of evolutionary "modelling" and "re-modelling" of vertebrate immune system can be called the failure of anti-tumor immunity.

As my intention has been to include only those critical points related to the origin as well as parallelism between immunoregulatory/suppressive mechanisms in pregnancy and tumor sufferers, I set out to write this publication presuming that the majority of readers are already familiar with the fundamentals of medicine, biology, immunology, immunopathology, oncology, mammalian reproduction and vertebrate evolution. To that effect, this publication is free of descriptions otherwise found in most textbooks. Naturally, the book is intended for all readers showing interest. If, upon reading this subject matter, the interest of readers grows into practical work in the fields of reproductive or tumor immunology, or something even greater, the author's satisfaction would be complete, and the objective of the entire publication fulfilled.

Finally, I recommend the articles cited throughout the text to all those readers who would like to expand their knowledge regarding the evolution of the immune system, alloreactivity, immune recognition, various forms of immune tolerance, reproductive and tumor immunology, as well as comparative immunology and oncology.

Author, 2004

ibubanovic@yahoo.com

CONTENTS

Chapter 1

The Vertebrate Immune System

1.1 THE VERTEBRATE IMMUNE SYSTEM EVOLUTION

Some evolutionary processes can be studied directly, while the evolution of the immune system and immunity cannot. The evolution of immunity is a macro-evolutionary process, which can be studied by examining the patterns in biological populations or species of related organisms and inferring process from pattern. Given the observation of micro-evolution and the knowledge that the earth is billions of years old - the evolution of the immune system could be postulated. However, this extrapolation, in and of itself, does not provide a compelling explanation of the patterns of biological diversity we see today. The evidence for the evolution of the immune system and immunity, or common ancestry and modification with descent, comes from several other fields of study. These include: comparative biochemical, immunological and genetic studies, comparative developmental biology and pathology, patterns of biogeography, comparative physiology, morphology and anatomy, as well as the fossil record.

One of the possible ways to carry out a detailed investigation into the issue of the evolution of the immune system is virtually by comparing the characteristics of the immune system across species and classes. In addition, the determination of the genes and molecules conserved throughout evolution is helpful in identifying the mechanisms of the immune system evolution, while variable genes as well as the emergence of new genes along evolution enable the identification of various evolutionary pressures, their duration and strength. This provides a better understanding of how the immune system works under physiological and pathological conditions. This approach, however, is not an easy one since only few percent of the total

1

number of species that have inhabited the planet to date are available for research. Notwithstanding this limiting factor, previous studies have produced sufficient findings for the construction of a hypothetical model of the evolution of the immune system in vertebrates, and also cleared the pathway towards reaching further relevant presumptions and conclusions.

The mechanisms for defending the host against foreign invaders are present in all multicellular organisms in different forms. For example, in all invertebrate species these mechanisms constitute innate immunity only. More specific and specialized defence mechanisms that constitute innate and adaptive immunity are found in vertebrates only, but many components of invertebrate immune systems have been preserved, and continue to play a crucial role in the vertebrate immune system.

The evolution of the immune system is a direct consequence of microbes-exerted selection pressure on multicellular organisms. The vertebrate immune system was to some extent inherited from invertebrates, whereas a part of it has advanced considerably in the course of its own evolution. Although certain vertebrate-specific properties such as immune recognition and immune memory have also been identified in invertebrates in rudimentary forms, it is particularly those qualities like progressive development of humoral and cellular adaptive immunity, Major Histocompatibility Complex (MHC), variable class I and class II genes, precise mechanisms of immune recognition and long-term immune memory that reflect the fundamental evolutionary advancement of the vertebrate immune system.

By tracing the evolution of the invertebrate immune system, it can be seen that it largely followed the "classical" model based on bi-directional "predator-prey" relationships. Also, the evolutionary emergence of MHC system and the mechanisms of immune recognition in vertebrates came as a direct result of a microbe-exerted selection pressure. Naturally, the new possibilities gave rise to new conveniences, but also brought about certain risks in the new forms, like auto-immunity, alloimmunity and reproductive efficacy.

The presumption that auto-immunity and, especially, alloimmunity and reproductive efficacy could have represented evolutionary new forms of strong selection pressures closely associated with the evolution of the mammalian immune system correlates with the unique control mechanisms of immune response verified in this vertebrate. In short, the evolution of the mammalian immune system has possibly undergone the pressures of at least four quite diverse factors: microbes, auto-immunity, alloimmunity/repro-ductive efficacy and tumors. However, it is unlikely that auto-immunity and tumors represent strong selection pressures, as most of these generally occur after leaving offspring. On the contrary, alloimmunity/reproductive efficacy might have been the source of a very strong selection pressure that greatly influenced the evolution of the mammalian immune system. The possible effects of such heterogeneous and complex evolutionary pressures are the evolutionary development of the mammalian immune system into one of the most complex, most organized and multilevel controlled system in the world of living beings.

The basic features of the adoptive immune system such as immune recognition, specific immune reactivity and immune memory, are based on the antigen-processing/presenting machinery, recognition mediated by the T cell receptor (TCR) and immunoglobulins (Ig), as well as high TCR and Ig variability as a result of the Recombination Activating Genes (RAG) activity (Abbas *et al.*, 2003). MHC class I and class II genes, RAG genes as well as TCR and Ig have only been identified in jawed vertebrates. No molecular evidence has been found that the adoptive immune system traces its origin straight from invertebrates, which leads us to a conclusion that this is one of the evolutionary advancements specific only for vertebrates (Greenhalgh *et al.*, 1993, 1995; Bernstein *et al*, 1996; Agrawal *et al*,. 1998).

1.1.1   Macrophages

Cells with macrophagal function have been discovered in almost all animals, from sponges to mammals. These cells are the carriers of the innate immunity, but they also play a significant role in initiating immune reaction by presenting peptides and communicating with the cells of the adoptive immune system. Various cells in invertebrates respond to microbes by enclosing these infectious agents within aggregates and destroying them. These responding cells resemble phagocytes and have been called phagocytic amebocytes in *acelomates*, haemocytes in *molluscs* and *arthropods*, coelomocytes in *annelids*, and blood leukocytes in *tunicates*. Invertebrates do not contain antigen-specific lymphocytes and do not produce immunoglobulin molecules or complement proteins. However, they express a number of soluble molecules that bind to and lyse microbes. These molecules include lectin-like proteins, which bind to carbohydrates on microbial cell walls and agglutinate the microbes, and numerous lytic and antimicrobial factors such as lysozyme, which is also produced by neutrophils in higher organisms. Phagocytes in some invertebrates may be capable of secreting cytokine-like molecules that resemble macrophage derived cytokines in the vertebrates. Thus, host defence in invertebrates is mediated by the cells and molecules that resemble the effector mechanisms of the innate immunity in higher organisms (Loker, 1994).

The evolutionary conserved functions such as phagocytosis and the NO mediated intracellular signal system have been equally identified in many evolutionary distant species (Ottaviani *et al.*, 1997). The macrophagal receptor (CD36) which participates in the phagocytosis of the apoptotic cells is evolutionary highly conserved, and is found in many species from *Drosophila* to mammalian species. The link discovered between apoptosis and metamorphosis in some animals, as well as the role of macrophages in the elimination of the apoptotic cells, indicates that the primordial function of macrophages might have been to mediate tissue and organ remodelling, acquiring the function of immune cells through later evolution (Franc *et al.*, 1996). Although macrophages belong to the innate immune cells with the non-specific antigen relationship as its basic feature, in vertebrates they play a key role in processing and presenting antigen peptides, and also in the controlling of final differentiation and proliferation of the adaptive immunity cells. These functions are essential in the activation of the normal and, more importantly, effective efferent immune mechanisms in vertebrates (Du Pasquer *et al.*, 1999; Abbas *et al.*, 2003).

1.1.2   Lymphocytes and Lymphoid Organs

Nemertea and *Annelidea* are evolutionary the oldest animals in which lymphocytes, or more precisely, lymphocyte-like cells have been identified. Although they resemble mammalian T lymphocytes, these cells do not respond to stimulation by Concanavalin A (ConA), Phytohemagglutinin (PHA), and bacterial Lipopolysacharide (LPS) (Roch *et al.*, 1975; Ottaviani *et al.*, 1997), and they do not display functional properties typical of vertebrate lymphocytes.

Unlike their ancestor, lymphocyte-like cells of *Echinodermata* and *Urochordata* participate in the rejection of transplanted tissue and react to the stimulation by PHA (Raftos *et al.*, 1987). Although lymphocyte-like cells have been also identified in other, evolutionary younger species of invertebrates, no molecules have been found that represents the possible precursor of TCR molecules. The lymphocytes containing variable (V) domain and TCR can only be found in vertebrates (Humphreys *et al.*, 1994).

Our knowledge of the immune system in the early vertebrates, the *Agnatha*, and the molecules involved in their immune reactions is fragmentary. *Agnatha* can reject allotransplant, but no presence of either real lymphocytes or thymus is identifiable in this vertebrate group. Lymphocyte-like cells can be identified as lymphoid accumulation and can function as the thymus (Matsunaga *et al.*, 1998; Du Pasquer *et al.*, 1999). Although in *Agnatha* lymphocyte-like cells are found in the blood and intestines, there are no defined lymphoid organs like spleen or lymph nodes (Nei *et al.*, 1997; Du Pasquer *et al.*, 1999).

Fish display a typical vertebrate adaptive immune system characterized by immunoglobulins, TCR, cytokines, and MHC molecules, but the immune system of fish is quite different in its efficiency and complexity from that of higher vertebrates. Cartilaginous fish express canonical B and T cell recognition genes, but their lymphoid organs and lymphocyte development have been poorly defined. This class of vertebrates represents evolutionary the oldest animals with the thymus, originated from pharyngeal pouches. Their thymus has the structure very similar to mammalian (cortex, medulla, thymocytes and a number of accessory cells). In sharks, lymphocytes have been found to respond to the stimulation by PHA and ConA, but not in Mixed Lymphocyte Reaction (MLR). Besides the thymus, cartilaginous fish also have the spleen and Gut-Associated Lymphoid Tissue (GALT). In rays and sharks, the lymphopoietic processes are placed in the liver, epigonadal tissue and kidney (Nei *et al.*, 1997, Du Pasquer *et al.*, 1999, Miracle *et al.*, 2001). It is conclusive that cartilaginous fish can be regarded as the oldest class of vertebrates with the immune system that has some organizational and functional properties characteristics of mammals.

Dramatic changes in the organization of the immune system that occurred in cartilaginous fish as compared to the *Agnatha* are not present when we compare the immune system of cartilaginous fish with that of bony fish. Like in other vertebrates, bony fish have a well-defined thymus and, for the first time in the evolution of vertebrates, the spleen with elements specific to mammals such as red and less developed white pulp. The red pulp contains mostly macrophages, whereas the white pulp contains lymphocytes that proliferate during the immune response (Petrie-Hanson *et al.*, 2000). Bony fish lymphocytes react to ConA, LPS and in MLR, but these reactions depend largely on the exterior temperature. The young thymectomized fish reject allotransplant very slowly or even fail to do so and also display poor antibody production which is a state very similar to mammals after receiving intrathymic injection of antigens. In bony fish, antibodies produced by B lymphocytes and function of T cells are closely associated with the functions of APCs. These data indicate the mechanisms of immune reaction in fish that follow the model of antigen processing and presenting, probably like complex MHC/peptide (Du Pasquer *et al.*, 1999; Petrie-Hanson *et al.*, 2000).

Non-specific cytotoxic lymphocyte-like cells have been isolated in bony fish and could be analogue to Natural Killer (NK) cells in mammals. These cells are involved in the non-specific lysis of the xenotransplant, but not in the rejection of allotransplant. There is another class of lymphocyte-like cells in bony fish which attacks viruses-infected cells and allogeneic cells. These lymphocytes could be the first evolutionary analogue to CTL in mammals (Petrie-Hanson *et al.*, 2000).

Frogs also have the thymus which develops from the dorsal epithelium of the visceral pouches. Lymphocytes begin to appear in the thymus of frogs at days 5 and 6 of embryonic development. Immediately after colonization, the process of clonal selection begins in the thymus, controlled by thymic class II positive, but class I negative dendritic cells.

Table 1.1. Comparative anatomic and functional features of vertebrate immune system.

Immune Features	Agn	Cho	Ost	Apo	Anu	Rep	Ave	Mam
Non-self recognition	+	+	+	+	+	+	+	+
Memory	+	+	+	+	+	+	+	+
MHC	-	+	+	+	+	+	+	+
PHA response	-	+	+	+	+	+	+	+
Mixed Lymphocyte response	-	-	+	+	+	+	+	+
Antibody	-	+	+	+	+	+	+	+
IgM	-	+	+	+	+	+	+	+
IgG	-	-	-?	-	+	+	-	+
IgN	-	-	+?	-	-	-	+	-
IgA, IgE, IgD	-	-	-	-	-	-	-	+
Secondary response	+	+	+	+	+	+	+	+
Thymus	-	+	+	+	+	+	+	+
Spleen	-	+	+	+	+	+	+	+
Bone Marrow	-	-	-	-	+	+	+	+
Lymph Nodes	-	-	-	-	-	-	-?	+
Bursa of Fabricius	-	-	-	-	-	-	+	-
Th1/Th2 functions	-	-	-	-	-	-	-?	+
Association TAP and LMP with class I or class II genes	-	I	I	I	I	I	I	II
Time of first set allotransplant rejection (days)	21-240	21-53	7-76	3-40	3-40	41-245	7-14	7-14
Time of second set allotransplant rejection (days)	7-252	15-22	6-26	8-19	8-19	21-32	1-7	1-7

Abbreviations: Agn-Agnatha, Cho-Cartilaginous fish, Ost-Bony fish, Apo-Apoda, Anu-Anura, Rep-Reptiles, Ave-Birds, Mam-Mammals

References for *Table 1.1.* (Saad *et al.,* 1984; Hughes, 1988; Kaufman *et al.,* 1995; Kasahara *et al.,* 1996a, 1996b; Du Pasquer *et al.,* 1990, 1999; Ohta *et al.,* 2000)

This is a unique evolutionary feature of amphibians that has not been examined enough, but it can be proved to be connected with the MHC class I expression on the tissues that will be remodelled via apoptotic mechanisms during metamorphosis. In addition, amphibians have the spleen with red and white pulp, GALT system and quite well-defined leukocyte accumulation (but not lymph nodes) in the liver, intestines and kidneys. Finally, to reach more complete conclusions on the characteristics of the immune system of amphibians is extremely difficult because of the variability in the morphology of lymphoid tissues from species to species (Arnall *et al.*, 1987).

Thymectomy in amphibians results in a complete absence of peripheral T lymphocytes, which points to activation of mechanisms of extrathymic lymphocytes maturation. Nevertheless, an extremely small number of T-like lymphocytes can still be detected in the frogs undergoing thymectomy at early stage. Also, there are some data suggesting that these can behave in the manner of NK cells. In reptiles, the early thymectomy causes tolerance to allotransplant and highly inhibits the synthesis of antibodies (Arnall *et al.*, 1987; Du Pasquer *et al.*, 1992).

Table 1.2. Evolution of immunocompetent cells in vertebrates.

Cartilaginous fish	Bony fish	Amphibians	Reptiles	Birds	Mammals
					Th3
					Th2
					Th1
			$CD4^+$	$CD4^+$	$CD4^+$
	$CD8^+$	$CD8^+$	$CD8^+$	$CD8^+$	$CD8^+$
	T cells	T cells	T cells	T cells	T cells
	B cells	B cells	B cells	B cells	B cells
	NK cells	NK cells	NK cells	NK cells	NK cells

In *Xenopus* species, thymocytes are only symbolically involved in helping and inducing the Graft-Versus-Host (GVH) reaction unlike the T lymphocytes from spleen that are good helpers and strong GVH inducers. Using *in vitro* chromium 51 release assays NK cells were detected in *Xenopus* species. Splenocytes effectors from early thymectomized one year old frogs lyse spontaneously allogeneic thymus tumor cell lines that lack MHC molecules expression. This ability is increased after the injection of the tumor cells, or after treating the splenocytes *in vitro* with mitogens. This suggests lymphokine activation of the killer mechanisms. Spontaneous killer activities also have been identified in *Rana* species (Du Pasquer *et al.*, 1992).

Urodeles do not have well-defined thymus and spleen. Lymphocytes of *Urodela* are less responsive to the stimulation by mitogens unlike the lymphocytes of *Xenopus* and mammals, especially when the lymphocytes are from larva or young animal. In adult stage, the lymphocytes of *Urodela* react much better to the stimuli such as PHA or allogeneic cells (Du Pasquer *et al.*, 1992).

All the studies performed on reptiles have shown that this vertebrate class has well defined lymphoid organs. The thymus is well differentiated into medulla and cortex, while the spleen shows a well defined demarcation line between white and red pulp. In *Python reticulatus* dendritic cells involved in immune complex trapping have been identified and could be the precursors of mammalian follicular dendritic cells. GALT develops later than the spleen during development and appears to be a secondary lymphoid organ; it does not seem to contain the equivalent of the bursa of *Fabricius* (Du Pasquer *et al.*, 1999; Petrie-Hanson *et al.*, 2000). There have been several reports of lymph node like anatomical defined structures, especially in snakes and lizards (Saad *et al.*, 1984; Du Pasquer *et al.*, 1999; Petrie-Hanson *et al.*, 2000). The functional characteristics of T and B lymphocytes in reptiles have not been investigated in greater detail, but it is known that after glass-wool filtration, non-adherent peripheral blood lymphocytes responded to PHA and not to LPS, whereas adherent cells could be stimulated by LPS, results analogous to those obtained in mammals (Saad *et al.*, 1984). Seasonal changes influencing the morphological and functional characteristics of the lymphoid tissue in reptiles greatly interfere with the investigation into the immune system of this vertebrate class (Nelson *et al.*, 1996).

In birds, the thymus is clearly defined into cortex and medulla, while the spleen is less clearly defined into red and white pulp in comparison with mammals. Although in some bird species lymphocyte accumulations were found that could resemble lymph nodes, the real lymph nodes as present in mammals have not been identified in birds. The effects of thymectomy, T and B cell cooperation and generation of MHC restricted helper and killer cells are similar to those of other warm-blooded vertebrates, such as mammals (Vandaveer *et al.*, 2001). In addition, bird lymphocytes react to non-specific stimulation by PHA, ConA and LPS, and also in MLR in a way that is very similar to that of mammalian lymphocytes (Du Pasquer *et al.*, 1999; Petrie-Hanson *et al.*, 2000).

The bursa of *Fabricius* is in birds a special, temporary lymphoid organ that arises at day 5 of development and involutes 4 weeks later. This organ, which is situated in the vicinity of the cloaca, is deputed to the maturation and production of B lymphocytes and it is, therefore, the major antibody forming organ. Indeed the study of this organ in the mid 1950's led to the first postulation of the existence of two major populations of lymphocytes - the T cells and the B cells. No homologous organ can be identified in mammals: this function is partially taken up by the spleen, but the bone marrow is also involved, at least, as far as the haemopoiesis of B cells is concerned (Du Pasquer *et al.*, 1999; Zekarias *et al.*, 2002).

1.1.2.1 Origin of TCR(s)

Diversity is a hallmark of the vertebrate immune system. Lymphocyte repertoires with millions of different specificities function in accord with a large diversity of cytokines, chemokines, and different types of APCs to protect vertebrates against infections and tumors. Different pathogens are handled by qualitatively different immune responses, varying from cellular to humoral responses, and varying in e.g. immunoglobulin isotype and cytokine expression. At the same time, unwanted immune responses against "self" peptides and innocuous antigens are typically avoided. Due to the polymorphism of MHC molecules, involved in antigen presentation to the vertebrate immune system, different individuals in a population typically respond differently to identical antigens (Arstila *et al.*, 1999).

Even though the presence of the so-called Ig-like gene has been established in several species belonging to evolutionary a very old group of *Parazoa* animals, no other animal group except for the vertebrate has been known to display the phenomenon of gene rearrangement (Benian *et al.*, 1989). These genes are involved in mechanisms of immune response in a number of invertebrates, yet they can also perform quite uncommon functions such as the synthesis of myosin-binding protein (Adema *et al.*, 1997).

The Ig domain comprises two major groups: the constant region (C) and the variable region (V). Primitive C2 region products capable of being identified in all invertebrates participate in the synthesis of many membranous proteins. On the other hand, the constant region (C1-type) which comprises the molecular basis for the adaptive immunity can only be found in the vertebrates which express MHC/TCR (Li *et al.*, 1997). The V domain included in the synthesis of various non-immune proteins has been identified in invertebrates, while in mammals it is primarily involved in the synthesis of the variable split immunoglobulin and TCR which gives an antigen-specific nature to these. The V domain in mammals possibly traces its origin from a non-arranging V domain of invertebrate. In *Arthropoda*, like in mammals, the molecules can be found containing the products of genes from a non-arranging V domain. In vertebrates, these are mostly so-called adhesion and co-stimulatory molecules such as CD2, CD80, CD86, etc. (Thompson, 1995; Du Pasquer *et al.,* 1999).

The structure of the TCR reveals both striking similarities with and fundamental differences from its functional counterpart, the antibody, in the humoral immune system. The conserved manner in which the TCR recognizes and interacts with its MHC/peptide ligand allows the TCR great latitude in its potential to form productive interactions with antigen-presenting cells that bear numerous ligands to which the TCR has not been previously exposed. This phenomenon of cross-, or alloreactivity arises from the combination of conserved structural features across all MHC molecules, both "self" and "non-self", and some degree of molecular mimicry. Non-classical MHC ligands presenting modified or specialized peptides, lipids, carbohydrates, or no ligand at all, are now thought to play increasingly important roles in cellular immunity (Arstila *et al.,* 1999; Abbas *et al.,* 2003)

The explosive evolution of the adaptive immunity in vertebrates is closely connected with mechanisms of TCR molecule rearrangement. Most models propose that the generation of somatically rearranging receptors occurred abruptly in evolution via the evolutionary emergence of the RAG machinery made of two lymphocyte-specific proteins, RAG1 and RAG2 (Bernstein *et al.,* 1996; Hansen *et al.,* 1996). The RAG1 and RAG2 genes encode nuclear proteins that directly mediate the mechanism V(D)J recombination process that occurs in T and B cells. The expression of RAG1 and RAG2 is required for the proper development of maturing lymphocytes (Peixoto *et al.,* 2000). Immunoglobulin and TCR genes are assembled from component gene segments in developing lymphocytes by a site-specific recombination reaction, V(D)J recombination. The proteins encoded by the RAG1 and RAG2, are essential in this reaction, mediating sequence-specific DNA recognition of well-defined recombination signals and DNA cleavage next to these signals. There is the theory that RAG1 and RAG2 were once components of a transposable element, and that the split nature of Ig and TCR genes derives from a germ line insertion of this element into an ancestral receptor gene soon after the evolutionary divergence of jawed and jawless vertebrates (Agrawal *et al,.* 1998).

No RAG homologs have been isolated from *Agnathans*, consistent with the absence of the antibody and TCR. The RAG genes have been found in cartilaginous fish, bony fish, amphibians, birds and mammals (Greenhalgh *et al.,* 1993, 1995; Bernstein *et al,* 1996; Hansen *et al.,* 1996; Agrawal *et al,.* 1998). The finding that RAG2 genes in sharks are approximately 50% identical with RAG2 genes from other vertebrates supports the thesis about evolutionary conservation of these genes (Schluter *et al.,* 2003). Three complete genes, RAG1, RAG2 and ACS (a possible homologue of the plant 1-amino-cyclopropane-carboxylate synthase gene) were identified in bony fish. The human ACS also was identified in a cosmid assigned to chromosome 11p11, which is close to the location of the RAGs (11p12). This indicates the conservation of the linkage between human and bony fish (Peixoto *et al.,* 2000). Also, the teleost RAG1 gene, unlike the mammalian one, contains a 666-bp intron in the trout and two in the zebrafish (Hansen *et al.,* 1996). Otherwise, the zebrafish RAG1 entire sequence is at least 78% identical to the human RAG-1 and 89% from positions 417 to 1042. The zebrafish RAG2 gene is 75% identical to the human RAG2 gene. Like in mammals, the RAG2 gene of fish is close to RAG1 and in reverse orientation. Finally, the untranslated region between the two genes increases in size from teleost to mammals (Greenhalgh *et al.,* 1993, 1995; Bernstein *et al.,* 1996; Hansen *et al.,* 1996; Agrawal *et al.,* 1998).

In *Xenopus* species, the two genes are also in opposite transcriptional orientation, separated by 6 kb sequence. *Xenopus* RAG1 is 71% similar to that of warm-blooded vertebrates and 88% between the positions 392 and 1012. RAG1 and RAG2 expression was detected in the thymus, liver, spleen, and even at a lower level in the kidneys. In adults, the thymus and bone marrow are the principal sites of expression (Greenhalgh *et al.,* 1993, 1995).

Only a small number of studies deal with the relationship between the RAG genes and reptiles. There is no firm evidence suggesting the similarity in the function of RAG genes between these and other vertebrates. Groth *et al.,* (1999) propose that RAG1-like genes have been identified in crocodiles and alligators, but it has not been established yet whether or not these participate in the TCR-like gene rearranging process. The poor function of RAG genes in reptiles may be responsible for a small TCR repertoire and, consequently, weak alloimmune reaction.

Although very little evidence is available regarding the role of RAG genes in birds, recent studies investigating their role in rearranging TCR of bird thymocytes bear out the evolutionary conserved function of RAG genes in vertebrates (Groth *et al.,* 1999). The RAG2 protein expression in the thymocytes of birds is most prominent in immature $CD4^-CD8^-$, a little less in intermediate $CD4^+CD8^+$ and the least such or even absent in mature $CD4^- CD8^+$ or $CD4^+CD8^-$ thymocytes (Ferguson *et al.,* 1994).

Previous findings about the evolution of class I and class II genes, RAG genes, Ig and TCR, suggest their evolution, from the ancestors of cartilaginous fishes to the present-day mammals. Microbes, the phenomenon of MHC gene diversification and amplification of their variability, increased the tempo of functioning of the rearranging machinery resulted in the expanding of the Ig and TCR repertoire, as well as in dramatic increase in the number of B and T lymphocyte clones. Also, there is the presumption that T cells co-expressing CD8 and NK markers (CD8 NK/T cells) are phylogenetically early mediators of cellular immunity.

1.1.3   Cytokines throughout Vertebrate Evolution

Chemokines are small, inducible, structurally related proteins that guide cells expressing the right chemokine receptors to the sites of immune response. They have been identified and studied extensively in mammals, but little is known about their presence in other vertebrate groups. There are described chemokines in bony fish and cartilaginous fish, as well as chemokine receptors in other jawed vertebrate. The phylogenetic analysis does not reveal any clear evidence of the orthology of lower vertebrates and human chemokines. Although the divergence of the subfamilies began before the fish-tetrapod split, much of the divergence within the subfamilies took place separately in the two vertebrate groups. The existence of chemokine receptors in the lamprey indicates that chemokines are apparently also present in the *Agnatha* (Kuroda *et al.,* 2003).

Many cytokines and their receptors, like most molecules of the immune system, tend to evolve rapidly, so that it has not been a simple task to isolate their ancestor genes. Although cytokine homologs are found within jawless, only a small number of cytokine or cytokine receptor genes have been sequenced in cartilaginous fish including newly-discovered IL-1β-like gene (Bird *et al.,* 2002a). Several cytokines, such as interferon(s) (IFN), IL-1, IL-2, IL-6, IL-8 and transforming growth factor (TGF)-β, have been identified in birds (Schneider *et al.,* 2001; Sijben *et al.,* 2003), reptiles (el Ridi *et al.,* 1987), amphibians (Franchini *et al.,* 1995) and bony fish (Inoue *et al.,* 2003; Fujiki *et al.* 2003). Albeit cytokines with vertebrate counterparts are likely to be present in invertebrates, no gene has been cloned that leaves a large gap in our knowledge of cytokine evolution. The functional analogues of vertebrate pro-inflammatory cytokines have been described in a variety of invertebrates. The analogy is based mainly on the crossreactivity of antibodies elicited against vertebrate cytokines, the sensitivity of invertebrate immunocytes to the action of vertebrate cytokines, and the responsiveness of vertebrate immune cells to invertebrate factors. However, without knowing the aminoacids or gene sequences of the putative invertebrate cytokine analogues, it has not been possible to demonstrate unequivocally a phylogenetic relationship between vertebrate cytokines and their invertebrate functional analogues. For example, a defence molecule from the earthworm *Eisenia foetida (Annelidea)* and the mammalian TNF-α perform similar functions, but they probably emerged independently during the evolution (Beschin *et al.,* 1999). In recent years, cytokines, which have been well characterised within mammals, have begun to be cloned and sequenced within non-mammalian vertebrates, with the number of cytokine sequences available from primitive vertebrates growing rapidly. The identification of cytokines, which are mammalian homologues, will give a better insight into where the immune system communicators arose and may also reveal the molecules, which are unique to certain organisms.

LPS-treated fish monocytes produce a cytokine which is the homologue of IL-1 in mammals. This cytokine displays similar effects on immune and other cells as well. The structural similarity between fish IL-1-like molecule and mammalian IL-1 is approximately 50%. In order to be able to demonstrate these analogies, it is necessary to know that the similarity between mammalian IL-1α and IL-1β is about 25% (Beschin *et al.*, 1999; Bird *et al.*, 2002a, 2002b). The LPS stimulated supernatant of both bird and amphibian monocytes, does not elicit IL-1-like activity, unlike mammalian lymphocytes. It is surpassingly due to the similarity of IL-1-like molecules in birds and mammals are approximately 60%.

These are convincing evidences corroborating the assumption that IL-1 and IL-1R probably emerged earlier than vertebrates and have been relatively conserved during the evolution:

1. IL-1 molecules of mammals and other vertebrate classes elicit interspecies crossreactivity with IL-1R;
2. Anti-IL-1 antibodies also display interspecies crossreactivity;
3. There is a relatively high similarity between mammalian and non-mammalian IL-1 like molecules;
4. Effects produced by IL-1 and IL-1-like molecules are identical or very similar on the cells of mammals, other vertebrates and even invertebrates (Bird *et al.*, 2002a, 2002b).

Unlike IL-1, IL-2-like molecules stimulate T cell growth in species-specific manner, though some evidence reveals certain inter-species cross-reactivity of anti-IL-2R antibodies. For example, mammalian IL-2 administered before or after allotransplantation, did not affect allograft rejection or regeneration in amphibians (Fahmy *et al.*, 2002). However, some studies have shown that intraperitoneal injection of human recombinant IL-2 can effectively modulate *in vivo* immune reactivity to thymus-dependent and thymus-independent type 2 immunogens in *Xenopus laevis*, but is less successful at affecting toad cells *in vitro* (Ruben *et al.*, 1994). Similar results were obtained by cultivating the lymphocytes of turtle with human IL-2. Eventually, IL-2 did not increase the fraction of turtle lymphocytes in mitosis even after days of repeated cultivation (Ulsh *et al.*, 2000).

Although IL-2-like molecules have been identified in higher vertebrate classes only, a better understanding of IL-2 and IL-2R would still require a more detailed comparative investigation. Nevertheless, we are currently able to conclude that evolution of IL-2 has been characterized by higher polymorphism, i.e. lower degree of conservation as unlike IL-1 genes (el

Ridi *et al.*, 1987; Franchini *et al.*, 1995). Bird IL-2-like molecule and IL-2R are most similar to mammalian, and are known to strongly stimulate proliferation of CD4$^+$ and CD8$^+$ of bird peripheral lymphocytes (Hilton *et al.*, 2002). In addition, the mammalian IL-15 analogue has been discovered in birds, displaying similar effects as IL-2 (Min et. al., 2002).

No firm evidence has been found consistent with the fact that mammalian analogue IL-4 participates in the regulation of immune response in non-mammalian classes of vertebrates, unlike the analogues of IL-6 superfamily genes which have been isolated in all classes of these, except cartilaginous fish. Leukaemia Inhibitor Factor (LIF) and IL-6 genes in bony fishes show a low degree of evolutionary conservation, since the similarity between these and the mammals are less than 20% (Fujiki *et al.*, 2003). IL-6-like genes have also been found in some species of amphibians *(Rana esculenta)*, but our knowledge remains scarce regarding the function of this cytokine family in amphibians (Franchini *et al.*, 1995). Bird IL-6-like molecule is known to be synthesized and secreted after stimulation of the lymphocytes, and also to participate in the up-regulation of lymphocyte proliferation. Finally, the similarity between bird and mammalian IL-6 is approximately 35% (Schneider *et al.*, 2001).

Inoue *et al.* (2003) found that some fish species have gene for IL-8. The same authors showed that the dogfish *(Triakis scyllia)* IL-8 sequence shared 50.5, 37.1 and 40.4-45.5% identity with the chicken, trout and mammalian IL-8 sequences, respectively.

Genes encoding anti-inflammatory cytokine IL-10 have not been identified in non-mammalian vertebrate classes, except for a small number of studies indicating presence of IL-10-like and IL-10R-like genes in birds, being clustered together with IFNR genes in the same way as in mammals (Reboul *et al.*, 1999).

To the surprise of many evolutionary immunologists, Yoshiura *et al.* (2003) found that comparative genomic analysis showed a conserved synthesis within the IL-12 regions between *Fugu-fish* and human, indicating that the *Fugu* genes are orthologues for mammalian IL-12 encoding genes, respectively. The deduced amino acid sequences of the *Fugu* IL-12 subunits showed homology with mammalian IL-12 subunits (50.4-58.0% similarity). These studies have been the only evidence of IL-12 presence in non-mammalian vertebrates so far.

IL-15 is cytokine that has been identified only in birds and mammals. In birds, IL-15 is secreted by splenocytes in response to LPS and performs the function of pro-inflammatory cytokines, similarly to mammals (Sijben *et al.*, 2003).

TGF-β genes are probably evolutionary older than vertebrates, so the TGF-β gene superfamily is clearly ubiquitous among vertebrates. This cytokine is a strong inhibitor of the adaptive immunity across classes and species. A wide interspecies presence of TGF-β, high functional similarity and least 50% similarity between TGF-β genes of phylogenetically distant species, all indicate the relative conservation of these genes. The evolutionary conservation of TGF-β genes is not likely to be the result of their involvement in the regulation of immune response, but probably the consequence of other, also very important biological mechanisms, such as cell differentiation, maturation of oocytes (Pogoda *et al.,* 2002), mechanisms controlling the formation of the embryonic axis in amphibians and probably in all classes of vertebrates (Stern, 1992). Finally, Paulesu *et al.,* (1997) studies have shown that the production of cytokines like TGF-β by the feto-placental unit is not limited to mammalian species, since TGF-β can be secreted by the placenta of such viviparous squamate reptiles as *Chalcides chalcides.* The finding of this parallelism between reptilian and mammalian reproduction suggests that immunological mechanisms, possibly mediated by the secretion of cytokines, played an important role in the evolution of viviparity.

Table 1.3. Diversification of cytokines in vertebrates.

Cartilaginous fish	Bony fish	Amphibians	Reptiles	Birds	Mammals
					IL-10
					IL-4
					IL-3
				IL-18	IL-18
				IL-15	IL-15
				IL-8	IL-8
				IL-6	IL-6
				IL-2	IL-2
	IFN-γ	IFN-γ	IFN-γ	IFN-γ	IFN-γ
	IFN type I	IFN type I	IFN type I	IFN type I	IFN type I
	TNF	TNF	TNF	TNF	TNF
TGF	TGF	TGF	TGF	TGF	TGF
IL-1	IL-1	IL-1	IL-1	IL-1	IL-1

Molecular cloning and expression analysis of TNF-α in all classes of vertebrates reveal its constitutive expression and ubiquitous nature. The protein sequence deduced from *(Sparus aurata L.)* TNF-α gene shows a high degree of homology with the *Japanese flounder* TNF-α (65.6% identity and 78.9% similarity) and, more important, it is more homologous to mammalian TNF-α (41.1-48.6% similarity) than to TNF-β (36.0-43.5% similarity) (Garcia-Castillo *et al.,* 2002). Although TNF superfamily plays important role in immune reaction in all vertebrates, the evolutionary conservation of the family, like in the case of TGF, is probably associated with the non-immune and co-immune mechanisms, like apoptosis, regeneration, tissue remodelling, metamorphosis etc.

The IFN family consisting of IFN-α, IFN-β, IFN-Ω, IFN-δ, IFN-κ, and IFN-τ is a large group of cytokines involved in the innate immune response against various microbes. While the IFN have not been found in invertebrates, the genes for IFN type I are expressed by almost vertebrates. Genes for IFN have been cloned from a variety of mammalian and avian species; however, IFN genes from cartilaginous fish have not been forthcoming. Despite the considerable advances in our understanding of teleost immunity, relatively few cytokine genes, including those for IFN, have been identified at the molecular level. In contrast, numerous studies have shown that following virus infection or exposure to double-stranded RNA, fish immune cells produce a soluble factor that is functionally similar to mammalian IFN. Also, there are reports considering the cloning and characterization of the IFN gene from the zebrafish (Altmann *et al.*, 2003). IFN type I comprise IFN-α secreted by leucocytes and IFN-β secreted by virus-infected cells, whereas interferon type II (IFN-γ) is secreted by the activated T cells and APCs. IFN-α and IFN-β are mainly involved in the inhibition of virus replication, while the role of IFN-γ in the immune response is the up-regulation of class I, class II and TAP/LMP genes transcription. IFN type II is much more difficult to identify in lower vertebrates. Although there are no consistent reports about the presence of this cytokine in fishes and reptiles, it has been found that the lymphocytes of many fish species react to human IFN-γ, with increased MHC expression, which is indicative of the presence of IFN-γ-like the factor which performs a similar function like in mammals. IFN-γ-like factor has been isolated from birds, and found to elicit similar interspecies crossreactivity with mammalian lymphocytes, although the similarity between bird and mammalian IFN type II is as low as 35% (Jensen *et al.*, 2002).

In conclusion, the evolution of the adaptive immunity is associated with the emergence of auto-immunity and alloimmunity/reproductive efficacy as by-products of "self/non-self" recognition. With auto-immunity and alloimmunity (especially in mammals) as new evolutionary forms of selection pressures, the emergence of strong control mechanisms of immune reaction is not surprising. Moreover, there are significant correlations between the strength of adaptive immunity and number of cytokines in different classes of vertebrates. The oldest classes of vertebrates are characterized by a small number of cytokines, probably tracing the origin from the regulatory factors whose primary role was probably not associated with immune mechanisms. During vertebrate evolution, this number grew dramatically, as well as the parallel growth of the function in the controlling of immune reaction. In lower vertebrates, cytokine network of the immune reaction consists mainly of the cytokines which are in mammals called the pro-inflammatory cytokines. Clearly defined, "specialized" and strong immunosuppressive cytokines have only been found in mammals, although the role of anti-inflammatory cytokines in lower vertebrates could be associated with the TGF superfamily. Even in birds, whose immune system is most similar to mammalian, cytokines like IL-4 and IL-10 have not been clearly identified (*Table* 1.3.).

1.2   MHC MOLECULES – NATURE, FUNCTION AND EVOLUTION

MHC molecules display antigenic peptides on the cell surface for surveillance by T lymphocytes. MHC class I molecules present peptides to CD8[+] CTL, whereas MHC class II molecules present peptides to CD4[+] Th cells. The current opinion is that antigens from the extracellular fluid enter the exogenous processing pathway by endocytosis and are partially degraded in acidic endosomal or lysosomal structures to yield peptides that bind MHC class II molecules. In the endogenous processing pathway intracellular proteins are degraded in the cytosol by the proteasome complex (LMP), generating peptides that are transported from the cytoplasm into the lumen of the endoplasmic reticulum (ER) by the TAP molecules, where they bind to nascent MHC class I heavy chain-β2-microglobulin (β2m) heterodimers. Fully assembled class I/peptide complexes exit the ER and are transported through the Golgi to the cell surface by the constitutive secretory route (Li *et al.*, 1997; German *et al.*, 1996; Abbas *et al.*, 2003).

Although endogenous antigens are presented by MHC class I molecules, and exogenous antigens are displayed at the cell surface by MHC class II molecules, accumulating evidence has shown that this dichotomy in presentation of antigen from endogenous and exogenous origin is not absolute. It was demonstrated that CTL responses can be primed *in vitro* and *in vivo* with exogenous antigen. At least, two fundamentally different pathways for the presentation of exogenous antigens by MHC class I molecules *in vitro* have been described: one involving access of exogenous antigen to the classical MHC class I loading pathway (TAP dependent) and another involving unconventional post-Golgi loading of MHC class I molecules (TAP independent). In the latter pathway the antigen presumably is processed in an acidic endosomal or lysosomal compartment. How peptides generated by endosomal/lysosomal degradation are loaded onto MHC class I molecules is unknown. The antigenic peptides either can be regurgitated followed by binding to peptide-receptive cell surface MHC class I or they can be captured by endocytosed MHC class I molecules, which then recycle back to the cell surface (Kropshofer *et al.*, 1997).

The evolutionary pressure of microbes on vertebrates and their ancestors gave rise to the emergence of the MHC molecules and adaptive immunity later on. A more precise and efficacious anti-microbe immune reaction with elements of immune memory contributed its own share in the expansion and diversification of vertebrates (Forsdyke, 1991). Generally, we know very little about the evolution of MHC system and its role in the immune reaction of lower vertebrates and the species that no longer exist. The research into MHC system has moved furthest in mammals and in only a few of the species like humans, mice and pigs. However, even in these there are some fundamental limitations. For example, the definite number of MHC gene alleles in humans and mice still needs to be determined. The other mammal species are also relatively unexplored in that regard.

In non-mammalian vertebrate classes, the organization of MHC system and its role in the mechanisms of immune reaction are still largely unknown. The MHC system has been partly explored in several species within each class of vertebrates: some species of shark as the representatives of cartilaginous fish (Kasahara *et al.*, 1993), zebrafish and carp as the representatives of bony fish (Sultman *et al.*, 1993), *Xenopus* as the representative of amphibians (Flajnak *et al.*, 1986), chickens as the representatives of birds (Plachy *et al.*, 1992), and humans and mice as the representatives of mammals (Du Pasquer *et al.*, 1999; Hughes, 1993).

1.2.1   MHC System in Mammals

1.2.1.1   The Structure
The class I and class II proteins have similar structures with subtle functional differences. Class I molecules are made up of one heavy chain (45 kD) encoded within the MHC and β2m (12 kD). Class II molecules consist of one α (34 kD) and one β chain (40 kD) both of which are encoded within the MHC (Steinmetz *et al.*, 1983; Castellino, 1997). The antigenic peptides of 8 to 10 amino acids bind to the cleft with low specificity but high stability. The a 3 domain contains a conserved seven amino acid loop (positions 223 to 229) which serves as a binding site for CD8[+]. Another site in this segment is also very important. The amino acid residue at position 227-300 is critical for the interaction of MHC class I molecule with the chaperon calreticulin, while class I heavy chain residues of 77 to 83 are important in NK cell recognition (Gumperz *et al.*, 1995).

In the class II molecule, generally both α and β chains are polymorphic. Hypervariable regions tend to be found in the walls of the cleft. Antigenic peptides of 12 to 24 amino acids long bind to the cleft and extend on either side (Castellino, 1997; Abbas *et al.*, 2003). A major CD4-binding site is contained within residues 241 to 255 in β2 domain. In the same domain, the polymorphic residues between positions 180 and 189 determine the quality of APCs-CD4$^+$ interaction (Steinmetz *et al.*, 1983; Castellino, 1997).

1.2.1.2 The Function

The MHC represents a group of highly polymorphic genes whose products are expressed on the membranes of all cells, except for the erythrocytes in mammals. The MHC molecules have a central role in the presentation of antigenic peptides to the immune cells. Following the synthesis, these molecules bind to intracytoplasmatic peptides from virus antigens, tumor-specific antigens or "self" molecules. The newly assembled MHC/peptide complex is then transported to the cell membrane and expressed as a membrane-associated molecule. From the conformation of the MHC molecule the immune cells are able to recognize virus-infected cells, tumor cells or mutated cells, and eliminate them. If the MHC molecules are expressed with "self" molecule peptides, immune cells will recognize the cell as "self" structure, which prevents their auto-immune cytotoxic action (German *et al.*, 1996; Li *et al.*, 1997; Abbas *et al.*, 2003).

The activation of T lymphocytes is a multi-step process involving the secretion of cytokines, proliferation of the activated clone, and various effectory activities. One of the first steps in the immune reaction is the recognition and binding of TCR CD4$^+$ and CD8$^+$ cells to the MHC/peptide complex. However, this is not the only link that is made between the cells and lymphocytes for, at the same time, the interaction occurs among different accessory molecules. Because of their significance in the lymphocyte activation and the final effects of the immune reaction, these are called the co-stimulatory and/or adhesion molecules (Bretscher, 1999). The most important molecules and their complementary receptors from this group are: B7.1/2 (CD80/86), CD28/CTLA-4, ICAM-1/2/3 (CD54/102/50), LFA-1 (CD11a/CD18), LFA-3 (CD58) and CD2. Their role is mainly the regulation of various intercellular signals like secretion of IL-2, and the mediation of clonal expansion of the activated lymphocytes (Chambers *et al.*, 1997; Croft *et al.*, 1997). The function performed by the co-stimulatory molecules is highly important, since the occurring of any error in their expression and/or the expression of complementary receptors, with simultaneous occupation of TCR, may result in anergy or even apoptosis of the T cell clones (Vidovic and Matzinger, 1988; Abbas *et al.*, 1997; Bretscher, 1999).

It is assumed that each cell expresses more than 250,000 MHC class I molecules (Parham *et al.*, 1996). These molecules are basically unstable and susceptible to recycling unless bound to antigenic peptides. However, once formed, the MHC/peptide complex becomes very stable and persists on the cell membrane with a half-time of 24 hours. Each moment, a healthy cell expresses about 1,000 different MHC/peptides complexes, though the number can be higher or even lower, depending mainly on the microenvironmental cytokine network and other regulatory factors (Le, 1994). Different cells express different number and kind of class I, in different ratios of class I A, B and C. In human, subclasses A and B comprise the majority of expressed class I molecules, whereas subclass C makes only 10% of the total expressed class I molecules (Le, 1994).

The MHC class I molecules are systematized into 2 groups, MHC Ia class and MHC Ib class. The MHC Ia group comprises the so-called classical MHC molecules like HLA-A, BLA-B and HLA-C, while the non-classical MHC Ib group comprises HLA-G, and HLA-F. MHC Ib molecules are much less polymorphic, the tissue distribution is more restrictive and their role in the immunological and immunopathological processes is completely different compared to class Ia. In normal circumstances, class Ib are basically engaged in the communication with the NK cells and regulation of their activity (Ljunggren, 1990).

MHC expression is very significant for NK cell function, which is an unusual quality of the NK cells, regardless of the fact that these cells are the representatives of innate immunity. Actually, the expression of MHC molecules is an inhibitory signal for NK cells; however, the poor expression of the molecules or a complete absence of the expression (viral infections and malignant alteration) triggers off killer mechanism in NK cells. This mechanism of killing the cells that have failed to express the MHC molecules probably is one of the most effective and quickest. The NK cells have different membranous receptors through which they bind to the MHC class I molecules. Each family of these interact with different MHC class I molecules. Since this interaction inhibits the NK cells' killing activities, it is called the Killer Inhibitory Receptor (KIR) (Lanier, 1998).

The MHC class II molecules mainly express on the immune cells and bind mostly extracellular molecules such as bacterial proteins and viral capsid proteins. The antigenic presentation within MHC class II is mainly assigned to CD4$^+$ Th cells (Abbas *et al.,* 2003). The tissue distribution of class II is more restrictive than in class I. The MHC class II only express the APCs such as B lymphocytes, macrophages, Langerhans and related dendritic cells as well as the activated T lymphocytes. The degree of expression of various MHC class II subtypes in humans can be simply described by the following non-equation: DR>DP>DQ. In addition, the degree of class II molecules expression depends from presence of microenvironmental factors such as cytokine, prostaglandine, viruses, hormones, cell detritus etc. These factors also have important influence on the ratio of expressed DR, DP and DQ class II molecules (Glimcher *et al.*, 1992; Guardiola *et al.*, 1993).

1.2.1.3   Non-Classical MHC Molecules (class Ib)

Class Ia (classical) and class Ib (non-classical) genes are found in all of the major groups of jawed vertebrates. In addition to being present in all the vertebrates, class Ia genes probably perform the same function and display the same characteristics in most of these. The ubiquitous manner of tissue expression, high polymorphism, constant and variable regions, particular evolutionary conserved gene sequences, participation in tissue recognition and antigenic peptide presentation are probably the mutual attributes of the class I in vertebrates (Kaufman *et al.*, 1992,1994).

Unlike class Ia, the class Ib genes display minimal polymorphism and express themselves in a tissue-specific fashion (Klein *et al.*, 1994). The sites of their expression are mainly the trophoblast cells, thymic epithelial cells and malignant cells. Low class Ib variability may be associated with the suppression of decidual NK cells in pregnancy and some other forms of immune tolerance (Yelavarthi *et al.*, 1991). Like in mammals, the presence of class Ib is also found in non-mammalian vertebrates (Kaufman *et al.*, 1992, 1994). Some studies point to the possible presence of non-classical MHC class II genes. So far, their presence has been only demonstrated in mammals, though the other classes as well are suspected to have these genes (Cho *et al.*, 1991; Kelly *et al.*, 1991).

From an evolutionary point of view, the most remarkable feature of class Ib loci is that they do not show orthologous relationships between mammals of different orders. Thus these loci have arisen independently in different mammalian lineages (Hughes *et al.*, 1989). Class Ib loci have evidently evolved as a result of the duplication of class Ia loci followed by a change in the expression pattern due to the mutation in the promoter region (Hughes *et al.*, 1989). The class Ia and class Ib genes thus do not represent mutually exclusive groups over evolutionary time. Dramatic evidence for this is the fact that the class Ia genes of *New World primates* are homologous to the human class Ib locus HLA-G (Cadavid *et al.*, 1997). This locus was evidently present in the common ancestor of *New World monkeys* and *Old World monkeys*, apes, and hominids, but in the former it has been duplicated to form separate class Ia loci. In spite of the independent origin of class Ib loci in different mammals, there is evidence that they can evolve similar functions convergently. For example the class Ib HLA-E molecule of humans and other primates has convergently evolved features of the PBR that are similar to those of the mouse class Ib molecule H2-Qa-1a, and there is evidence that these molecules may bind similar peptides (Hughes *et al.*, 1989; Cadavid *et al.*, 1997).

1.2.2   The MHC System in Non-mammals

Studies of the cartilaginous fish MHC showed that the nurse shark class II genes (Kasahara *et al.*, 1993) and hound-shark class Ia (Okamura *et al.*, 1997) genes are highly polymorphic and under Darwinian positive selection similar to their mammalian homologues. Some data suggest that the MHC consisted of class Ia and class II genes ever since the common ancestor of cartilaginous fish and all other vertebrates existed between 460 and 540 million years ago (Caroll, 1988). These data strongly suggest that one gene duplicated from the other in *cis* and the two loci subsequently remained closely linked in most vertebrate taxa. In contrast to all other vertebrates, every bony fish species examined so far carries unlinked class Ia and class II genes (Klein *et al.*, 1993). It was suggested that this unusual feature might have been the primordial condition, with class I and class II arising on two paralogous chromosomes and being "brought together" in a tetrapod ancestor. Even if recent mitochondrial DNA studies suggesting that cartilaginous fish have a terminal position in the piscine phylogenetic tree are true, then the class II and class I genes would have had to come together rapidly in a "functional cluster" twice in the vertebrate evolution. It is much more likely, and simpler to imagine, that class I and class II genes were linked in the primordial MHC typified by the shark. The "non-linkage" in bony fishes may have occurred by differential silencing of MHC genes after genome-wide chromosomal duplication events proposed for a bony fish-specific ancestor (Ohta *et al.*, 2000).

The function of MHC molecules in "self-non-self" recognition, presentation of small antigenic peptides, restriction of the interaction between T cells and APCs (i.e. MHC restriction), processing of antigen by professional APCs and T cell "education" in the thymus, are capable of being identified in most vertebrate taxa. However, the lymphocyte activity non-mediated by MHC molecules is only evidenced in cartilaginous fish and the evolutionary older species. The verification of the MHC class I and II and the phenomenon of TCR rearranging (Rast *et al.*, 1994) in cartilaginous fish, also indicate the presence of some characteristics of the adaptive immunity in these (Ohta et. al., 2000).

Like in cartilaginous fish, *Urodela* amphibians show weak immune response, which was accounted for by the poor expression and low polymorphism of MHC genes, even their absence (Charlemagne, 1987; Kaufman et. al., 1995). However, the MHC classes I and II have still been found in *Urodela* and some researches indicate that polymorphism they display is higher than it was expected (Sammut *et al.*, 1997).

The three-dimensional structure of MHC classes I and II is very similar in all vertebrate classes (Brown *et al.*, 1993). Certain parts of MHC genes are evolutionary conserved, showing the identical organizational model in all vertebrates, whereas only a small portion of molecules related to peptide presentation and TCR binding differs from class to class, and from species to species (Bjorkman *et al.*, 1987; Brown *et al.*, 1993). The conserved MHC gene sequences suggest the importance of these molecules in the evolution of the MHC system. In addition, the differences within MHC genes point to a possible divergent evolution of the MHC molecules.

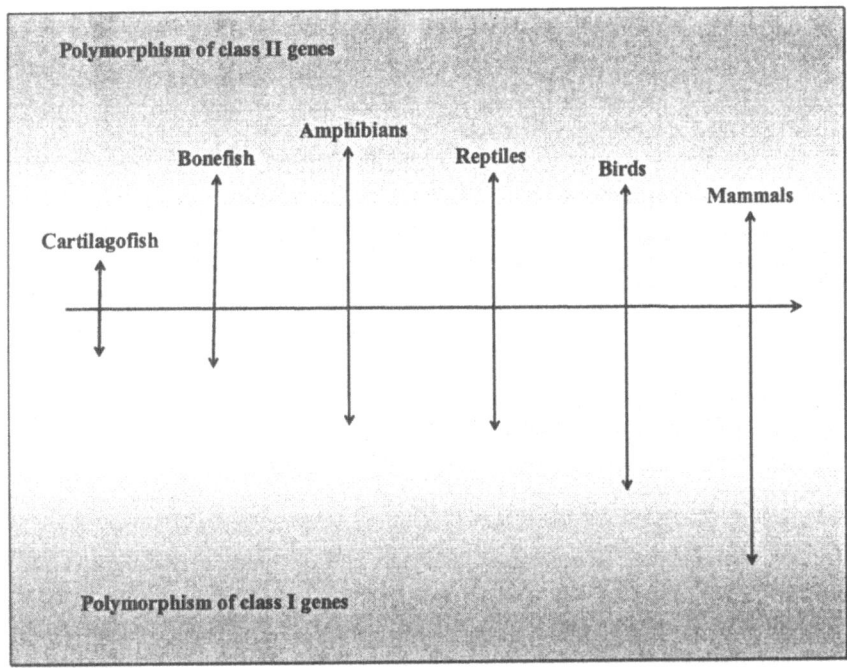

Figure 1.1. Degree of polymorphism of class I and II genes in different classes of vertebrates (ratios are not given in real proportions).

Another comparison to be made with bony fish and all other vertebrates is the relative stability of the class I and class II genes. In mammals, class I genes are plastic, fluctuating in numbers and function in different species, whereas class II genes are more stable (Margulies, 1999). In bony fishes, the situation is reversed, with class II being plastic and found on multiple chromosomes whereas class I appears to be more conserved (Kasahara *et al.*, 1993; Ohta *et al.*, 2000) (*Figure* 1.1.).

Also, class II gene number in the teleostean cichlid fishes differs greatly from haplotype to haplotype. The close association of class Ia to the TAP and proteasome genes in bony fishes, birds and amphibians indicates the co-evolution of such genes and may account for the relative stability of class I in these species whereas bony fish class II, probably being linked to no other genes involved in MHC peptide presentation, is capable of rapid diversification (Deverson *et al.*, 1990; Trowsdale *et al.*, 1990). In the nurse shark, both class II and class I genes seem rather stable, although studies in other species must be done; tight linkage to the other MHC genes probably maintains this stability in both classes. In addition, with the little data Ohta *et al.*, (2000) have obtained so far, cartilaginous fish class Ia and class Ib lineages seem old, suggesting little expansion and the contraction of class Ia-like genes as is found in mammals.

1.2.3   Ontogeny of MHC Molecules Expression

In mammals and birds, the class I and II molecules give the same manner of expression and tissue distribution on embryonic and adult tissues. The MHC expression on .all the tissues (except for the trophoblast) at the embryonic stage in mammals and birds is quite a uniform process and depends mainly on the general regulatory mechanisms of MHC expression (Jaffe *et al.*, 1991; Kaufman et. al., 1992, 1994).

In carp, class I and class II were isolated from one-day old embryo, showing the tendency towards the linear increase in function up to the thirteenth week of life. Some differences have been detected in the structure of MHC molecules between adult animals and the animals undergoing early development, which indicates that one type of MHC molecules is expressed in the developmental stage, whereas the other type is expressed in the adult stage (Kaufman *et al.*, 1990).

The greatest divergence in MHC expression during ontogeny is evident in the species whose development is related with the phenomenon of metamorphosis. The larva of *Xenopus* species express the class II molecules in greater measure than in adult frogs, whereas class I expression is either extremely poor or completely absent. Class I expression in the *Xenopus* larva is characterized by a restricted tissue distribution. These molecules are only expressed on the skin, mucous membrane of intestines and the tissues that are in contact with the external environment. Albeit, class I expression provides immune protection for the larvae, the astonishing fact is, that it only occurs on the organs undergoing massive destruction and apoptotic remodelling during metamorphosis (Du Pasquer *et al.*, 1990; Rollins-Smith *et al.*, 1990; Kaufman *et al.*, 1990).

Axolotl class II molecules are also regulated differentially during ontogeny, being expressed in young animals on B cells and then expanding to all haematopoietic cells, including erythrocytes later in life. The changes in MHC expression are not correlated with cryptic metamorphosis in *axolotls*, but class II expression by erythrocytes is correlated to the switch from larval to adult globins. Finally, class I transcripts isolated so far are expressed early in ontogeny, from hatching onward (Kaufman *et al.*, 1990).

1.2.4   The Evolution of MHC Molecules

The innate immune system is the only defence mechanism to be found in invertebrates, but in vertebrates it is only a part of the immune system. The second, probably the most important part of vertebrate immune system is adaptive immunity. Due to the lack of molecular evidence that vertebrates inherited the adaptive immunity from invertebrates, there is the presumption that this part of the immune system was developed as an effective advancement of the innate immunity (Rittig *et al.*, 1996).

Agnatha do not have genes for MHC molecules, although they have the ability to reject the transplants. Cartilaginous fish are evolutionary the earliest vertebrate group displaying clearly defined MHC class I and class II genes, which indicates that some of their ancestors must have been the precursor of the MHC system. This suggests that class I and class II genes are older than 450 million years (Ohta *et al.*, 2000).

Geneticists and molecular biologists still have not identified the gene likely to be the evolutionary precursor of MHC genes. While some authors propose that class II genes are evolutionary older than class I, others support the idea that class I genes are the earliest molecules of the tissue compatibility. Also, there is a real presumption that the precursors of class I molecules are actually Heat Shock Proteins (HSP) (Lawlor *et al.*, 1990; Flajnik *et al.*, 1991; Hughes *et al.*, 1993; Klein *et al.*, 1993).

Absence of MHC genes in jawless fish or invertebrates, suggesting that the MHC arose rather abruptly in a jawed vertebrate ancestor, probably a *placoderm* (Kasahara *et al.*, 1992; Okamura *et al.*, 1997; Matsunaga *et al.*, 1998). One hypothesis suggests that genome-wide duplications played a role in the emergence of the MHC and the entire adaptive immune system (Kasahara *et al.*, 1996a), as genes linked to class I and class II are found in four paralogous clusters in mammalian genomes (Boyson *et al.*, 1996; Kasahara *et al.*, 1997). In all tetrapod species examined to date, including several primates, the bird *Gallus* (Gyllensten *et al.*, 1989), and the amphibian *Xenopus* (Moriuchi *et al.*, 1985; Kasahara *et al.*, 1992), class I and class II genes are closely linked. However, among older taxa, in all investigated bony fish species, including the zebrafish (Karr *et al.*, 1986), carp, salmon

(Nei *et al.*, 1997), and trout (Parham *et al.*, 1996), classical class I and class II genes are not linked and even are found on different chromosomes. It was proposed that one of two scenarios occurred in vertebrate evolution (Karr *et al.*, 1986; Parham *et al.*, 1997): (i) class I and class II genes arose on different paralogous chromosomes in a jawed vertebrate ancestor and "clustered" together in a tetrapod ancestor, or (ii) the genes were originally in the same linkage group but were rent apart in a recent teleost ancestor and now lie on different chromosomes in this single vertebrate lineage (Hughes *et al.*, 1988) (*Figure 1.2.*).

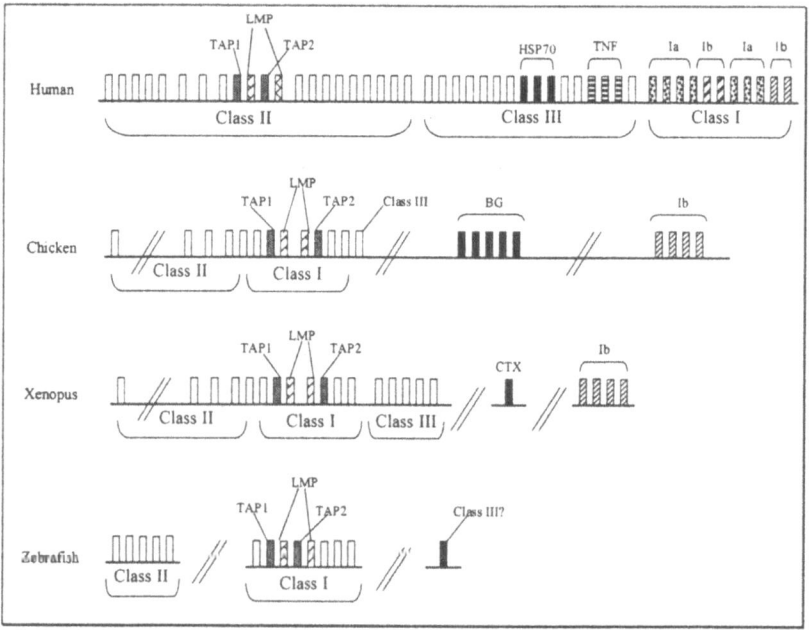

Figure 1.2. Map of the linkage between genes in human, chicken, *Xenopus* and zebrafish. Only a small number of the most important genes are shown, while the rest of the genes have been omitted, as well as a relative distance between them. Backslashes on line means that the genes are on the same chromosome, but on different clusters. Backslashes between lines means that the genes are on different chromosomes (Moriuchi *et al.*, 1985; Kasahara *et al.*, 1992, 1996a, 1996b, 1997).

Regarding the tendency towards the linking and clustering of class I and class II genes along the evolution of vertebrates, it seems logical to ask: what is the nature of the selection pressure that directed the development of this phenomenon? Almost every argument on the evolution of the immune system puts microbes into the foreground as the source of the strongest evolutionary pressure that modelled the vertebrate immune system. However, could microbes really be considered the only factor of evolutionary pressure that could have led to the clustering of class I and class II in higher vertebrates?

Linking and clustering of the class I and class II genes is not the only phenomenon clearly detected in the evolution of the immune system from cartilaginous fish to mammals. In birds (Kaufman *et al.*, 1995), bony fishes (Karr *et al.*, 1986) and reptiles (Moriuchi *et al.*, 1985; Kasahara *et al.*, 1992), but not in mammals, the genes responsible for determination of LMP and TAP molecules are linked with the class I genes. This phenomenon is most striking in zebrafish, because the class I gene complex, LMP and TAP are found on the same chromosome, while class II genes are found on a quite different one. It is presumed that MHC III genes were inserted into the MHC in the later evolution of vertebrates, but some data suggesting that the genes which determine the C4 component of complement and HSP70 in reptiles and mammals are linked with the class I, also propose that this phenomenon occurred in the common ancestors of reptiles and mammals (Kasahara *et al.*, 1996a, 1997). This assumption is in corollary with the hypothesis that the class I and II genes appear to have the common precursor gene. Meanwhile, these data also indicate the likely scenario of the MHC gene evolution, where MHC class I and class II had quite independent evolutionary pathways, whereas classes I and III probably had the common evolution.

From the gene arrangement on the chromosome map (*Figure* 1.2.) it is clearly noticeable that in all non-mammalian classes the TAP and LMP genes are highly conserved within class I genes region, which suggests that the processing and expressing of genes take place on the same cluster. In mammals, TAP and LMP genes are highly evolutionary conserved within class II genes region (Kasahara *et al.*, 1996b).

Available evidences suggest that the genetic content and level of MHC complexity are comparable in all mammalian species. However, several intriguing peculiarities have been identified in equine and cattle. Recent genetic mapping suggests that MHC of horses may be disrupted even more than it is in chickens. This would be the first example in mammals where MHC sequences are located on different chromosomes (Fraser *et al.*, 1998). Surprisingly, horse homologue to TAP2 is conserved in MHC class II region. Also, TAP2 genes and class II surrounding genes in horses order seems to be fairly well conserved with the human class II organization (Personal communication with Dr. Antczack, J.A. Baker Institute for Animal Health College of Veterinary Medicine Cornell University Ithaca, NY 14853).

1.2.5   Mechanisms of MHC Molecules Variability

Although both MHC molecules and TCR are known for their extreme degrees of diversity, the underlying mechanisms are fundamentally different. Whereas TCR owe their diversity to special somatic diversification processes, MHC molecules have mutation rates similar to those of most other genes (Parham *et al.*, 1995). An explanation for the high degree of MHC polymorphism cannot be sought in vertebrate allograft rejections, as these are experimental artefacts and thus not naturally involved in evolutionary selection, probably until the emergence of viviparity (De Boer, 1995; Bubanovic *et al.*, 2004). One of possibility is that the vertebrate MHC polymorphism is a "relict" of ancestor genes polymorphism (Buss *et al.*, 1985). Alternatively, the selection pressure for MHC diversity may be due to peptide presentation to the immune system. Several most commonly held views are that MHC polymorphism is due to selection favouring MHC heterozygosity and evolutionary accumulation of MHC molecule diversity (Doherty *et al.*, 1975) or due to the selection for hosts with rare MHC molecules (Bodmer, 1972).

Regarding the role of MHC molecules in the anti-microbe defence and antigen-presentation to the immune system, the number of MHC genes expressed per individual is surprisingly small. For example, each human individual expresses maximally six different classical MHC class I genes, and twelve different MHC class II molecules (Abbas *et al.*, 2003). One would expect evolution to favour the expression of many MHC genes per individual. A solution to this paradox has been sought in "self-non-self" discrimination. A widely accepted argument is that an excessive expression of MHC molecules leads to the depletion of the T cell repertoire during "self" tolerance induction (Abbas *et al.*, 2003). In addition, the highly polymorphic MHC genes control immunological "self-non-self" recognition; therefore, the polymorphism may function to provide "good genes" for an individual's offspring. There are three adaptive hypotheses for MHC dependent survives under evolutionary pressure of microbes (especially viruses): (i) the consequence of high MHC genes variability is mainly MHC-heterozygous offspring that may upgrade anti-microbe immune response. Although this hypothesis is not supported by tests of single microbe infection, MHC heterozygotes may be resistant to multiple viruses or other microbes; (ii) MHC variability enables hosts to provide a "moving target" against rapidly evolving microbes that escape immune recognition. Such viruses are suspected to drive MHC diversity through rare allele advantage. Thus, the two forms of viruses-mediated selection thought to drive MHC diversity, heterozygote and rare allele advantage, will also favour MHC variability and (iii) the diversification of MHC genes may also function to avoid inbreeding; a hypothesis consistent with other evidence that MHC genes play a possible role in kin recognition or even bifurcation of species from subspecies and emergence of a new species.

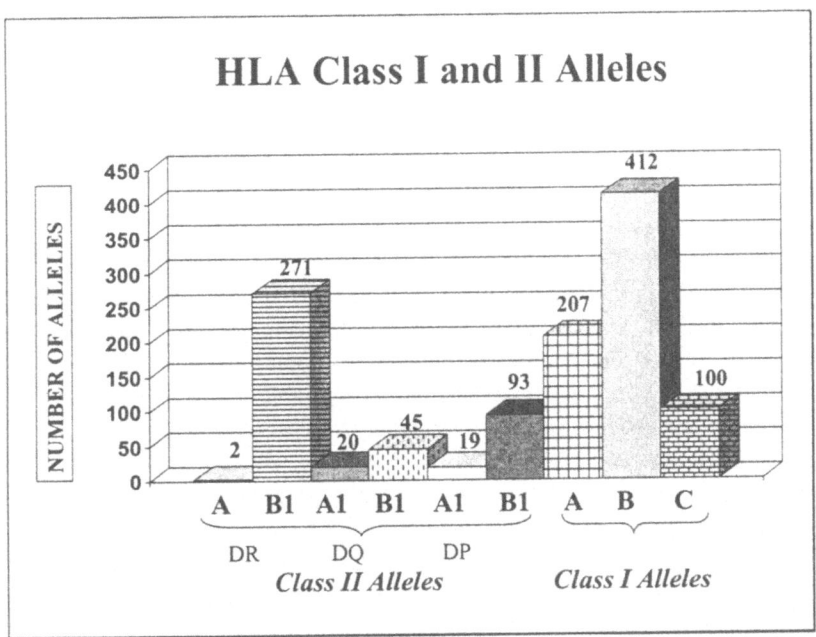

Figure 1.3. Number of different class I and class II alleles in humans.

The MHC loci are known to be highly polymorphic in humans, mice and certain other mammals, with heterozygosity as high as 80-90%. Six different hypotheses have been considered to explain this high degree of polymorphism (*Figure* 1.3.):

1. A high mutation rate with gradual accumulation of spontaneous mutational substitution over evolutionary time. The main source of the variability in the MHC genes sequences is a point mutation but the mutation rate is by no means higher in the MHC than elsewhere in the genome (Lawlor *et al.,* 1988; Parham *et al.,* 1995). Because of transspecies polymorphism, the accumulation of point mutations over evolutioanry times (millions of years) results in the extensive polymorphism.
2. Gene conversion, interlocus genetic exchange or periodic intragenic (interallelic) and more rarely, intergenic, recombination within the class I genes.

3. The selection against mutational divergence in the regions of the class I molecule involved in T cell receptor interaction and also in certain regions that interact with common features of antigens.
4. Positive selection pressure in favour of the persistence of MHC polymorphism and heterozygosity at the antigen recognition site.
5. The negative selection of the MHC alleles associated with tolerance to microbes.
6. Microbes-exerted the negative selection of low polymorphic or monomorphic MHC genes.

1.2.5.1 Viruses like Inductors of MHC Variability

Generally, all microbes could have influenced the evolution of the MHC system. Although class I and class II molecules participate in the immune reaction against extracellular microbes, the intracellular microbes are thought to be the most important source of evolutionary pressure that has governed the MHC gene evolution and their variability.

MHC alleles differ in their resistance to viruses and susceptibility to auto-immune diseases (Apanius *et al.*, 1997), so why does natural selection not eliminate all but the most resistant allele? Several lines of evidence indicate that the antigen-binding site of MHC molecules is under balancing selection for long periods of evolutionary time (Hughes *et al.*, 1995; Apanius *et al.*, 1997). Because the MHC plays such a pivotal role in the immune reaction, the diversity of MHC alleles is generally assumed to be maintained by microbes, especially viruses (Clarke *et al.*, 1966; Hughes *et al.*, 1995; Parham *et al.*, 2000) (*Figure* 1.4.). It is often assumed that the MHC diversity is maintained because it "provides broad immunological protection for the species as a whole" during catastrophic epidemics and as "a strategy to keep viruses from spreading through the entire population" (Klein *et al.*, 1994). This argument implies that MHC polymorphisms are maintained because populations with high MHC diversity have a better chance of survival than populations with low diversity. Such group selection may be favouring the MHC diversity (Apanius *et al.* 1997); however, the problem is that if directional selection and drift are eliminating MHC diversity within populations in the short term, then there will be no diversity for selection to act on among populations in the long term. Therefore, there must be some other explanation besides group selection for the evolutionary preservation of MHC diversity. There are two non-mutually hypotheses for how viruses can preserve MHC diversity within host populations: (i) selection can preserve MHC polymorphisms if certain MHC allele or MHC heterozygotes are more resistant to viruses than homozygotes (Hughes, 1992) and (ii) if MHC alleles are under negative frequency-dependent selection from viruses (Clarke *et al.*, 1966; Slade *et al.*, 1992). There is strong evidence suggestive of the universal correlation between the degree of MHC expression on the infected cell and the host's resistance to viral infection (Kaufman *et al.,* 1995; Sammut *et al.,* 1997). For example, in chickens there is a strong link between some diseases and MHC molecules expression. Namely, the percentage of survival of the animals with haplotype B12, infected with Marek's virus is over 95%, while the other haplotypes show different levels of resistance to the virus, but still significantly lower in comparison with the haplotype B12 (Plachy *et al.,* 1992).

1.2.5.2 The Heterozygote Advantage Hypothesis

There are data that MHC heterozygotes are more resistant to microbes than homozygotes, also MHC heterozygotes present a wider diversity of antigens to the immune system than homozygotes, but they have significant smaller variety of TCR repertoire (Doherty *et al.,* 1975); however, there is surprisingly little evidence from population surveys and experimental infections to support the heterozygote advantage hypothesis. A recent study on feral sheep found no MHC-heterozygote advantage against a nematode microbe (Paterson *et al.* 1998), and a large survey study on malaria in humans found that MHC heterozygotes had a disadvantage (Hill *et al.,* 1991). Despite numerous experiments in the laboratory with mice and chickens, which used a wide range of infectious agents, MHC heterozygotes do not show any general resistance compared with homozygotes (Apanius *et al.,* 1997). There are several possible reasons why the MHC-heterozygote advantage hypothesis has not been adequately tested:

1. There is hypothesis that high MHC variability and MHC heterozygote advantage may be observed only on the level of population or species under special conditions such as big epidemics. Namely, antiviral immune reaction is in close relation with the expression of viral antigen epitops presented by MHC class I, recognition of the infected cell and activation of the effectory cytotoxic mechanisms. The survival of individuals from the virus affected population may be in close link with the quality and quantity of the immune reaction that is developed by each individual against the virus. An inadequate immune reaction will reduce chances of surviving of the virus-infected individuals. Thus, a weak immune reaction may be as undesirable as a strong one (Kleinet *et al.,* 1993, 1994). The high polymorphism of MHC genes and high percent of MHC heterozygotes within one species allow for a great number of combinations of the single antigenic epitops expression, resulting in a proportionally high number of qualitative and quantitative variants of the immune reaction. This phenomenon may be highly relevant with respect to favouring the "optimal" immune response within a species, especially

if one has in mind that the aggressive immune response can be as fatal for the individual as an extremely weak one. High MHC variability and the diversity of qualitative and quantitative types of immune reaction guarantee the survival of some individuals of one species in the case of catastrophic epidemics caused by intracellular parasites. Therefore, the evolution and polymorphism of MHC could be a direct consequence of the variability and adaptation of the intracellular parasites, and the need of the immune system for the recognition between "self" and "non-self" structures, precise intercellular communication, as well as fine adjustments of the quality and quantity of the immune reaction on the level of species (*Figure* 1.4.).

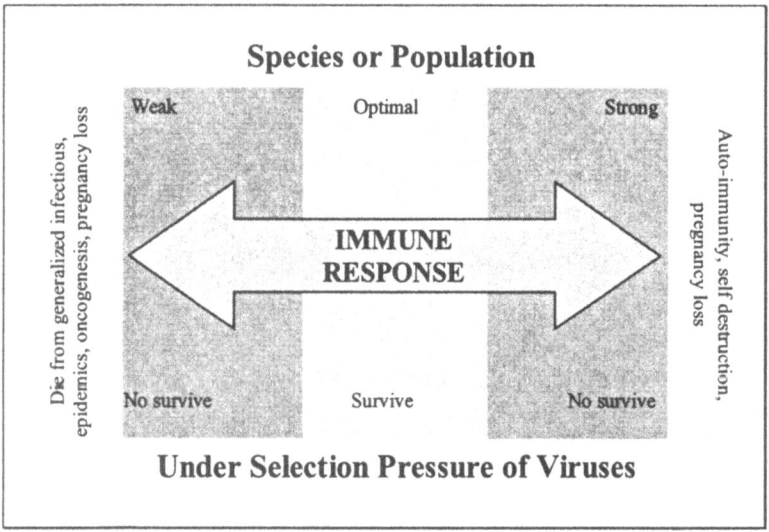

Figure 1.4. Different models of selection pressure related to viral infection, variability of MHC molecules and quality and quantity of immune reaction.

2. The protective effect of high MHC variability and MHC heterozygosity may only occur when individuals are infected with multiple microbes (strains or species) as occur in the wild. Also, the protective effect may only occur when individuals are infected with intracellular parasites such as viruses. Hughes *et al.* (1992) suggest that MHC heterozygotes are protected against multiple microbes because they recognize a wider array of antigens than homozygotes. However, if these arguments were correct, then MHC heterozygotes should also be resistant to single, as well as multiple microbes. A stronger reason to expect that MHC heterozygotes are protected against multiple infections is that MHC alleles conferring

resistance to one microbe increase susceptibility to others (Apanius *et al.*, 1997). Contrary to Hughes *et al.* (1992), such trade-offs in resistance are common among MHC alleles (Apanius *et al.*, 1997). Thus, if resistance to infection is generally dominant or semidominant to susceptibility, then MHC heterozygotes should have an advantage over homozygotes. A recent survey of humans in *West Africa* found that individuals heterozygous at a class II MHC locus are resistant to *hepatitis B* (Thrusz *et al.*, 1997).

3. Since MHC heterozygotes can potentially recognize a given microbe in more ways than homozygotes, a successful evasion of immune recognition may be more difficult. This is supported by the study that viral escape variants emerged more easily in MHC homozygous compared with heterozygous mice (Weidt *et al.*, 1995).

4. MHC heterozygote advantage may have been overlooked if functional MHC homozygotes have been misclassified as heterozygotes. Most human MHC alleles belong to only a few supertypes based on similarities in their peptide-binding properties (Sidney *et al.*, 1996). If MHC-heterozygote advantage only occurs when individuals are heterozygous for MHC functional super-types, then classifying individuals as "heterozygotes" based on allelic differences may fail to detect a true heterozygote advantage. Some evidence suggests that the allelic distribution of MHC super-types is more uniform than allelic differences indicating that selection is operating on super-types. This suggests that some other form of selection is operating on MHC subtypes besides their ability to bind to foreign antigens. Thus, the discovery of MHC super-types may have important implications for understanding how MHC genes variability influences anti-microbe protection (Penn *et al.*, 1998a).

5. MHC heterozygote advantage may be overlooked if the benefit of heterozygosity lies in reduced immunopathology rather then increased immune responsiveness (Carter *et al.*, 1992). Immune responses can be too strong as well as too weak; however, experimental and evolutionary studies have generally ignored immunopathology even though it is probably the most important price of immunological defences (Wakelin 1997; Gemmill *et al.*, 1998; Penn *et al.*, 1998b). One problem with this "optimal immunity" hypothesis is that experimental evidence from mice indicates that MHC heterozygotes respond more aggressively to infection and consequently probably suffer more immunopathology than homozygotes (Doherty *et al.*, 1975) (*Figure* 1.4.). The relationship of the phenomenon of MHC functional super-types and immunopathology is mostly unexplored, particularly regarding evolutionary aspects of population immunology and "optimal immunity" during catastrophic epidemics.

6. MHC heterozygote advantage may be overlooked if the optimal number of MHC molecules expressed in an individual's immune system is less than complete heterozygosity. Individuals with more heterozygous MHC loci present more antigens to the immune system; however, they probably have smaller T cell repertoires (because of thymic selection), that is, there is a pleiotropic trade-off between maximizing the number of different antigens presented by MHC and the number recognized by T cells. The finding that tetraploid *Xenopus* frogs have silenced half of their MHC genes (Du Pasquer *et al.,* 1989) suggests that there is a cost of having too many MHC genes expressed. This optimal MHC-heterozygosity hypothesis is consistent with evidence that MHC heterozygotes sometimes have an advantage but other times have no advantage or a disadvantage (Apanius *et al.,* 1997).

1.2.5.3   The Moving Target Hypothesis

Another way that viruses can maintain MHC diversity is through a frequency-dependent, co-evolutionary "arms race" between hosts and viruses. If MHC alleles have different susceptibilities to a particular microbe, then the most resistant allele will be favoured and spread through the population or species (Hill *et al.,* 1991, 1992). However, a resistant MHC allele will not necessarily go to fixation because, when the resistant allele becomes common, this increases selection on viruses to evade recognition by this common allele. Any microbe that escapes recognition will spread and impose selection against the common host MHC allele. This co-evolutionary "arms race" is suspected to create cycles of frequency-dependent selection that preserve MHC polymorphisms for an indefinite period (Clarke *et al.,* 1966; Slade *et al.,* 1992). If MHC diversity is maintained by rapidly evolving viruses, then MHC genes variability preferences will provide a moving target to viruses that evade immune recognition. The most important viruses driving MHC diversity are suspected to be vertically transmitted (Klein *et al.,* 1994). As viruses adapt to their host's MHC genotype, then genes variability preferences will enable hosts to render viruses adaptations obsolete in their offspring. The MHC genes variability preferences may function to produce offspring that are MHC dissimilar from their parents and other members of the population rather than heterozygous per se. In conclusion, the moving target hypothesis suggests that MHC genes variability and high incidence of heterozygous, as an adaptation mechanism in evolutionary "arms race", provides a moving target "strategy" against rapidly evolving viruses (Hamilton *et al.,* 1990; Ebert *et al.,* 1996).

1.2.5.4 Shifting MHC Presentation Holes

Co-dominant and high variable MHC genes will produce offspring that can present a different set of antigens than other members of population and should therefore recognize viruses that have evaded, for example, their parent's MHC presentation. Because MHC molecules bind to small peptides at only two to three critical amino-acid anchor positions; therefore, substitutions at these positions should enable microbes to evade presentation (Potts *et al.*, 1995). For example, a strain of Epstein-Barr (EBV) virus that infects people in *New Guinea* has an amino acid substitution that prevents presentation of peptides normally recognized by a class I allele (De Campos *et al.*, 1993). This particular MHC allele is uncommon except in *New Guinea*, suggesting that the common allele has favoured the viral escape variant.

1.2.5.5 Shifting T-Cell Recognition Holes

Another way that viruses can evade MHC-dependent immunity is by escaping T cell recognition. Since the MHC shapes an individual's TCR repertoire during thymic selection, different parental MHC alleles will alter the T cell repertoire of an individual's offspring and their ability to recognize foreign antigens. Viruses can evade T cell recognition through several mechanisms: by single amino-acid substitutions in antigens recognized by T cells (Pircher *et al.*, 1990; Price *et al.*, 1997), by punching a "hole" in their host's T cell repertoire by inactivating or anergy antigen specific T cell clones (Bertoletti *et al.*, 1994; Klenerman *et al.*, 1994), and by resembling host antigens (molecular mimicry) viruses take advantage of the holes in their host's T cell repertoire (Hall, 1994).

Molecular mimicry creates a particularly important challenge to the immune system if cross-reactivity triggers auto-immunity (Benoist *et al.*, 1998). Common MHC alleles will tend to accumulate an auto-immune load as a result of molecular mimicry, and this auto-immune load may create a negative frequency-dependent selection on MHC genes (Apanius *et al.*, 1997). The moving target hypothesis assumes that viruses are able to adapt to their host's MHC genotypes and that the high polymorphism of MHC molecules will alter the immune system of an individual's offspring. Finally, if high rate of heterozygous provides a moving target, then altering MHC at each generation should retard the rate of viral adaptation to a host's MHC. Thus, viruses-mediated selection on MHC genes, both MHC heterozygote advantage and frequency-dependent selection, would favour the evolution of MHC genes variability. If viruses maintain the diversity of MHC genes, through heterozygote or rare-allele advantage, then MHC variability and the high number of MHC alleles can function to create heterozygous offspring or to provide a moving target against rapidly evolving viruses.

1.2.6   Variability of TAP, LMP and Other Immunological Genes

In mammalian the class II region of MHC, four genes have been described implicated in processing of MHC class I presented peptides. Two of these are TAP1 and TAP2 code for ER membrane transporters proteins and the other two are LMP2 and LMP7 for proteasome subunits. These genes are polymorphic, although much less so than classical MHC class I and II genes. There is a controversy concerning the possible functional implications of this variation.

The molecular and functional analyses of rat and primate TAP2 homologues indicated major differences in gene diversification patterns and selectivity of peptides transported. The sequence analysis of the TAP2 cDNAs from gorilla EBV virus-transformed B-cell lines revealed four alleles with a genetic distance of less than 1%. The diversification of the locus appears to have resulted from point substitutions and recombinational events. Evolutionary-rate estimates for the TAP2 gene in gorilla and human closely approximate those observed for other hominoid genes. The amino acid polymorphisms within the gorilla molecules are distinct from those in the human homologues. The absence of ancestral polymorphisms suggests that gorilla and human TAP2 genes have not evolved in a trans-species fashion but rather have diversified since the divergence of the lineages (Loflin *et al.*, 1996). Polymorphism within these genes could alter the level of the immune response, a phenomenon relevant to the development of auto-immune diseases. For example, Moins-Teisserenc *et al.* (1995) investigate that TAP2 gene polymorphism contributes to the genetic susceptibility to multiple sclerosis. Similarly with previous citation, Martinez-Laso *et al.* (1994) found that TAP2 genes are placed within the HLA complex, have limited genetic variability and encode two main groups of TAP, the so-called TAP2*01 alleles, with a short ATP-binding domain, and the TAP2*0201 allele with a long domain. The shorter TAP2*01 alleles are present in 99% of diabetics and 90% of controls.

TAP and LMP genes are undoubtedly polymorphic, but if microbes maintain MHC diversity, then why are other genes that influence disease resistance not as polymorphic as the MHC? The MHC is widely cited as an example of genetic diversity driven mainly by viruses, yet the largest survey on MHC and disease resistance found evidence for directional selection (Hill *et al.*, 1991), which reduces genetic diversity. There are several possible reasons for this inconsistency:

1. Disease resistance genes are generally polymorphic, but the variation is hidden and will require molecular techniques to uncover. This explanation seems unlikely because most major immune system genes,

such as TCR and Ig genes are not particularly polymorphic (Kurth *et al.*, 1993).

2. MHC genes are unusually polymorphic because their role in the immune reaction is qualitatively different from other genes such as TCR, Ig, TAP, LMP and RAG. The absence of high TAP/LMP gene variability can be explained by such a stable intracellular antigen processing machinery.

3. In mammals, the TAP/LMP genes are located within the class II region but they control antigen presentation associated with class I molecules (Powis *et al.*, 1996) therefore, microbe evasion should provide a similar selective force on TAP/LMP and MHC genes. Accordingly, it is likely that the evolutionary pressure of microbes as well as the other selection pressures may also have acted to favour high MHC variability and relative evolutionary conservation of TCR, Ig, TAP, LMP, and the RAG genes.

1.2.7   Regulation of MHC Genes Transcription

Synthesis and expression of MHC molecules is a multileveled controlled biological process which is determined by several factors including genetic and microenvironmental factors such as intercellular interactions, cytokine network, presence of microbes, etc.

The constitutive expression of MHC molecules is genetically determined and occurs independently from external influences. This mechanism is possibly involved in the regulation of tissue distribution of MHC class I and class II molecules. Even though class I MHC proteins are found on the surface of almost all nucleated cells, the level of expression varies between different cell types. Lymphocytes probably express the highest levels of class I while hepatocites have only a low level expression. There is no detectable cell surface class I MHC expression on brain cells, germline cells, trophoblast (except class Ib) and undifferentiated embryonic carcinoma cells (Garrido *et al.*, 1995).

Class II molecules are mainly expressed by the immune cells (APCs, lymphocytes), with the level of their expression being proportional to the degree of activation and differentiation of the immune cells. MHC class II molecule expression changes with maturation; immature APCs concentrate MHC class II molecules intracellularly, whereas maturation increases surface expression of MHC class II and costimulatory molecules to optimize antigen presentation (Setterblad *et al.*, 2003).

The regulation of frequency of MHC gene transcription is one of the most significant factors in the regulation of MHC molecules expression. Pre-transcriptional signals, such as cytokines, are the most important factor of the regulation of MHC expression. However, there are other post-transcriptional mechanisms that regulate the MHC expression. In cells of the central nervous system that do not express class I genes, the crucial promoter/regulatory sequences of class I and β2m genes are largely unoccupied *in vivo* although factors that bind to the regulatory regions are present in the nuclear extracts of the cells (Burke *et al.*, 1989; Howcroft *et al.*, 2003). This suggests the existence of mechanisms that determines accessibility of regulatory nuclear binding factors to relevant *cis*-acting elements during development. The expression of MHC gene products can also be regulated at the post-transcriptional level. It is reported that the 3' untranslated region of HLA-DR mRNA interacts with at least two compartmentalized proteins, which seem to participate in mRNA partitioning in the nucleus and in the cytoplasm, respectively. The transfection of cDNA from the crucial 3' untranslated region results in an mRNA that is preferentially released in the cytoplasm, where it mainly associates with ribosomes (Accolla *et al.*, 1995; Howcroft *et al.*, 2003). The post-transcriptional regulation of the MHC expression can also be mediated by cytokines, intracellular microbes (especially viruses) and TAP/LMP molecules. One of the efficient mechanisms of the post-transcriptional suppression of the MHC expression is the inhibition of TAP genes and failure of MHC molecules to bind to peptides. Empty MHC molecule does not express on the cell surface but is recycled instead, so that the cell is under such circumstances restricted to expressing a very small number of MHC molecules. Many viruses employ such "craftiness" in order to prevent the presentation of viral peptides and destruction of the host cell, though there are some groups of viruses, such as *flaviviruses*, which enhance class I expression on infected cells (Lobigs *et al.*, 2003).

Corticosteroids, sex hormones and prostaglandins have long been known to be immunosuppressive factors. It turns out that they are able to decrease the expression of class II molecules, while class I molecules expression is also influenced by various viruses. Specifically, *cytomegalovirus* (CMV), *hepatitis B* and *adenovirus-12* (Ad-12) can decrease class I molecules expression. CMV express a viral peptide which binds to β2m preventing class I assembly and transport to the plasma membrane, while Ad-12 causes a pronounced decrease in the transcription of TAP1 and TAP2 which are required to transport processed peptides from the cytoplasm into the ER. The down-regulation of class I molecules helps these viruses to evade the immune response since viral peptides are presented to CD8[+] cells by class I molecules (Zhao *et al.*, 2003).

The strategy employed by EBV is a completely different one. EBV expresses the gene which determines the synthesis of IL-10-like molecule (EBV IL-10) with low affinity for the known IL-10 receptor, an accessory chain to that receptor is expressed in certain cells and that EBV IL-10 binds the accessory receptor with a high affinity (Liu *et al.* 1997; Zdanov *et al.* 1997). Among the cells on which EBV IL-10 has an effect are macrophages/monocytes and B cells. In the former, EBV IL-10 down-regulates the expression of class II molecules, thereby suppressing a specific host immune response latter, EBV IL-10 enhances proliferation, which is presumably also to the virus's advantage, since B cells are its primary host cell type (Liu *et al.* 1997). Furthermore, the anti-apoptotic effect of EBV IL-10 might be associated with the resistance of the infected cells to the cytotoxic activity and apoptotic signals.

Cytokines are key mediators in the regulation of gene transcription, determining the synthesis of classes I and II MHC molecules. As immune cells are the primary source of cytokines, this mechanism can be regarded as a trigger for the control of immune reaction. Pro-inflammatory cytokines generally enhance MHC molecules expression, whereas anti-inflammatory cytokines suppress it. Pro-inflammatory cytokines, such as IFN-α, β and γ as well as TNF-α, intensify class I expression on most cell types. The mechanism of action of these cytokines is based on amplification of MHC gene transcription. Most cell types expressing class II MHC molecules enhance expression primarily through mediation by IFN-γ; however, these effects are antagonized by IFN-α, β and IL-10. Only a minority of cell types expressing class II MHC molecules, such as B lymphocytes, enhance MHC expression in the presence of IL-4 and IL-10, while suppressing it in the presence of IFN-γ (Wolfgang *et al.,* 1999).

1.2.7.1 Cytokines which modulate expression of MHC class I molecules

IFN-α and β were originally described as anti-viral and anti-proliferative agents which were subsequently also shown to have immunoregulatory properties. IFN-α and β can antagonize the effects of IFN-γ on MHC class II expression on murine macrophages and endothelial cells (Inaba *et al.,* 1986). IFN-γ is produced by NK cells, Th1 and CD8$^+$ lymphocytes on activation by antigens or mitogens (Fong *et al.,* 1990). Although IFN-γ displays most of the biologic activities that have been described to the other IFNs, it has 10-100 fold lower specific antiviral activity than either IFN-α or IFN-β. On the other hand, IFN-γ is 100-10,000 times more active as an immunmodulator than are other types of IFNs (Pace *et al.,* 1985).

All three types of INFs enhance the expression of class I MHC molecules on a number of immunologically important cell types including mononuclear phagocytes, epidermal cells, endothelial cells, B lymphoid cells and macrophages (Chang *et al.,* 1994), while IFN-γ is somewhat more efficient in bringing about this effect. Cells with enhanced MHC I expression may become better targets for CTLs recognizing viral, tumor or auto-antigens present in such cells. Following IFN treatment, the increase of HLA class I antigens on human cells is rapid (detectable after one hour), stable for at least 24 h and ranges from 2 to 10 times the initial levels (Rosa *et al.*, 1984). Northern blot analysis of mRNA encoding HLA class I heavy chain has shown that IFN treatment increases the steady state level of mRNA for HLA-A, -B, -C in every cell type studied (Fellous *et al.*, 1992). IFN-γ could induce *de novo* expression of class I MHC molecules in K562 human leukaemia cells which have no basal expression of these antigens (Rosa *et al.*, 1984). IFN thus appears to be both a modulator, as well as an inducer of MHC genes transcription.

TNF-α increases mRNA levels and surface expression of HLA-A and B molecules in normal (untransformed) human vascular endothelial cells and fibroblasts *in vitro*. This effect plateaus in several days and is sustained in the presence of TNF. This cytokine has also been shown to up-regulate constitutively expressed HLA genes in different tumor cell lines (Pfizenmaier *et al.*, 1987) and several neoplastic cell lines of carcinoma and leukaemia origin (Scheurich *et al.*, 1986). TNF-α by itself is unable to induce *de novo* expression of HLA-A, B, C genes but acts as an enhancer of constitutive or IFN-γ induced HLA gene expression (Pfizenmaier *et al.*, 1987).

1.2.7.2 Cytokines which modulate expression of MHC class II molecules

The constitutive expression of class II molecules is restricted to professional APCs, such as dendritic cells, B cells, macrophages, and thymic epithelium. However, class II MHC expression can be induced both *in vivo* and *in vitro* upon exposure to IFN-γ on a wide variety of cell types. The aberrant expression of class II MHC molecules has been described in auto-immune disorders such as multiple sclerosis and has been linked to the progression of auto-immune neurological disease development (Benichou *et al.*, 1991; Jonathan *et al.*, 2000).

The regulation of class II gene expression occurs primarily at the transcriptional level. A coordinated expression occurs through conserved *cis*-acting regions, termed W (Z, S, or H), X (X1 and X2), and Y elements, within the proximal promoter of most class II MHC genes (Bradley *et al.*, 1997). The optimal expression requires the cooperative binding of several constitutively expressed trans-acting factors to the W, X, and Y boxes of the class II promoter (Brown *et al.*, 1998). Although the presence of DNA binding proteins is necessary, it is not sufficient for class II transcription. The transcription of class II MHC genes (both constitutive and inducible) occurs only in the presence of the recently described class II transactivator (CIITA) (Chang *et al.*, 1995).

Many reports have found that placing CIITA in an array of cell types can result in not only the induction of class II MHC promoters but also the expression of cell surface class II MHC proteins (Chang *et al.*, 1995; Otten *et al.*, 1998). Further, expression of class II MHC is controlled quantitatively by CIITA (Otten *et al.*, 1998). In CIITA$^{-/-}$ mice, class II MHC is missing in almost all tissues and cells (Chang *et al.*, 1996). Most cytokines which alter class II MHC expression, such as IFN-γ, TNF-α, TGF-β, IL-1, IL-4, and IL-10, either up- or down-regulate CIITA and class II MHC accordingly (Chang *et al.*, 1995; Sims *et al.*, 1997, 1999). Immunosuppressive cytokines such as TGF-β and IL-10 generally have an inhibitory effect on IFN-γ induced class II MHC expression (Lee *et al.*, 1997).

The CIITA gene is an attractive target for the cytokine-mediated inhibition of class II MHC expression. In this regard, we and others have recently shown that TGF-β suppress IFN-γ induced class II MHC expression by inhibiting the expression of CIITA mRNA (Lee *et al.*, 1997; Sims *et al.*, 1997, 1999). The inhibitory effect of TGF-β on CIITA mRNA expression was mediated at the transcriptional level, suggesting that the CIITA promoter may be targeted by TGF-β (Lee *et al.*, 1997).

IL-1, a cytokine with predominantly pro-inflammatory properties, has been shown to inhibit IFN-γ induced class II MHC expression in many cells (Smith *et al.*, 1993). The molecular mechanisms underlying IL-1β-mediated inhibition of class II MHC expression are associated with IL-1β mediated inhibitory effect by suppressing IFN-γ induced CIITA mRNA expression. IL-1β inhibition of CIITA mRNA expression results from the ability of this cytokine to inhibit IFN-γ activation of the type IV CIITA promoter. Thus, IL-1 may contribute to the regulation of immunological events by reducing class II MHC gene expression. In addition to cytokine-regulated class II MHC expression, the in situ expression of CIITA is also tightly linked to class II MHC gene expression (Sims *et al.*, 1997, 1999). The evolutionary conservation of W-, X-, and Y-containing promoters in mammals, birds (Riegert *et al.*, 1996), amphibians (Kobari *et al.*, 1995), and fish (Sultmann *et al.*, 1994) suggests that CIITA may be extremely old and very important factor of immune reaction control.

1.3   AUTO-IMMUNITY - A BY PRODUCT OF ADAPTIVE IMMUNITY

In all vertebrates, the functional immune system is able to react against foreign antigens while remaining irresponsive to "self"-antigens. This "self"-tolerance is acquired and maintained by the combination of central and peripheral tolerances. In addition to clonal selection, immune specificity is regulated by the receptor selection in T and B lymphocytes at different stages of their differentiation (Parkin *et al.*, 2001). Through various specific processes such as positive and negative selections, a peripheral repertoire that is depleted from auto-reactive lymphocytes is generated (Nemazee *et al.*, 2000; Parkin *et al.*, 2001). To establish irresponsiveness to "self" during negative and positive selection, the selected thymocytes are expected to increase their activation threshold to "self" peptides, thereby peripheral T cells are properly activated by only foreign high affinity peptides. These cells can be regarded as high affinity T cells. However, "self"-reactive lymphocytes can be found in the blood and tissues of healthy individuals (Burns *et al.*, 1983; Arnold *et al.*, 1993). These findings imply that central and peripheral tolerances are somehow defective. The activation of these potential auto-reactive lymphocytes is expected to be controlled by various mechanisms of antigen-induced tolerance such as apoptosis, anergy, T regulatory cells and immune deviation/modulation (Arnold *et al.*, 1993; Maloy *et al.*, 2001). In auto-immune diseases, "self"-reactive T and B cells become aggressive and cause tissue injury, in some cases leading to severe injuries. How the mechanisms of tolerance mentioned are dysregulated in patients with auto-immune diseases is not well understood.

One of the biggest threats to survival is infection, so that the immune system is under strong evolutionary pressure to be highly responsive. The evolutionary emergence of the MHC has enabled a more effective defence from intracellular parasites, such as viruses. However, the whole complex of processing/presenting/recognizing of antigens could be closely related to the auto-immunity as a by-product of the evolution of MHC system and adaptive immunity. There is a large body of data that many auto-immune diseases are a characteristic of vertebrates and that they are associated with MHC molecules. In fact, there is no firm evidence that would suggest the existence of auto-immune phenomena in invertebrates (Rittig *et al.*, 1996; Ohta *et al.*, 2000). Naturally, the phenomenon of auto-immunity is well explored in mammals, especially in humans. For example, susceptibility to type 1 diabetes mellitus is particularly associated with HLA-DR3, 4 and associated DQ2, 8 alleles and this is well documented in genetic association studies. These molecules play an important role in the presentation of peptide antigens after intracellular processing to $CD4^+$ lymphocytes (Wong *et al.*, 2003). The presumably MHC molecules of aberrant target cells, TCR and APCs need to interact abnormally before auto-immune disease can fully develop. In this abnormal interaction, additional aberrancies in other regulatory systems may play a role in a further exacerbation of the "self"-directed immune response, such as defects in the hormone and cytokine synthesis and secretion. The various aberrancies are partly genetically determined by a variety of separate genes, particularly MHC and related genes like TAP/LMP, but they may also be environmentally induced by viruses, chemicals, drugs or injuries (Lam-Tse *et al.,* 2002).

In evolutionary new condition of strong (adaptive) immunity, the survival advantage imposed by an extremely reactive immune system is jeopardized if that system turns against the host and causes "self" destruction. Thus, evolutionary pressures selecting for a hyperactive immune system must be combined with similar pressures optimizing self-tolerance. Accordingly, the mechanisms which evolved in response to the auto-immunity-imposed evolutionary pressure or, more precisely, co-evolved with the phenomenon of auto-immunity, are related to various forms of immune tolerance, strong and multileveled control immunomodulatory and suppressive mechanisms like sex hormones, IL-10, TGF-β, Th2 cells, apoptosis and/or anergy of "self"-reactive clones, blood-barrier sequestration of "self" molecules, cells, tissues and organs (Cua *et al.,* 1995; Dalal *et al.,* 1997).

Surprisingly, auto-immunity is not a feature of a young immune system, when the immune network functions at its prime. Instead, the risk of developing auto-immune disease increases with age. In general, auto-immunity manifests in hosts who have passed the apex of their reproductive years and in whom evolutionary pressures towards prompt immune responsiveness are declining. The ageing of the immune system should be associated with the loss of function, and the likelihood of developing auto-immunity should progressively decrease. The traditional paradigm interprets auto-immunity as an aberrant response of the adaptive immune system to "self" molecule(s), consistent with the view that auto-immunity is a result of overreacting. It has been proposed that T lymphocytes specific for such "self" molecule(s) induce a memory response, which is relatively resistant to natural immuno-suppressive mechanisms. Tissue destruction has been understood as the after-effects of persistent immunocompetent cells. This model ignores that the risk for auto-immunity is inversely related to the functionality of the adaptive immune system throughout a lifetime. The new evolutionary concept of auto-immunity proposes that the accelerated immunity and failure of control mechanisms after reproductive time might be the primary risk factor for auto-immunity. From the evolutionary point of view, the immune system based on adaptive immunity has been made into a more complex and advanced defence system, developed under a strong selection pressure of microbes during the vertebrate evolution. Such model of vigorous immunity in vertebrates produced a new form of selection pressure, known nowadays as auto-immunity. Because the positive selection pressure of the adaptive immunity was probably stronger than the negative pressure of auto-immunity, the selection pressure of auto-immunity gave rise to the emergence of the control immunomodulatory and immunosuppressive mechanisms and to the "deferring" of the emergence of auto-immune diseases until post-reproductive age.

1.3.1   Sex Hormones and Auto-immunity

Recent data have provided evidence of a feed-back loop between reproductive hormones, mainly estrogens, and the expression, distribution and activity of cytokines. *In vitro* studies using mice cell cultures showed that while androgens decreased the production of IFN-γ, IL-4 and IL-5, estrogens enhanced IFN-γ production by murine lymphoid cells. Moreover, estrogens treatment of macrophages from male mice increased IL-1 secretion. In CD4$^+$ cell clones from auto-immune patients, both IL-10 and IFN-γ production were increased in the presence of estradiol (Cua *et al.*, 1995; Dalal *et al.*, 1997; Bebo Jr. *et al.*, 1998).

In general, females have a more responsive immune system than males. Females have a greater humoral response, as evidenced by higher serum Ig concentrations than males (Butterworth, 1967) and a greater antibody response to various antigens after immunization (London, 1977). In addition, females reject skin allograft faster and have a reduced incidence of tumors, indicating that they also have a greater cellular immune response (Hilgert, 1981; Enosawa, 1989). This difference in immune response is thought to be responsible for the greater susceptibility of females to the auto-immune diseases such as multiple sclerosis, rheumatoid arthritis, and systemic lupus erythematosus. This gender difference has also been observed in animal models of auto-immune disease in NZBxNZW mice (Cua *et al.*, 1995).

A protective effect of testosterone is thought to underlie why males are less susceptible to auto-immune disease than females. This is based on studies that include removing testosterone from male mice via castration as well as by treatment of female mice with testosterone. For example, the castration of male non-obese diabetic mice resulted in an increased prevalence of diabetes (Fitzpatrick *et al.*, 1995), and the castration of male mice increased the incidence of auto-immunity (Bebo Jr. *et al.*, 1998). Conversely, female non-obese diabetic mice implanted with testosterone pellets had a lower incidence of diabetes and less incidence of auto-immune disease, respectively, compared with those implanted with placebo pellets (Dalal *et al.*, 1997). The same studies have indicated that gender differences in susceptibility may be due to gender differences in cytokine production upon auto-antigen-specific stimulation. In males, compared with females, greater Th2 and less Th1 cytokine production has been observed. The balance between cytokines produced by Th1 and Th2 lymphocytes is considered central to the development of auto-immune disease. Th1 lymphocytes produce IFN-γ, IL-2, and TNF-α. Th2 lymphocytes secrete IL-

4, IL-5, IL-6, IL-10, and IL-13. These two cell types are mutually inhibitory, and their development occurs under very specific conditions. If a naive T lymphocyte is initially stimulated with antigen in the presence of IL-12, the immune response is skewed toward Th1. However, if a naive T lymphocyte is initially stimulated with antigen in the presence of IL-4, the immune response is skewed toward Th2 (Cua *et al.*, 1995; Dalal *et al.*, 1997). The same and other studies have collectively shown that immune cells under male sex hormones produce more IL-4 and IL-10, and less IFN-γ and IL-12, supporting the conclusion that the male immune system is shifted toward Th2 immunity (Dalal *et al.*, 1997; Bebo Jr. *et al.*, 1998). The mechanisms underlying why there is a sex hormones difference in cytokine production remain unknown. Many possibilities exist such as differences in the levels of male sex hormones, differences in female sex hormones, and differences in genes located on sex chromosomes.

Similar to previous studies, Stephanie *et al.* (2001) found that levels of the Th2 cytokines IL-4 and IL-10 were higher and the IL-12 level was lower in splenocytes from males compared with females. Also, splenocytes from female mice implanted with testosterone pellets, like splenocytes from male mice, secreted more IL-10 and less IL-12. However, the treatment with testosterone did not cause increased IL-4 production. This clearly indicates that testosterone does not recapitulate all the cytokine differences seen in male versus female mice, and that the increase in IL-4 must be due to gender differences in other sex hormones and/or genes found on sex chromosomes (Stephanie *et al.*, 2001). The finding of increased IL-10 production is equally as important as the finding of decreased IL-12 production upon testosterone treatment. Numerous studies have shown that IL-10 is essential in down regulation of cellular immune reaction. Specifically, the treatment of auto-immune patients with IL-10 has been shown to ameliorate disease (Rott *et al.*, 1994; Cannella *et al.*, 1996), whereas the administration of anti-IL-10 antibodies has exacerbated disease (Cannella *et al.*, 1996). Although the treatment of auto-immune patients with IL-4 also ameliorated disease (Shaw *et al.*, 1997), studies of IL-4- and IL-10-deficient mice and IL-4 as well as IL-10 transgenic mice have shown that IL-10 may play a more critical role in the protection from auto-immunity. Indeed, IL-10$^{-/-}$ mice developed more severe auto-immune disease compared with wild-type mice, and overexpression of IL-10 rendered mice resistant to auto-immunity (Bettelli *et al.*, 1998). Because IL-10 has been shown to play a protective role and IL-12 a disease-promoting role in auto-immunity, and because testosterone increases IL-10 and decreases IL-12, testosterone would appear to play an important role in susceptibility to auto-immunity and immune reaction control. Although many cells within spleen express the testosterone receptor (TR), testosterone probably can act directly upon CD4$^+$ T lymphocytes to increase IL-10 expression during stimulation with anti-CD3. The PCR analysis showed that CD4⁺ lymphocytes express the TR, supporting the possibility of direct action of testosterone on these cells. However, the TR is also expressed by CD8⁺ lymphocytes and macrophages. Thus, an indirect action of testosterone mediated through these cells was also possible. *In vitro* stimulation of CD4⁺ T lymphocytes in the presence of testosterone and in the absence of other cells resulted in increased IL-10 production (Stephanie *et al.*, 2001).

Estrogens modulate its effect by binding to estrogens receptors (EsR) present in the immune target cells. The EsR is a nuclear transcription factor that regulates gene expression. Some of the genes regulated by estrogens are progesterone receptor, *bcl-2* apoptosis inhibitor, *FasL* and other growth-related genes responsible for estrogen's effects on cell death and proliferation. Different authors have shown EsR in human peripheral blood mononuclear cells, thymocytes, spleen cells and APCs. The recent discovery of a second estrogens receptor, EsRb, presents new possibilities for control of immune targets by different selective estrogens receptor modulators (Cua *et al.*, 1995; Dalal *et al.*, 1997; Bebo Jr. *et al.*, 1998; Stephanie *et al.*, 2001).

A shift toward Th2 cytokine production has been demonstrated during pregnancy and high dose estrogens therapy and is thought to be the primary mechanism by which estrogens suppress the cellular immune response. However, a low dose estrogens treatment is equally suppressive in the absence of a significant shift in cytokine production. Estrogens treatment in cytokine-deficient and wild type mice up-regulate Th2 cytokine production. Also, estrogens effectively suppress the development of experimental auto-immunity in both, IL-4/IL-10 knockout mice and in auto-antigen-immunized wild type mice (Cua *et al.*, 1995; Dalal *et al.*, 1997; Bebo Jr. *et al.*, 1998; Stephanie *et al.*, 2001).

1.3.2   Evolutionary Pressure of Auto-immunity

Unlike invertebrates, the evolution of the vertebrate immune system involved several complex processes, the details of which have not been entirely cleared up. Nevertheless, the evolutionary new mechanisms of anti-microbe defence, such as adaptive immunity, probably caused the emergence of auto-immunity as a new form of evolutionary pressure.

Notwithstanding the possibility that auto-immunity might be the by-product of the evolution of adaptive immunity, this phenomenon could have been one of the factors significantly influencing the course of the evolution of vertebrate immune system and development of the mechanisms for the control of immune reaction. The evolutionary modelling of the vertebrate immune system under the influence of microbes and auto-immunity, did not probably result in the weakening of the killer mechanisms efficacy and MHC genes variability, but could have been followed by the introduction and co-evolution of evolutionary new mechanisms for the control of immune reaction, that could have restrained the potentially "self"-destructive power of the adaptive immunity.

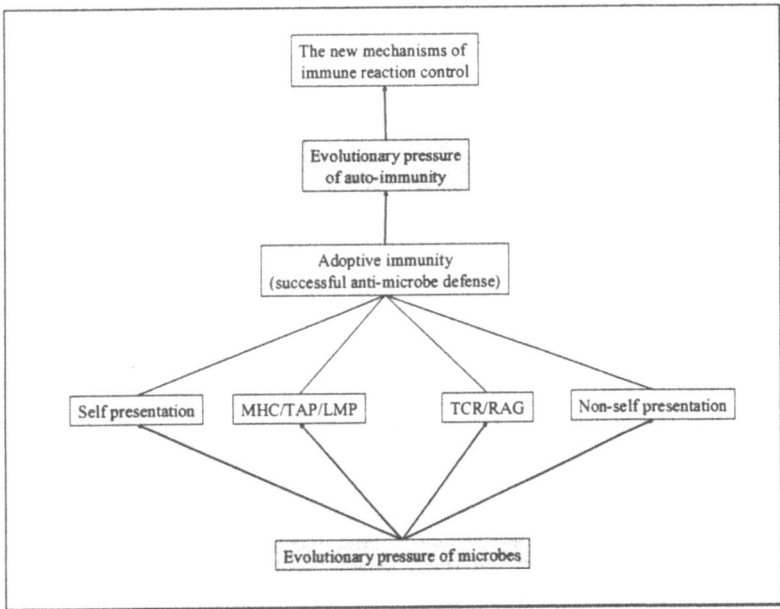

Figure 1.5. Evolution of auto-immunity as a by-product of adaptive immunity and evolutionary emergence of the new mechanisms of immune reaction control.

The evolution of the vertebrate immune system from cartilaginous fish to mammals is characterized by several processes featuring clearly perceivable evolutionary trends:

1. The grouping and clustering of MHC genes;
2. Associating of TAP/LMP genes with a less variable class of MHC genes (class II in mammals);
3. The emergence of auto-immunity like a form of selection pressure;
4. The sophisticated mechanisms of self-tolerance mechanisms;
5. The sophisticated mechanisms of immune reaction control mediated by cells of innate immune system;
6. The increasing of the number of the immune cell subtypes involved in the control of immune reaction (Th1, Th2, APCs, DCs);
7. The regulation of immunoreactivity mediated by sex hormones;

8. The increasing of the number of cytokines;
9. The regulation of immunoreactivity mediated by a complex cytokine network;

These phenomena could be associated with a better and more precise control of the immune reaction. However, it is very difficult to answer the question how big the contribution of the selection pressure of auto-immunity on the evolutionary development of the vertebrate immune system is. Judging from the pathogenesis of auto-immune diseases which are clearly associated with the basic features of the adoptive immunity like MHC, and also from the high incidence of auto-immune diseases in mammals, this phenomenon could be a significant factor of the selection pressure and evolutionary modelling, and permanent re-modelling, of the vertebrate immune system and their control mechanisms.

REFERENCES

Abbas K.A., Lichtman H.A., Pober S.J. (2003). The MHC. In Cellular and Molecular Immunology. W.B. Saunders Co., London, pp.: 96-115.

Accolla R.S., Adorini L., Sartoris S. *et al.* (1995). MHC: orchestrating the immune response. *Immunol Today.* 16:8-11.

Adema C.M., Hertel L.A., Miller R.D. *et al.* (1997). A family of fibrinogen-related proteins that precipitates parasite-derived molecules is produced by an invertebrate after infection. *Proc Natl Acad Sci USA.* 94:8691-8696.

Agrawal A., Eastman Q.M., Schatz D.G. (1998). Transposition mediated by RAG1 and RAG2 and its implications for the evolution of the immune system. *Nature.* 394:744-751.

Altmann S.M., Mellon M.T., Distel D.L. *et al.* (2003). Molecular and functional analysis of an IFN gene from the zebrafish, *(Danio rerio). J Virol.* 77:1992-2002.

Apanius V.D., Penn P., Slev L.R., *et al.* (1997). The nature of selection on the MHC. *Crit Rev Immunol.* 17:179-224.

Arnall J.C., Horton J.D. (1987). *In vivo* studies on allotolerance perimetamorphically induced in control and thymectomized Xenopus. *Immunology.* 62:315-319.

Arnold B., Schonrich G., Harmmerling G.J. (1993). Multiple levels of peripheral tolerance. *Immunol Today.* 14:12–14.

Arstila T.P., Casarouge A., Baron V. *et al.* (1999). A direct estimate of the human αβ TCR diversity. *Science.* 286:958–961.

Bebo B.F.Jr., Zelinka-Vincent E., Adamus G., *et al.* (1998). Gonadal hormones influence the immune response to PLP 139–151 and the clinical course of relapsing experimental auto-immune encephalomyelitis. *J Neuroimmunol.* 84:122-130.

Benian G.M., Kiff J.E., Neckelmann N. *et al.* (1989). Sequence of an unusually large protein implicated in regulation of myosin activity in *C elegans. Nature.* 342:45-50.

Benichou B., Strominger J.L. (1991). Class II-antigen-negative patient and mutant B-cell lines represent at least three, and probably four, distinct genetic defects defined by complementation analysis. *Proc Natl Acad Sci USA.* 88:4285-4288.

Benoist C., Mathis D. (1998). The pathogen connection. *Nature.* 394:227-228.

Bernstein R.M., Schluter S.F., Bernstein H. *et al.* (1996). Primordial emergence of the RAG1: sequence of the complete shark gene indicates homology to microbial integrases. *Proc Natl Acad Sci USA.* 93:9454-9459.

Bertoletti A., Sette A., Chisari A.V. *et al.* (1994). Natural variants of cytotoxic epitopes are T-cell receptor antagonists for antiviral cytotoxic T-cells. *Nature.* 369:407–410.

Beschin A., Bilej M., Brys, L. *et al.* (1999). Convergent evolution of cytokines. *Nature.* 400:627– 628.

Bettelli E., Das E.P., Howard E.D, *et al.* (1998). IL-10 is critical in the regulation of auto-immune encephalomyelitis as demonstrated by studies of IL-10- and IL-4-deficient and transgenic mice. *J Immunol.* 161:3299-330.

Bird S., Wang T., Zou J. *et al.* (2002a). The first cytokine sequence within cartilaginous fish: IL-1 β in the small spotted catshark *(Scyliorhinus canicula).* *J Immunol.* 68:3329-3340.

Bird S., Zou J., Wang T. *et al.* (2002b) Evolution of IL-1β. *Cytokine Growth Factor.* 13:483-502.

Bjorkman P.J., Saper M.A., Samraoui, B. *et al.* (1987). Structure of the human class I histocompatibility antigen, HLA-A2. *Nature.* 329:506-512.

Bodmer W.F. (1972). Evolutionary significance of the HL-A system. *Nature.* 237:139-145.

Boyson J.E., Shufflebotham C., Cadavid L.F. *et al.* (1996). The MHC class I genes of the rhesus monkey. Different evolutionary histories of MHC class I and II genes in primates. *J Immunol.* 156:4656-4660.

Bradley M.B., Fernandez J.M., Ungers G. *et al.* (1997). Correction of defective expression in MHC class II deficiency (bare lymphocyte syndrome) cells by retroviral transduction of CIITA. *J Immunol.* 159:1086-1095.

Bretscher A.P. (1999). A two-step, two-signal model for the primary activation of precursor helper T cells. *Proc Natl Acad Sci USA.* 96:185–190.

Brown J.A., Rogers E.M., Boss J.M. (1998). The MHC class II transactivator (CIITA) requires conserved leucine charged domains for interactions with the conserved W box promoter element. *Nucleic Acids Res.* 26:4128-4136.

Brown J.H., Jardetzky T.S., Gorga J.C., *et al.* (1993). Three-dimensional structure of the human class II histocompatibility antigen HLA-DR1. *Nature.* 364:33-39.

Bubanovic I., Najman S. (2004). Anti-tumor Immunity Failure in Mammals - Evolution of the Hypothesis. *Acta Biotheor.* 52:57-64.

Burke P.A., Ozato K. (1989). Regulation of major histocompatibility complex class I genes. *Year Immunol.* 4:23-40.

Burns J., Rosenzweig A., Zweiman B. *et al.* (1983). Isolation of myelin basic protein-reactive T cell lines from normal human blood. *Cell Immunol.* 81:435-440.

Buss L.W., Geen D.R. (1985). Histoincompatibility in vertebrates: the relict hypothesis. *Dev Comp Immunol.* 9:191-201.

Butterworth M., McClellan B., Allansmith M. (1967). Influence of sex in immunoglobulin levels. *Nature.* 214:1224-1226.

Cadavid C., Shufflebotham F.J., Ruiz M. *et al.* (1997) Evolutionary instability of the MHC loci in New World primates. *Proc Natl Acad Sci USA.* 94:14536-14541.

Cannella B., Gao Y.L., Brosnan C. *et al.* (1996). IL-10 fails to abrogate experimental auto-immune encephalomyelitis. *J Neurosci Res.* 45:735-741.

Carroll R.L. (1988). Vertebrate Paleontology and Evolution. (Freeman, New York), pp.: 62-83.

Carter R., Schofield L., Mendis K. (1992). HLA effectsin malaria: increased parasite-killing immunity or reduced immunopathology. *Parasit Today.* 8:41-42.

Castellino F. (1997). Antigen presentation by MHC class II molecules: invariant chain function, protein trafficking, and the molecular basis of diverse determinant capture. *Hum Immunol.* 54:159-169.

Chambers C.A., Allison J.P. (1997). Co-stimulation in T cell responses. *Cur Op Immunol.* 9:396-404.

Chang C.H., Furue M., Tamaki K. (1994). Selective regulation of ICAM-1 and MHC class I and II molecule expression on epidermal Langerhans cells by some of the cytokines released by keratinocytes and T cells. *Eur J Immunol.* 24:2889-2895.

Chang C.H., Flavell R.A. (1995). Class II transactivator regulates the expression of multiple genes involved in antigen presentation. *J Exp Med.* 181:765-767.

Chang C.H., Guerder S., Hong S.C. *et al.* (1996). Mice lacking the MHC class II transactivator (CIITA) show tissue-specific impairment of MHC class II expression. *Immunity.* 4:167-178.

Charlemagne J. (1987). Antibody diversity in amphibians. Non-inbred axolotls used the same unique heavy chain and a limited number of light chains for their anti-2,4-dinitrophenyl antibody responses. *Eur J Immunol.* 17:421-424.

Cho S.G., Attaya M., Monaco J.J. (1991). New class II-like genes in the murine MHC. *Nature.* 353:573-576.

Clarke B., Kirby R.S. (1966). Maintenance of histocompatibility polymorphisms. *Nature.* 211:999-1000.

Croft M., Dubey C. (1997). Accessory molecule and costimulation requirements for CD4 T cell response. *Crit Rev Immunol.* 17:89-118.

Cua D.J., Hinton D.R, Stohlman S.A. (1995). Self-antigen-induced Th2 responses in experimental allergic encephalomyelitis (EAE)-resistant mice: Th2-mediated suppression of auto-immune disease. *J Immunol.* 155:4052-4057.

Dalal M., Kim S., Voskuhl R.R. (1997). Testosterone therapy ameliorates experimental auto-immune encephalomyelitis and induces a T helper 2 bias in the autoantigen-specific T lymphocyte response. *J Immunol.* 159:3-7.

De Boer R.J. (1995). The evolution of polymorphic compatibility molecules. *Mol Biol Evol.* 12:494-502.

De Campos P.O., Gavioli R., Zhang Q.J. *et al.* (1993). HLA-A11 epitope loss isolates of Epstein-Barr virus from a highly A111 population. *Science.* 260:98-100.

Deverson E.V., Gow I.R., Coadwell W.J. *et al.* (1990). MHC class II region encoding proteins related to the multidrug resistance family of transmembrane transporters. *Nature.* 348:738-741.

Doherty P.C., Zinkernagel R.M. (1975). Enhanced immunological surveillance in mice heterozygous at the H-2 gene complex. *Nature.* 256:50-52.

Du Pasquier L., Schwager J., Flajnik M.F. (1989). The immune system of *Xenopus. An Rev Immunol.* 7:251-275.

Du Pasquier L., Flajnik M. (1990). Expression of MHC class II antigens during Xenopus development. *Dev Immunol.* 1:85-95.

Du Pasquier L., Robert J. (1992). *In vitro* growth of thymic tumor Cell lines from Xenopus. *Dev Immunol.* 2:295-307.

Du Pasquer L., Flajnik M. (1999). Origin and evolution of the vertebrate immune system. In Fundamental Immunology, 4th (ed W.E. Paul) Lippincot-Raven Publishers. Philadelphia. pp. 605-650.

Ebert D., Hamilton W.D. (1996). Sex against virulence: the coevolution of parasitic diseases. *Trend Ecol & Evol.* 11:79-82.

el Ridi R., Wahby A.F., Saad A.H. *et al.* (1987). Concanavalin A responsiveness and IL-2 production in the snake Spalerosophis diadema. *Immunobiol.* 174:177-189.

Enosawa S., Hirasawa K. (1989). Sex-associated differences in the survival of skin grafts in rats: enhancement of cyclosporine immunosuppression in male compared with female recipients. *Transplantation.* 47:933-936

Fahmy G.H., Sicard R.E. (2002). A role for effectors of cellular immunity in epimorphic regeneration of amphibian limbs. *In vivo.* 16:179-184.

Fellous M., Nir U., Wallach D. *et al.* (1992). IFN-dependent induction of mRNA for the MHC antigens in human fibroblasts and lymphoblasts and lymphoblastoid cells. *Proc Natl Acad Sci USA.* 79:3082-3086.

Ferguson S.E., Accavitti M.A., Wang D.D. *et al.* (1994). Thompson CB. Regulation of RAG-2 protein expression in avian thymocytes. *Mol Cell Biol.* 14:7298-7305.

Fitzpatrick F., Lepault F., Homo-Delarche F. *et al.* (1991). Influence of castration, alone or combined with thymectomy, on the development of diabetes in the nonobese diabetic mouse. *Endocrinol.* 129:1382-1390.

Flajnik M.F., Kaufman J.F., Hsu E. *et al.* (1986). MHC-encoded class I molecules are absent in immunologically competent *Xenopus* before metamorphosis. *J Immunol.* 137:3891-3899.

Flajnik M.F., Canel C., Kramer J. *et al.* (1991). Which came first, MHC class I or class II? *Immunogen.* 33:295-230.

Fong T.A., Mosmaun T.R. (1990). Alloreactive murine CD8⁺ T cell clones secrete the Th1 pattern of cytokines. *J Immunol.* 144:1744-1752.

Forsdyke D.R. (1991). Early Evolution of MHC Polymorphism. *J Theor Biol.* 150:451-456.

Franc N.C., Dimarcq J.L., Lagueux *et al.* (1996). Croquemort, a novel *Drosophila* hemocyte/macrophage receptor that recognizes apoptotic cell. *Cell Immunol.* 4:431-443.

Franchini A., Ottaviani E., Franceschi C. (1995). Presence of immunoreactive pro-opiomelanocortin-derived peptides and cytokines in the thymus of an anuran amphibian *(Rana esculenta)*. *Tissue Cell.* 27:263-267.

Fraser D.G., Bailey E., Swinburne J. *et al.* (1998). Two MHC class II DQA for the horse. *Animal Genetics.* 29(S1):29-34.

Fujiki K., Nakao M., Dixon B. (2003). Molecular cloning and characterisation of a carp *(Cyprinus carpio)* cytokine-like cDNA that shares sequence similarity with IL-6 subfamily cytokines CNTF, OSM and LIF. *Dev Comp Immunol.* 27:127-136.

Garcia-Castillo J., Pelegrin P., Mulero V. *et al.* (2002). Molecular cloning and expression analysis of TNF-α from a marine fish reveal its constitutive expression and ubiquitous nature. *Immunogen.* 54:200-207.

Garrido F., Cabrera T., Lopez-Nevot M.A. *et al.* (1995). HLA class I antigens in human tumors. *Adv Cancer Res.* 67:155-195.

Gemmill A.W., Read F.A. (1998). Counting the cost of disease resistance. *Trend Ecol & Evol.* 13:8-9.

German R.N., Castellino F., Han R. *et al..* (1996). Processing and presentation of endocytically acquired protein antigens by MHC class II and class I molecules. *Immunol Rev.* 151: 5-30.

Glimcher L.H., Kara C.J. (1992). Sequences and factors: a guide to MHC class II transcription. *An Rev Immunol.* 10:13-49.

Greenhalgh P., Olesen C.E., Steiner L.A. (1993). Characterization and expression of recombination activating genes (RAG-1 and RAG-2) in Xenopus laevis. *J Immunol.* 151:3100-3110.

Greenhalgh P., Steiner L.A. (1995). Recombination activating gene 1 (Rag1) in zebrafish and shark. *Immunogen.* 41:54-55.

Groth J.G., Barrowclough G.F. (1999). Basal divergences in birds and the phylogenetic utility of the nuclear RAG-1 gene. *Mol Phylogenet Evol.* 12:115-123.

Guardiola J., Maffei A. (1993). Control of MHC class II gene expression in auto-immune, infectious, and neoplastic diseases. *Crit Rev Immunol.* 13:247-268.

Gumperz J.E., Parham P. (1995). The enigma of the natural killer cell. *Nature.* 378:245-248.

Gyllensten, U.B., Erlich, H.A. (1989). Ancient roots for polymorphism at the HLA-DQ α locus in primates. *Proc Natl Acad Sci USA.* 86:9986.

Hall R. (1994). Molecular mimicry. *Advan Parasit.* 34:81-133.

Hamilton W.D., Axelrod R., Tanese R. (1990). Sexual reproduction as an adaptation to resist parasites (a review). *Proc Natl Acad Sci USA.* 87:3566-3573.

Hansen J.D., Kaattari S.L. (1996). The recombination activating gene 2 (RAG2) of the rainbow trout Oncorhynchus mykiss. *Immunogen.* 44:204-211.

Hilgert I., Pokorná Z., Singh K. *et al.* (1981). Different efficiency of mercurascan in allograft survival prolongation in male and female mice. *Folia Biol.* 27:379-383.

Hill A.V.S., Allsopp D., Dwiatkowski N.M. *et al.* (1991). Common West African HLA antigens are associated with protection from severe malaria. *Nature.* 352:595-600.

Hill A.V.S., Elvin J. Willis A.C. *et al.* (1992). Molecular analysis of the association of HLA-B53 and resistance to severe malaria. *Nature.* 360:434-439.

Hilton L.S., Bean A.G., Kimpton W.G. *et al.* (2002). IL-2 directly induces activation and proliferation of chicken T cells *in vivo*. *J Interferon Cytokine Res.* 22:755-763.

Howcroft T.K., Raval A., Weissman J.D. *et al.* (2003). Distinct transcriptional pathways regulate basal and activated MHC class I expression. *Mol Cell Biol.* 23:3377-3391.

Hughes A.L., Nei M. (1988). Pattern of nucleotide substitution at MHC class I loci reveals overdominant selection. *Nature.* 335:167-168.

Hughes A.L., Nei M. (1989). Evolution of the MHC: independent origin of nonclassical class I genes in different groups of mammals. *Mol Biol Evol.* 6:559-579.

Hughes A.L. (1992) Maintenance of MHC polymorphism. *Nature.* 335:402-403.

Hughes A.L., Nei M. (1993). Evolutionary relationships of the classes of MHC genes. *Immunogen.* 37:337-341.

Hughes A.L., Hughes M.K. (1995). Natural selection on the peptide-binding regions of MHC complexmolecules. *Immunogen.* 42:233-243.

Humphreys T., Reinherz E.L. (1994). Invertebrate immune recognition, natural immunity and the evolution of positive selection. *Immunol Today.* 15:316-320.

Inaba K., Kitaura M., Kato T. *et al.* (1986). Contrasting effects of α/β and γ-IFNs on expression of macrophage Ia antigens. *J Exp Med.* 163:1030-1035.

Inoue Y., Haruta C., Usui K. *et al.* (2003). Molecular cloning and sequencing of the banded dogfish *(Triakis scyllia)* IL-8 cDNA. *Fish Shellfish Immunol.* 14:275-281.

Jaffe L., Robertson E.J., Bikoff E.K. (1991). Distinct patterns of expression of MHC class I and β 2-microglobulin transcripts at early stages of mouse development. *J Immunol.* 147:2740-2749.

Jensen I., Larsen R., Robertsen B. (2002). An antiviral state induced in Chinook salmon embryo cells by transfection with the double-stranded RNA poly I:C. *Fish Shellfish Immunol.* 13:367-378.

Jonathan A.H., Ting J.P.Y. (2000). Class II Transactivator: Mastering the Art of MHC Expression. *Mol Cell Biol.* 20:6185-6194.

Karr R.W., Gregersen P.K., Obata F. *et al.* (1986). Analysis of DR β and DQ β chain cDNA clones from a DR7 haplotype. *J Immunol.* 137:2886-2890.

Kasahara M., Vazquez M., Sato K. *et al.* (1992). Evolution of the MHC: isolation of class II A cDNA clones from the cartilaginous fish. *Proc Natl Acad Sci USA.* 89:6688-6695.

Kasahara M., McKinney E.C., Flajnik M.F. *et al.* (1993). The evolutionary origin of the MHC: polymorphism of class II α chain genes in the cartilaginous fish. *Eur J Immunol.* 23:2160-2165.

Kasahara M., Hayashi M., Tanaka K. *et al.* (1996a). Chromosomal localization of the proteasome Z subunit gene reveals an ancient chromosomal duplication involving the MHC. *Proc Natl Acad Sci USA.* 93:9096-9100.

Kasahara M., Kandil E., Salter-Cid L. *et al.* (1996b). Origin and evolution of the class I gene family: why are some of the mammalian class I genes encoded outside the MHC? *Res Immunol.* 147:278-285.

Kasahara M., Nakaya J., Satta Y. *et al.* (1997). Chromosomal duplication and the emergence of the adaptive immune system. *Tren Gen.* 13:90-95.

Kaufman J., Skjoedt K., Salomonsen J. (1990) The MHC molecules of nonmammalian vertebrates. *Immunol Rev.* 113:83-117.

Kaufman J., Anderson R., Avila D., *et al.* (1992). Different features of the MHC class I heterodimer have evolved at different rates. *J Immunol.* 148:1532-1546.

Kaufman J., Salomonsen J., Flajnik M. (1994). Evolutionary conservation of MHC class I and class II molecules different yet the same. *Semin Immunol.* 6:411-424.

Kaufman J., Volk H., Wallny H.J. (1995) A minimal essential Mhc and an nonrecognized Mhc-two extremes in selection for polymorphism. *Immunol Rev.* 143:63-68.

Kelly A.P., Monaco J.J., Cho S.G. *et al.* (1991). A new human HLA class II-related locus, DM. *Nature.* 353:571-573.

Klein J., O'huigin C. (1993). Composite origin of MHC genes. *Cur Opin in Gen & Develop.* 3:923-930.

Klein J., O'huigin C. (1994). The conundrum of nonclassical MHC genes. *Proc Natl Acad Sci USA.* 91:6251-6252.

Klenerman P., Rowland-Jones S., McAdam J. *et al.* (1994). Cytotoxic T-cell activity antagonized by naturally occurring HIV-1 gag variants. *Nature.* 369:403-407.

Kropshofer H., Arndt S.O., Moldenhauer G. *et al.* (1997). HLA-DM acts as a molecular chaperone and rescues empty HLA-DR molecules at lysosomal pH. *Immunity.* 6:293-302.

Kobari F., Sato K., Shum B.P. *et al.* (1995). Exon-intron organization of *Xenopus* MHC class II β chain genes. *Immunogen.* 42:376-385.

Kuroda N., Uinukool T.S., Sato A. *et al.* (2003). Identification of chemokines and a chemokine receptor in cichlid fish, shark, and lamprey. *Immunogen.* 54:884-895.

Kurth J.H., Mountain J.L., Cavalli-Sforza L.L. (1993). Subclustering of human immunoglobulin kappa light chain variable region genes. *Genomics.* 16:69-77.

Lam-Tse W.K., Lernmark A., Drexhage H.A. (2002). Animal models of endocrine/organ-specific auto-immune diseases: do they really help us to understand human auto-immunity? *Springer Semin Immunopathol.* 24:297-321.

Lanier L.L., Follow A. (1998) NK cell receptors for classical and nonclassical MHC class I. *Cell.* 92:705-707.

Lawlor D.A., Ward F.E., Ennis P.D. *et al.* (1988) HLA-A and B polymorphisms predate the divergence of humans and chimpanzees. *Nature.* 335:268-270.

Lawlor D.A., Zemmour J., Ennis P.D. *et al.* (1990). Evolution of class I MHC genes and proteins: from natural selection to thymic selection. *An Rev Immunol.* 8:23-30.

Le Bouteiller P. (1994) HLA class I chromosomal region, genes, and products: facts and questions. *Crit Rev Immunol.* 14: 89-129.

Lee Y.J., Han Y., Lu H.T. *et al.* (1997). TGF-β suppresses IFN-γ induction of class II MHC gene expression by inhibiting class II transactivator messenger RNA expression. *J Immunol.* 158:2065-2070.

Li S., Sjogren H.O., Hellman U. (1997). Cloning and functional characterization of a subunit of the transporter associated with antigen processing. *Proc Natl Acad Sci USA.* 94:8708-8713.

Liu Y., de Waal Malefyt R., Briere F., *et al.* (1997). The EBV IL-10 homologue is a selective agonist with impaired binding to the IL-10 receptor. *J Immunol.* 158:604-613.

Lobigs M., Mullbacher A., Regner M. (2003). MHC class I up-regulation by flaviviruses: Immune interaction with unknown advantage to host or pathogen. *Immunol Cell Biol.* 81:217-223.

Loflin P.T., Laud P.R., Watkins D.I. *et al.* (1996). Identification of new TAP2 alleles in gorilla: evolution of the locus within hominoids. *Immunogen.* 44:161-169.

Loker E.S. (1994). On being a parasite in an invertebrate host: a short survival course. *J Parasitol.* 80:728-747.

London W.T., Drew J.S. (1977). Sex differences in response to hepatitis B infection among patients receiving chronic dialysis treatment. *Proc Natl Acad Sci USA.* 74:2561-2564.

Ljunggren H.G., Karre K. (1990). In search of the 'missing self': MHC molecules and NK cell recognition. *Immunol Today.* 11:237-244.

Maloy K.J., Powrie F. (2001). Regulatory T cells in the control of immune pathology. *Nat Immunol.* 2:816–822.

Margulies D.H. (1999). Fundamental Immunology, (ed. Paul, W. E.) Lippincott, Philadelphia, pp.: 263-285.

Martinez-Laso J., Martin-Villa J.M., Alvarez M. *et al.* (1994). Susceptibility to insulin-dependent diabetes mellitus and short cytoplasmic ATP-binding domain TAP2*01 alleles. *Tis Antigens.* 44:184-188.

Matsunaga T., Rahman A. (1998). What brought the adaptive immune system to vertebrates? *Immunol Rev.* 166:177-182.

Min W., Lillehoj H.S., Li G. *et al.* (2002). Development and characterization of monoclonal antibodies to chicken IL-15. *Vet Immunol Immunopathol.* 88:49-56.

Miracle A.L., Anderson M.K., Litman R.T. *et al.* (2001). Complex expression patterns of lymphocyte-specific genes during the development of cartilaginous fish implicate unique lymphoid tissues in generating an immune repertoire. *Int Immunol.* 13:567-580.

Moins-Teisserenc H., Semana G., Alizadeh M. *et al.* (1995). TAP2 gene polymorphism contributes to genetic susceptibility to multiple sclerosis. *Hum Immunol.* 42:195-202.

Moriuchi J., Moriuchi T., Silver J. (1985). Nucleotide sequence of an HLA-DQ α chain derived from a DRw9 cell line: genetic and evolutionary implications. *Proc Natl Acad Sci USA.* 82:3420-3425.

Nei M., Gu X., Sitnikova T. (1997). Evolution by the birth-and-death process in multigene families of the vertebrate immune system. *Proc Natl Acad Sci USA.* 94:7799-7804.

Nelson R.J., Demas G.E. (1996). Seasonal changes in immune function. *Q Rev Biol.* 71:511-548.

Nemazee D. (2000). Receptor selection in B and T lymphocytes. *An Rev Immunol.* 18:19–51.

Ohta Y., Okamura K., McKinney C. *et al.* (2000). Primitive synteny of vertebrate MHC class I and class II genes. *Proc Natl Acad Sci USA.* 97:4712-4717.

Okamura K., Ototake M., Nakanishi T. *et al.* (1997). The most primitive vertebrates with jaws possess highly polymorphic MHC class I genes comparable to those of humans. *Immunity.* 7:777-782.

Ottaviani E., Franceschi C. (1997). The invertebrate phagocytic immunocyte: clues to a common evolution of immune and neuroendocrine systems. *Immunol Today.* 18:169-174.

Otten L.A., Steimle V., Bontron S. *et al.* (1998). Quantitative control of MHC class II expression by the transactivator CIITA. *Eur J Immunol.* 28:473-478.

Pace J.L., Russell S.W., LeBlanc P.A. *et al.* (1985). Comparative effects of various classes of mouse IFNs on macrophage activation for tumor cell killing. *J Immunol.* 134:977-981.

Parham P., Adams E.J., Arnett K.L. (1995). The origins of HLA-A,B,C polymorphism. *Immunol Rev.* 143:141-147.

Parham P., Ohta T. (1996). Population biology of antigen presentation by MHC class I molecules. *Science.* 272:67-74.

Parham P., Arnett K.L., Adams E.J. *et al.* (1997). Episodic evolution and turnover of HLA-B in the indigenous human populations of the Americas. *Tis Antigens.* 50:219-225.

Parkin J., Cohen B. (2001). An overview of the immune system. *Lancet.* 357:1777–1789.

Paterson S., Wilson K., Pemberton J.M. (1998). MHC variation associated with juvenile survival and parasite resistance in a large unmanaged ungulate population *(Ovis aries L).* *Proc Natl Acad Sci USA.* 95:3714–3719.

Paulesu L. (1997). Cytokines in mammalian reproduction and speculation about their possible involvement in nonmammalian viviparity. *Microsc Res Tech.* 38:188-194.

Peixoto B.R., Mikawa Y., Brenner S. (2000). Characterization of the recombinase activating gene-1 and 2 locus in the Japanese pufferfish, *Fugu rubripes.* *Gene.* 246:275-283.

Penn D., Potts W. (1998a). How do MHC genes influence odor and mating preferences? *Advan Immunol.* 69:411-435.

Penn D., Potts W. (1998b). Chemical signals and parasite-mediated sexual selection. *Trend in Ecol & Evol.* 13:391-396.

Petrie-Hanson L., Ainsworth A.J. (2000). Differential cytochemical staining characteristics of channel catfish leukocytes identify cell populations in lymphoid organs. *Vet Immunol Immunopathol.* 25;73:129-144.

Pfizenmaier K., Scheurich P., Schluter C. *et al.* (1987). Tumor TNF enhances HLA-A, B, C and HLA-DR gene expression in human tumor cells. *J Immunol.* 138:975-980.

Pircher H., Moskophidis D., Rohrer U. *et al.* (1990). Viral escape by selection of cytotoxic T-cell-resistant virus variants *in vivo.* *Nature.* 346:629-632.

Plachy J., Pink J.R., Hala K. (1992). Biology of the chicken MHC (B complex). *Crit Rev Immunol.* 12:47-49.

Pogoda H.M., Meyer D. (2002). *Zebrafish* Smad7 is regulated by Smad3 and BMP signals. *Dev Dyn.* 224:334-349.

Potts W.K., Slev P.R. (1995). Pathogen-based models favoring MHC genetic diversity. *Immunol Rev.* 143:181-197.

Powis S.J., Young L.L., Joly E. *et al.* (1996). The cim effect: TAP allele dependent changes in a class I MHC anchor motif and evidence against C-terminal trimming of peptides in the ER. *Immunity.* 4:159-165.

Price D.A., Goulder P.J.R. Klenerman P. *et al.* (1997). Positive selection of HIV-1 cytotoxic T lymphocyte escape variants during primary infection. *Proc Natl Acad Sci USA.* 94:1890-1895.

Raftos D.A., Tait N.N., Briscoe D.A. (1987). Allograft rejection and alloimmune memory in the solitary urochordate, Styela plicata. *Dev Comp Immunol.* 11:343-351.

Rast J.P., Litman G.W. (1994). T-cell receptor gene homologs are present in the most primitive jawed vertebrates. *Proc Natl Acad Sci USA.* 91:9248-9252.

Reboul J., Gardiner K., Monneron D. *et al.* (1999). Comparative genomic analysis of the IFN/IL-10 receptor gene cluster. *Genome Res.* 9:242-250.

Riegert P., Andersen R., Bumstead N., *et al.* (1996). The chicken β 2-microglobulin gene is located on a non-MHC microchromosome: a small, G⁺C-rich gene with X and Y boxes in the promoter. *Proc Natl Acad Sci USA.* 93:1243-1248.

Rittig M.G., Kuhn K.H., Dechant C.A. *et al.* (1996). Phagocytes from both vertebrate and invertebrate species use pooiling-phagocytosis. *Dev Comp Immunol.*20:393-306.

Roch P., Valembois P., Du Pasquier L. (1975). Response of earthworm leukocytes to concanavalin A and transplantation antigens. *Adv Exp Med Biol.* 64:45-55

Rollins-Smith L., Blair P. (1990). The expression of class II MHC antigens on adult T Cells in *Xenopus* is metamorphosis dependent. *Dev Immunol.* 1:97-104.

Rosa F., Fellous M. (1984), The effect of γ-IFN on MHC antigens. *Immunol Today.* 9:261-262.

Rott O., Fleischer B., Cash E. (1994). IL-10 prevents experimental allergic encephalomyelitis in rats. *Eur J Immunol.* 24:1434-1438.

Ruben L.N., Rak J.C., Johnson R.O. *et al.* (1994). A comparison of the effects of human rIL-2 and autologous TCGF on *Xenopus laevis* splenocytes *in vitro*. *Cell Immunol.* 157:300-305.

Saad A.H., El Ridi R. (1984). Mixed leukocyte reaction, graft-versus-host reaction, and skin allograft rejection in the lizard, Chalcides ocellatus. *Immunobiol.* 166:484-493.

Sammut B., Laurens V., Tournefier A. (1997). Isolation of Mhc class I cDNAs from the axolotl Ambystoma mexicanum. *Immunogen.* 45:285-294.

Scheurich P., Kronke M., Schluter C. *et al.* (1986). Noncytocidal mechanisms of action of tumor TNF-α on human tumor cells: Enhancement of HLA gene expression synergistic with IFN-γ. *Immunobiol.* 172:291-300.

Schluter S.F., Marchalonis J.J. (2003). Cloning of shark RAG2 and characterization of the RAG1/RAG2 gene locus. *FASEB J.* 17:470-472.

Schneider K., Klaas R., Kaspers B. *et al.* (2001). Chicken IL-6. cDNA structure and biological properties. *Eur J Biochem.* 268:4200-4206.

Setterblad N., Roucard C., Bocaccio C., *et al.* (2003). Composition of MHC class II-enriched lipid microdomains is modified during maturation of primary dendritic cells. *J Leukoc Biol.* 74:40-48.

Shaw M.K., Lorens J.B., Dhawan A. *et al.* (1997). Local delivery of IL-4 by retrovirus-transduced T lymphocytes ameliorates experimental auto-immune encephalomyelitis. *J Exp Med.* 185:1711-1718.

Sidney J., Grey H.M., Kubo R.T. *et al.* (1996). Practical, biochemical and evolutionary implications of the discovery of HLA class I supermotifs. *Immunol Tod.*17:261-266.

Sijben J.W., Klasing K.C., Schrama J.W. *et al.* (2003). Early *in vivo* cytokine genes expression in chickens after challenge with *Salmonella typhimurium* lipopolysaccharide and modulation by dietary n-3 polyunsaturated fatty acids. *Dev Comp Immunol.* 27:611-619.

Sims T.N., Goes N.B, Ramassar V., *et al.* (1997). *In vivo* class II transactivator expression in mice is induced by a non-IFN-γ mechanism in response to local injury. *Transplantation.* 64:1657-1664.

Sims T.N., Halloran P.F. (1999). MHC class II regulation *in vivo* in the mouse kidney. *Microbes Infect.* 1:903-912.

Slade R.W., McCallum H. (1992). Overdominant vs. frequency-dependent selection at MHC loci. *Genetics.* 132:861-862.

Smith M.E., McFarlin D.E., Dhib-Jalbut S. (1993). Differential effect of IL-1β on Ia expression in astrocytes and microglia. *J Neuroimmunol.* 46:97-102.

Steinmetz M., Hood L. (1983). Genes of the MHC in mouse and man. *Science.* 222:727-733.

Stephanie M., Liva R., Rhonda R.V. (2001). Testosterone Acts Directly on CD4$^+$ T Lymphocytes to Increase IL-10 Production. *J Immunol.* 167:2060-2067.

Stern C.D. (1992) Mesoderm induction and development of the embryonic axis in amniotes. *Trends Genet.* 8:158-163.

Sultmann H., Mayer W.E., Figueroa F. *et al.* (1993). Zebrafish MHC class II α chain-encoding genes: polymorphism, expression, and function. *Immunogen.* 38:408-420.

Sultmann H., Mayer W.E., Figueroa F. *et al.* (1994). Organization of MHC class II B genes in the zebrafish *(Brachydanio rerio). Genomics.* 23:1-14.

Thompson C. (1995). New insights into V(D)J recombination and its role in the evolution of the immune system. *Immunity.* 3:531-539.

Thrusz M.R., Thomas H.C., Greenwood B.M. *et al.* (1997). Heterozygote advantage for HLA class II type in hepatitis B virus infection. *Nat Gen.* 17:11-12.

Trowsdale J., Hanson I., Mockridge I. *et al.* (1990). Sequences encoded in the class II region of the MHC related to the ABC-superfamily of transporters. *Nature.* 348:741-744.

Ulsh B.A., Congdon J.D., Hinton T.G. *et al.* (2000). Culture methods for turtle lymphocytes. *Methods Cell Sci.* 22:285-297.

Vandaveer S.S., Erf G.F., Durdik J.M. (2001). Avian T helper one/two immune response balance can be shifted toward inflammation by antigen delivery to scavenger receptors. *Poult Sci.* 80:172-181.

Vidovic D. and Matzinger P. (1988). Unresponsiveness to a foreign antigen can be caused by self-tolerance. *Nature.* 336:222-225.

Wakelin D. (1997). Parasites and the immune system. *Bio-Science.* 47:32-40.

Weidt G., Deppert W., Utermohlen O. *et al.* (1995). Emergence of virus escape mutants after immunization with epitope vaccine. *J Virol.* 69:7147-7151.

Wolfgang R., Ping L., Yuanshu D. *et al.* (1999). IL-1β Inhibits IFN-Induced Class II MHC Expression by Suppressing Transcription of the Class II Transactivator Gene1. *J Immunol.* 162:886-896.

Wong F.S., Wen L. (2003) The study of HLA class II and auto-immune diabetes. *Curr Mol Med.* 3:1-15.

Yelavarthi K.J., Fishback L., Hunt J.S. (1991). Analysis of HLA-G mRNA in human placental and extraplacental membrane cells by in situ hybridization. *J Immunol.* 146:2847-2851.

Yoshiura Y., Kiryu I., Fujiwara A. *et al.* (2003). Identification and characterization of *Fugu* orthologues of mammalian IL-12 subunits. *Immunogen.* 55:296-306.

Zdanov A., Schalk-Hihi C., Menon S., *et al.* (1997). Crystal structure of Epstein-Barr virus protein BCRF1, a homologue of cellular IL-10. *J Mol Biol.* 268:460-467.

Zekarias B., Songserm T., Post J. *et al.* (2002). Development of organs and intestinal mucosa leukocytes in four broiler lines that differ in susceptibility to malabsorption syndrome. *Poult Sci.* 81:1283-1288.

Zhao B., Hou S., Ricciardi R.P. (2003). Chromatin repression by COUP-TFII and HDAC dominates activation by NF-kappaB in regulating MHC class I transcription in adenovirus tumorigenic cells. *Virol.* 306:68-76.

Chapter 2

Alloimmunity and Pregnancy

2.1   ALLOIMMUNITY

Many studies have shown that invertebrates are capable of rejecting foreign tissue transplants, or allografts. If the sponges *(Porifera)* from two different colonies are parabiosed by being mechanically held together, they become necrotic in the period of 1 to 2 weeks, whereas the sponges from the same colony become fused and continue to grow. Earthworms *(Annelids)* and starfish *(Echinoderms)* also reject tissue grafts from other species of the phyla. These rejection reactions are mediated mainly by phagocyte-like cells. They differ from the graft rejection in vertebrates in that the specific memory for the grafted tissue either is not generated or is difficult to demonstrate. Nevertheless, such results indicate that even invertebrates must express cell surface molecules that distinguish "self" from "non-self", and such molecules may be the precursors of the MHC molecules in vertebrates.

Alloreactivity is the ability of the immune system to recognize and reject the grafted tissues or cells of the individuals of the same species. Allorecognition occurs via interaction of the MHC molecules expressed on transplant cells and complementary receptors on the recipient's immune cells. This mechanism was first detected in mammals only to be discovered later that other vertebrate classes also may reject grafts. To what degree mechanisms of alloimmunity in non-mammals (especially in lower vertebrates) coincide with those in mammals has not been precisely established yet. Because vertebrates are not naturally presented with situations of receiving transplants from the same or different species, it is still not clear from the evolutionary point of view why the immune system possesses very well defined mechanisms capable of rejecting allografts quite effectively and relatively quickly. Several evolutionary theories are currently available about the origin of alloreactivity; however, the most logical seems to be the assumption that alloreactivity did not have a direct evolution. It also might have been the accidental consequence of the "self-non-self" immune recognition, the type of communication between immune cells and the system of immune surveillance. Some authors also call the phenomenon of alloimmunity "self-near-self" recognition (Forsdyke, 1991). The absence of evolutionary pressure that may be directly linked with alloreactivity, necessarily defines it as *"Homo Laboratory"* induced phenomenon.

Some reproductive immunologists will not agree with this claim, citing pregnancy as a natural alloimmune phenomenon. However, the MHC gene variability and alloreactivity are most distinctive in mammals in comparison with other vertebrates (Du Pasquer *et al.,* 1999, Hughes, 1993). This corroborates the assumption that pregnancy, being the factor of evolutionary pressure, could not have significantly influenced the development of alloreactivity, diversification of MHC genes and mechanisms of immune reaction mainly based on MHC molecules. We are left with the possibility to regard alloreactivity as an accidental consequence of the evolutionary search of the vertebrate immune system for better solutions regarding "self-non-self" discrimination and anti-microbe defence.

2.1.1   Allorecognition

The immune reaction to donor MHC molecules of the graft is known to represent the main obstacle to the successful allotransplant engraftment (Snell, 1981; Dausset, 1981). The allorecognition elicits a potent immune reaction resulting in rapid elimination of donor cells and the rejection of the graft. Traditionally, allorecognition was thought to occur via only one mechanism: a direct allorecognition in which T cells recognize determinant peptides on the intact donor MHC molecules displayed on the surface of transplanted cells. There are, however, two ways in which T cells can recognize donor MHC molecules in the course of alloreaction. The first way is a direct allorecognition without previous co-recognition by recipients' APCs and the presentation of antigen peptides. Such a direct activation of effectory mechanisms is unique and can only be detected within the immune reaction to allogeneic MHC molecules. A direct alloresponse to intact donor MHC molecules is ensured by T cells, which are polyclonal and directed toward a variety of antigens. Another pathway is by the classical activation of effector mechanisms with prior ingestion, processing and expression of alloantigens via recipients' APCs. An indirect alloresponse is oligoclonal and involves a few dominant antigen peptides on donor MHC (Auchincloss *et al.,* 1993; Shoskes *et al.,* 1994).

The donor APCs probably do not participate in alloreaction in a classical way, due to the phenomenon of MHC restriction, as well as their being the target of alloreactive CTL. Nevertheless, the donor APCs are very important for the starting of the alloimmune reaction. This route for the sensitization of the recipient to a graft seems to involve donor APCs leaving the graft and migrating via the lymph to the regional lymph nodes. Here they can activate those host T cells that bear the corresponding TCR. The activated alloreactive effector T cells are then carried back to the graft, which they attack directly. Indeed, if the grafted tissue is depleted of antigen-presenting cells by the treatment with antibodies or by a prolonged incubation, the rejection occurs only after a much longer time. Also, if the site of grafting lacks lymphatic drainage, no response to the graft results (Auchincloss *et al.,* 1993; Shoskes *et al.,* 1994).

The direct and indirect alloreaction itself is a consequence of the mechanism of the antigene peptide recognition presented as an associated MHC/peptide complex. The MHC molecules which bind "self" peptide will be recognized as a "self" structure, while the allogeneic MHC molecules will be identified as the "self" MHC molecules binding a "non-self" peptide. The result of these mechanisms is such that the recipient's immune system recognizes the allogeneic MHC molecules as modified "self" molecules, while identifying the allogeneic cells as virus-infected, mutated or tumor cells (Matzinger, *et al.,* 1977; Liu *et al.,* 1992). As the number of the epitopes of allogeneic MHC molecules and allogeneic cells themselves is very high, so the number of Th and CTL clones being involved will be proportionally high. A much-supported thesis is that alloreaction is, in fact, a cross-reaction between a high number of Th and CTL and the proportionally high number of allogeneic epitops. Some authors call this a "self-near-self" mechanism, because Th and CTL identify allogeneic cells as "self", only with "slightly" modified MHC molecules, similarly as when virus-infected cell expresses MHC/virus-peptide complex.

2.1.1.1 Dichotomy of Direct and Indirect Allorecognition

Alloreactive T cells that directly interact with donor MHC/peptide complexes (direct pathway) are characterized by their high precursor frequency and the diversity of their TCR specificities. In contrast, T cells recognizing processed forms of donor antigens associated with the recipient MHC molecules (indirect pathway) are directed to a single or a few dominant determinant(s) on the donor MHC molecules and they display a limited TCR Vb gene usage (Liu *et al.,* 1992). While the direct allorecognition has been known for years to contribute to the graft rejection, recently, Auchincloss *et al.* (1993) have clearly demonstrated that the indirect pathway is also an essential element of the rejection process. Using donor cells from MHC class II knocked out mice, it was observed that while the direct allorecognition was abrogated, an indirect type of alloresponses were preserved and capable of mediating the graft rejection.

The recent demonstration of the involvement of the indirect allorecognition in the graft rejection prompted to readdress the mechanisms associated with the physiology of the graft rejection. It is clear that both direct and indirect types of allorecognition participate in T cell responses to donor antigens and contribute to the actual rejection of an allograft. This dichotomy raises the question of how much the direct pathway contributes to the initiation of alloreaction, compared to the indirect one. No precise answers have been defined yet; however, it is obvious that these mechanisms, in particular, make alloreaction a complex, potent and rapid one. The initiation of T cell responses to transplanted cells occurs via the recognition of alloantigens presented on donor passenger leukocytes (bone marrow-derived, MHC class II positive: macrophages, dendritic cells) infiltrating the recipient's lymphoid organs. Following alloantigen recognition, activated alloreactive T cells infiltrate the graft and secrete lymphokines. For example, IFN-γ induces MHC class II expression on the endothelial and epithelial cells of the graft (Botazzo *et al.*, 1983; Hayry *et al.*, 1984). After the passenger leukocytes have left the graft, these endothelial and epithelial cells appear to be the only donor MHC class II-expressing cells in the transplant. Several studies have demonstrated that T cells recognizing donor MHC molecules on these "amateur", defective APCs become anergized and fail to induce proliferation of T cells and IL-2 production (Gaspar *et al.*, 1988; Lo *et al.* 1989).

The recognition of native allo-MHC molecules is only critical for the initiation of the alloresponse but it might not be the main mechanism for ensuring the rejection process. Alternatively, the indirect pathway can be triggered by the presentation of donor MHC peptides by the recipient-derived "professional" APCs which infiltrate the graft. The induction of donor class II MHC molecules by IFN-γ during the secondary inflammatory reactions at the site of the graft should also enhance the processing of donor MHC molecules. Consequently, MHC class II positive inflammatory graft-infiltrating APCs of recipient origin, by processing donor MHC molecules and presenting donor peptides, may provide help for cytotoxic T cell activation and the production of donor-directed antibodies by alloreactive B cells. In conclusion, while the indirect alloresponse may not be the driving force in the initial *in vivo* sensitization of T cells to alloantigens, it may however play a key role in the actual rejection process at the site of the graft (Benichou *et al.*, 1990, 1994, 1998).

2.1.1.2 Secondary (auto-immune-like) Alloimmune Response

While some knowledge has been accumulated with regards to the cellular and molecular mechanisms eliciting an early acute rejection of allografts, very little is known about later immunological events involved in the long-term and chronic rejection process. Based upon the principle that rapidly after transplantation, grafted tissues become devoid of "professional" bone-marrow-derived APCs, there are hypotheses that the long-term and chronic types of rejection were mediated by T cells recognizing antigens on the recipient APCs (Benichou *et al.*, 1992). Although this model is still hypothetical, some recent experimental evidence has been provided indicating that the indirect allorecognition represents a driving force both in the late rejection and in the chronic rejection of allografts (Ciubotariu *et al.*, 1998). In one study, it has been shown that the chronic rejection process in patients with heart transplants is actually associated with the diversification or antigen-spreading of indirect T cell alloresponses to newly presented determinants on donor MHC proteins (Benichou *et al.*, 1998). Interestingly, there is an observation that T cell responses can also spread to determinants present on "self"- i.e. the recipient-derived MHC molecules themselves (Fedoseyeva *et al.*, 1996). In this case, the alloresponse led to the disruption of immunological tolerance to a "self"-antigen. Further supporting the potential involvement of this phenomenon in the graft rejection, is a study by Fedoseyeva *et al.* (1999) in which is shown that the following heart transplantation, "self"-tolerance to a cardiac tissue-specific antigen, cardiac myosin, is disrupted. Same authors showed that this phenomenon requires an initial alloimmune T cell response to donor MHC molecules of the graft. However, once this auto-immune response is engaged it is sufficient on its own to cause the rejection of the graft. Interestingly, cardiac myosin has been previously described as the auto-antigen that causes a heart auto-immune disease, auto-immune myocarditis. These data showed that allotransplantation-induced auto-immunity to cardiac myosin in transplanted mice leads to a histopathology that is indistinguishable to that observed in mice with auto-immune myocarditis. This study suggests that the secondary breakdown of "self" tolerance to key tissue antigens displayed on the graft may be an essential component of the long term rejection of allotransplanted tissues and organs (Fedoseyeva *et al.*, 1999).

2.1.2 Xenoreactivity

Xenoreactivity is the ability to recognize and reject transplanted cells or tissues of different species. Although xenoreactivity may cause the rejection of the transplant within shorter time (a few hours) than alloreaction, its mechanisms are fundamentally different from alloreaction. The mechanisms are mainly based on the humoral immunity and NK cells mediated cytotoxicity, and the least on cellular immunity (Lanza *et al.*, 1997). The hyperacute rejection of xenotransplant is most likely to be the result of a cross-reaction between the pre-existing antibodies of the recipient and the molecules of endothelial cells of the graft's blood vessels, resulting in the activation of the classical pathway of complement (Rose *et al.*, 1996). Following the transplantation of a pig organ into a human, or the extracorporeal perfusion of human blood through a pig organ, there is a marked increase in the titer of human anti-pig antibodies, specific for *galactosea 1-3 galactose (a Gal)* epitopes on the surface of pig vascular endothelium (Sandrin *et al.*, 1993; Galili, 1993; Cooper, 1998). There is a belief that anti-*a Gal* antibodies develop during the first few weeks of life through the exposure to certain microbes that colonize the gastrointestinal tract and which also express *a Gal* structures on their cell membranes (Galili *et al.*, 1988). At birth, anti-*a Gal* IgG can frequently be detected in the plasma, presumably passively transferred from the mother, but not IgM. As it is predominantly IgM binding that initiates the hyperacute rejection, a pig organ transplanted into a neonatal baboon is not rejected hyperacutely, but does undergo an acute vascular rejection over the next few days (Minanov *et al.*, 1997).

In this case, there is practically no immune recognition directed towards the MHC molecules or the activation of specific effectory mechanisms. The mechanism for rejecting xenotransplant is left to randomness being the hallmark of every cross-reaction. The evidence that the hyperacute rejection of xenotransplants is sometimes dependent on complement-independent humoral mechanisms indicated the importance of CD16[+] positive NK cells and monocytes. The rejection of porcine hearts by neonatal baboons, which have very low levels of performed IgM xenoantibody, was significantly delayed compared to the rejection that has been observed in adult baboons. The dominant findings in hearts rejected by neonatal baboons were perivascular and interstitial mononuclear cell infiltrates, contrasting with the intravascular thrombosis typically observed in hyperacute rejection in adult baboons. The cellular infiltrates were predominantly composed of host macrophages and NK cells (Intescu *et al.*, 1998).

NK cells are involved in the graft rejection in two ways. One is antibody-dependent, while the other is a direct cytotoxicity, focussed on the cells not expressing MHC of the recipient's species. Due to this, NK cells play only a minor role in the mechanisms of alloreaction. It was earlier said that the MHC expression was an inhibitory signal for NK cells and would thus be able to unmistakably identify and effectively eliminate cells not expressing the MHC molecules. Although the cells of xenotransplant express the MHC molecules, the NK cells of the recipient do not recognise them as the MHC molecules, which directly trigger the activation of NK cells (Intenscu *et al.*, 1998).

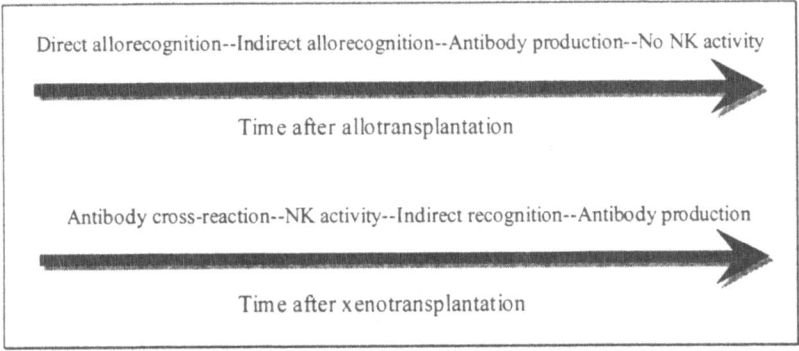

Figure 2.1. Order of activation of the immune mechanisms in alloreaction and xenoreaction.

CTL becomes involved in the mechanism of xenoreaction only in the cases when the transplant has survived long enough for the mechanisms analogue to the indirect alloreaction to be activated (1-2 weeks, or more). Sometimes, the mechanism of xenoreaction is analogue to the direct alloreaction. In fact, xenoreaction is much less predictable than alloreaction and depends on the phylogenetic linkage of the species included in the transplantation. In phylogenetically closer species, xenoreaction follows the model of alloreaction, whereas in distant species it is mediated mainly by antibodies, NK cells, monocytes, and the least by CTL (Lanza *et al.*, 1997).

2.1.3   Alloimmunity in Vertebrates

The comparative analysis of alloreactivity in different vertebrate classes may be helpful in identifying evolutionary details concerning some important mechanisms such as the presentation of antigen peptides, immune recognition and communication of immune cells. However, there are some basic problems concerning the mechanisms of alloreactivity:

1. MHC system was discovered only 40 years ago, so it is largely unexplored, especially, in lower vertebrates.
2. Lower vertebrates display less MHC gene variability and a proportionally less number of heterozygote offspring related to mammals.
3. The majority of the species has been known to survive only 1-2 million years, so the number of species available for the research comprises only

a tiny percent, compared to the total number of species participating in the evolution of vertebrates.

4. The types of expression and tissue distribution of class I and class II molecules are significantly different across vertebrate classes.

5. Some vertebrates display great variability of the MHC expression during the ontogenic development, such as metamorphosis in amphibians or seasonal changes in reptiles (Nelson *et al.*, 1996).

6. The mechanisms of immune cell communication and significance of MHC molecules for this communication vary across vertebrate classes and species.

7. The number of the immune cell subpopulations involved in the communication with the MHC molecules and regulation of the immune reaction grows progressively from cartilaginous fish towards mammals, which may be the source of great discrepancy in the mechanisms of alloimmunity.

8. Non-immune factors, such as external temperature, influence the speed and intensity of alloimmune reaction in some vertebrates like fish, amphibian and probably reptiles.

These as well as other factors largely impede the construction of a reliable experimental model and interpretation of experimental findings. This area of comparative immunology can thus be said to be very complex and relatively unexplored.

After allografting in *Agnatha*, heavy leukocyte infiltration becomes perceivable in the graft tissue. From 10 to 60 days after the allotransplantation, the infiltrate is almost entirely composed of polymorphonuclear (PMN) leucocytes and eosinophilic granulocytes, while macrophagal infiltration occurs between day 40 and day 60. Lymphocyte-like cells comprise only a minor part of the infiltrate, and are assumed to perform an insignificant role in transplant rejection. Based on these data, PMNs and macrophages are presumed to play a key role in transplant rejection in *Agnatha* (Hughes *et al.*, 1993).

Cartilaginous fish reject the primary allotransplant very slowly, following the model of chronic alloreaction. The rejection of the secondary transplant is also performed according to the model of chronic rejection, only within a much shorter time interval. These data indicate that lymphocytes of cartilaginous fish are not capable of direct alloimmune recognition, but the chronic rejection of the graft is mediated through the indirect allorecognition. This assumption is evidenced by the fact that lymphocytes of sharks do not proliferate in MLR; however, the stimulation of the lymphocytes with BCG, PHA or ConA results in their proliferation (Hashimoto *et al.*, 1992).

In bony fish, the primary graft is rejected very slowly following the model of the acute rejection. The secondary allotransplantation causes a strong reaction against the allograft, and the developing of a persistent cell memory. The mechanism of allograft rejection in bony fish is very similar to that in higher vertebrates, including lymphocyte infiltration, the destruction of capillary endothelial cells and other transplant cell destruction. Although alloreaction is elicited via the model of acute rejection, it is not clear yet whether these correspond to the direct allorecognition in mammals. MLR has been identified, but only in some bony fish species, thus causing even greater confusion regarding the mechanism of the direct allorecognition. With regard to this, some authors propose that MHC mediated intercellular communication in bony fish significantly differs from mammalian (Nakanishi *et al.*, 1999). Thus, Mc Kinney *et al.*, (1992) observed that lymphocyte-alloimmunized tropical marine teleost *(Pomacentrus partitus)* responded to allogeneic pronephros between days 8 and 10, but fishes immunized with epithelial cells showed accelerated MLR, with optimal responses occurring between days 4 and 6. Interestingly, no reactivity toward epithelial cells was observed in the absence of immunization, suggesting that accessory cell functions of damselfish pronephros and epithelial cells differ (McKinney *et al.*, 1992). In addition, GVH reaction and delayed hypersensitivity reaction (DTH) were observed in bony fish. These data strongly suggest the presence of cytotoxic T cells and MHC class I polymorphism in fishes (McKinney *et al.*, 1992; Nakanishi *et al.*, 1999).

The allograft rejection in amphibians may be very rapid (days 3-40). The process takes place via the acute rejection, accompanied by lymphocyte infiltration and the destruction of endothelial cells of the blood vessels. Whether or not amphibians undergo the acute rejection has not been evidenced yet. In these, the secondary graft is rejected more rapidly than the primary one (days 8-19). On the other hand, the phenomenon of a prolonged rejection of the secondary graft has been observed, even the establishing of specific tolerance after several successive transplantations. Amphibian lymphocytes respond in MLR and GVH reactions, and also when stimulated by PHA or ConA (Di Marzo *et al.*, 1982; Maeno *et al.*, 1987; Arnall *et al.*, 1987).

Surprisingly for many immunologists, reptiles represent a gap in the evolutionary development of alloreactivity. If a progressive increase in the complexity of the adaptive immunity is observable from *Cyclostomata* to amphibians, along with a tendency towards intensifying the alloreactivity, this evolutionary trend is disrupted abruptly in reptiles. We can currently only speculate the possible reasons for the occurrence of this phenomenon.

The rejection of allograft in reptiles is effected via mechanism of the chronic rejection, and the speed of rejecting (months and years) is almost completely independent from external temperature, unlike fish and amphibians. The rejection of secondary and tertiary allotransplant is much more rapid than the primary one, which indicates the presence of the immune memory (Saad *et al.,* 1984). Reptile lymphocytes proliferate in MLR, but the intensity and general efficacy of their immune system significantly vary during seasonal changes (Nelson *et al.,* 1996).

The immune system of birds displays the greatest similarity to the mammalian, so allografts in these are most likely rejected via acute alloreaction, and probably accompanied by a component of direct alloreactivity. Birds, like mammals, have a highly polymorphic MHC that determines a strong allograft rejection. Kaufman *et al.,* (1997) found that certain common chicken MHC haplotypes express only one class I molecule at high levels. The selection on a single MHC gene should be strong, in contrast to the situation in mammals. Like mammals, birds express strong GVH reaction which is supported by $CD4^+$ and $CD8^+$ $\alpha\beta TCR^+$, as well as $\gamma\delta TCR^+$ cells (Tsuji *et al.,* 1996). In chicken, MLR to MHC (B-complex) disparity was first detected in the thymus of 16 day old embryo and regularly detectable from the 18th embryonal day on. MLR to class I (B-F) molecules disparity was not detectable before hatching. Exogenous IL-2 containing supernatant added to the cultures had only a minor effect on MLR of embryonal thymocytes. In the spleen, MLR was first detectable in some strain combinations on day 3 and regularly found on the 7th day post hatching, indicating the colonization of peripheral lymphoid organs by the thymus derived mature cells. These data suggest that the ontogeny of alloreactivity in birds (like in mammals) is associated with the appearance of TCR, CD4 and CD8 bearing cells in the thymus and peripheral lymphoid organs (Lehtonen *et al.,* 1989).

2.2   ALLOIMMUNITY AND VIVIPARITY

Despite the fact that some examples of viviparity exist in invertebrates, fish, amphibians and reptiles, this manner of reproduction is a relatively new acquisition in evolution. The establishment of full, intimate bi-directional connection between mother and embryo is a characteristic of mammals only. Pregnancy is a new strategy in reproduction which allows the development of a small number of embryos under the protection of maternal organism. This became possible with the establishment of a functional uterus and immune tolerance to embryonic tissues. Although viviparity is an evolutionary advantage, it has one important limitation: the absolute necessity of synchronisation many events between the embryonic and maternal relationship.

Successful mammalian pregnancy depends upon the tolerance of a genetically incompatible fetus and especially placenta by the maternal immune system. When tolerance is not achieved, pregnancies fail, so that repeated pregnancy failures can be associated with reproductive efficacy of species. Antigenically the fetus and placenta is like a tissue allograft in the maternal uterus as half the MHC genes of the conceptus are of paternal origin. Although viviparity and placentation are advantageous for the nourishment and growth of offspring, it involves an immunological risk for the allogeneic fetus. The prolonged exposure of embryonic tissues bearing paternal antigens to maternal uterine tissues can cause the rejection of the embryo and decrease the reproductive efficacy of viviparous way of reproduction. The mechanism why it survives is still largely unknown. Several theories have been proposed and all seem to combine to enhance the survival of the fetus. Originally Sir Peter Medawar proposed that (a) the mother was immunologically tolerant with respect to the fetal antigens; (b) the fetus did not express foreign (paternal) antigens; (c) the uterus was an immunologically privileged site; (d) the placenta acts as a barrier to prevent maternal immune attack on the fetus (Bainbridge, 2000).

Many reproductive immunologists define pregnancy as a model of successful allotransplantation, as allogeneic fetus goes through a successful development throughout the pregnancy, without the activation of the maternal alloimmune response. However, the definition of allograft as a tissue expressing incompatible MHC class I and II molecules and functioning in the immediate surroundings of the intact immunocompetent cells of the host, rather relativizes the definition of pregnancy as a model of classical allotransplant. While it is true that fetal tissues express paternal MHC molecules from the earliest stage of its development, the maternal immune system is not in a direct contact with them. It simply does not recognize the allogeneic fetus due to the placental barrier. On the other hand, trophoblast cells, which are in direct communication with maternal immune cells, do not express allogeneic MHC molecules except for the class Ib molecules. Therefore, trophoblast tissue, though being of fetal origin, could not be defined as a classical allograft. Finally, we can conclude this argument, by stating that neither fetus nor placenta meets conditions set by this definition because the fetus is invisible to the maternal immune system, and also because placental tissues do not express the MHC molecules class Ia and II.

More recently the theories about the maternal immune tolerance of the feto-placental unit include:

1. General non-specific immunosuppression mediated by the elevation of pregnancy associated steroid hormones mostly produced by the placenta.

2. Reduced fetal immunogenicity by the reduction or alteration of expression of fetal MHC class Ia and class II molecules by trophoblast cells.
3. The activation of KIR on NK cells by placental class Ib molecules expression.
4. Inhibition of CTL by placental class Ib molecules expression.
5. Specific "placental graft" enhancement mediated by either blocking antibody or antigen-antibody complexes.
6. The specific activation of decidual cells to produce immunosuppressive factors like prostaglandine and cytokines which result is activation of Th2 and suppression of the Th1 immunity.
7. The expression of apoptotic ligand, such as *Fas* ligand, by trophoblast cells.
8. The activation of extrathymic lymphocyte maturation in local (decidual) compartment.
9. The expression of receptors on trophoblastic cells for anti-apoptotic factors like TGF-β.

Pregnancy and pregnancy-like model of reproduction probably has evolved many times during the long evolution of viviparity in vertebrates. The evolutionary development of the whole mechanism of immunotolerance of feto-placental unit in pregnancy, especially the loss of MHC molecules on trophoblast tissue, must have undergone a strong selection pressure of reproductive efficacy. Therefore, we can be highly certain that the placental tissues of primordial mammals expressed MHC molecules class Ia and class II. In addition, during the evolution certain regulatory mechanisms developed, resulting in the inhibition of the MHC molecules expression on placental tissues. These processes certainly caused pregnancy to lose its attribute of being an allotransplant, and increased the reproductive efficacy in mammals. The evidence for this can be found in the fact that the more primitive and less effective forms of placenta, such as the superficial placenta in equine which does not make a direct contact with maternal blood, express much higher number of the MHC molecules than a more effective and advanced placenta, such as the interstitial placenta in primates. The infiltrating nature of the trophoblast of interstitial placenta as well as a direct contact with maternal blood, contributed to the placental efficacy in respiratory, nutritional, excretory and secretory respect. On the other hand, in this type of placenta the risk of immune rejection was increased many times. Still, the evolutionary solution was found in a number of immunomodulatory and immunosuppressive mechanisms, and also in the inhibition of the MHC molecules expression, (except for the monomorphic or minimal-polymorphic class Ib molecules) on the trophoblast of the interstitial placenta.

2.2.1   Evolution of Viviparity

Viviparity probably has evolved many times within vertebrates, but the attendant morphological modifications and the sequence in which they occur remain unclear. Although this phenomenon is present in all vertebrate classes except birds, the evolutionary emergence of viviparity in different vertebrates is probably not connected with the common evolutionary ancestor and represents an example of convergent evolutionary phenomenon (Rothchild, 2003). Viviparity may involve the formation of a placenta, a structure formed by the apposition of extra-embryonic membranes and maternal tissues. Among vertebrates, placental viviparity is present in mammals (except monotremes), some squamate reptiles and cartilaginous fishes (Bainbridge, 2000). Probably the first placental structures can be inferred from the fossil evidence of *Ichthyosaurus*, more than 170 million years ago. The origins of the placenta lie in the reptilian (amniote) type of egg. Whether it represents one of the first stages in placental evolution or whether the placenta of mammals was "invented" several times in the evolutionary history of mammals is unknown (Blackburn 1995), but placenta is an organ that probably has co-evolved with the mammals, and therefore shows a much wider diversity of form and function than other organs.

2.2.1.1   Viviparity in Cartilaginous fish

Cartilaginous fish are the oldest extant jawed vertebrates and the oldest line to have placenta-like organ. Their pivotal evolutionary position makes them attractive models to investigate the mechanisms involved in the maternal-fetal interaction. Among cartilaginous fish, placental viviparity occurs only in some species of sharks. Unlike amniotes in which four basic membranes develop, the yolk sac is the only extra-embryonic membrane present in anamniotes. In all species of chondrichthyans, regardless of the reproductive mode, embryos are initially nourished by yolk stores in the yolk sac. Initially yolk is solubilized and transported across the endoderm to the yolk sac vessels. Additionally, yolk is transported to the fetal gut by ciliary action of the ductus vitellointestinalis in the yolk stalk. In most placental sharks, as yolk stores are depleted, the distal aspect of the modified yolk sac directly contacts the tertiary egg envelope that intervenes between the fetal and maternal tissues throughout gestation. At midgestation and in concert with the diminution of yolk stores, yolk sac tissues differentiate into a definitive yolk sac placenta-like organ. In all cases, the placenta-like organ is epitheliochorial, non-invasive and persists until term. During the ontogenetic modification of the yolk sac, a concomitant change occurs in the yolk stalk. It elongates to become an umbilical cord-like organ (Hamlett *et al.*, 1985, 1993).

The uterine epithelium is generally separated from the egg envelope by a discernable space. The yolk sac epithelial ectoderm, on the other hand, is in a close association with the egg envelope along most of its extent. The yolk sac endoderm is characterized by large cells that line the now empty former yolk sac cavity. As in the uterus, fetal vessels are separated from the surface ectodermal epithelium by a scant connective tissue. The vessels are in close proximity to the simple to bilayered squamous epithelium. Paraplacental uterine sites are characterized by a simple columnar epithelium. The epithelium rests on a connective tissue zone with the modest vascularity (Hamlett *et al.*, 1985, 1993).

The immune maternal-placental relationship in cartilaginous fish is still largely unexplored. However, there are some data suggesting the similar role played by certain cytokines in the process of placentation in these and mammals. For example, Cateni *et al.* (2003) proved a similar role of IL-1 in placentation in some species of sharks and mammals. There are few data regarding the possibility of developing alloimmune reaction on the paternal antigens (if they exist) on placenta-like organs in sharks. In addition, relatively low variability of the MHC molecules in cartilaginous fish, and a lower ability for alloimmunity, may minimize the impact of the maternal immune response on the reproductive efficacy in this class of vertebrates.

2.2.1.2   Viviparity in Reptiles

A large number of evolutionary biologists agree with the assumption that eggshell represents one of the greatest evolutionary steps made by vertebrates. This "invention" belongs to reptiles or their ancestors. The egg with a relatively hard shell protecting it from drying enabled these animals to brood out of water and be relatively independent from the habitat nearer the water. Consequently, the process of colonising dry areas, and ultimately the conquest of the inland territories began. However, this evolutionary advancement was impeded by harsh conditions in waterless habitats, variable temperatures, uncertain sources of food and a large number of "nest depredators". In response to these and similar evolutionary pressures, reptiles developed the adaptation mechanism of egg retention until the conditions were suitable for their laying (Guillette, 1993; Qualls, 1996). In some reptiles, the retention of eggs is a variable category, regarding the time. Different species of reptiles lay eggs in different phases of embryo's development, while there is a large number of species which retain eggs until the intrauterine hatching of offspring. In viviparous reptiles, however, the eggshell significantly reduces oxygenation of the fetus, and could limit the reproduction, which is not the case with oviparous reptiles. The reduction of thickness of eggshell, or its complete absence, is a next adaptation mechanism which developed only in viviparous reptiles (Qualls 1996;

Blackburn *et al.*, 1995, 2002) when conditions were set for the occurrence of the first immunoreproductive phenomena in the context of the maternal-fetal relationships. Viviparous reptiles having soft-shelled eggs are probably the first vertebrates to have been exposed to the new form of selection pressure, such as reproductive efficacy based on the immune mechanisms.

The evolution of reptilian viviparity and placentation has held the interest of biologists for over half a century (Weekes, 1927; Blackburn *et al.*, 1982, 2002; Guillette, 1993; Qualls, 1996), and the sequence of events culminating in these features has been an ongoing point of discussion (Weekes 1935; Blackburn *et al.*, 2002). Most oviparous reptiles produce flexible-shelled eggs that lose water rapidly if oviposited in dry substrates (Ackerman *et al.*, 1984). Thus, unlike the rigid-shelled eggs of many crocodilians, some chelonians, and some lizards, flexible shelled eggs, in general, are not suited for the protection against desiccation. Nonetheless, the shells of flexible shelled eggs may exhibit a lower permeability to water vapor under the conditions of low ambient humidity than high ambient humidity. This raises the possibility that under adversely dry conditions, the structure of the shell may mechanically adjust in such a way that limits water loss from the egg (Feder *et al.*, 1982).

The evidence to support the idea that thinning of the shell occurs concurrently with longer periods of retention, is limited. First, no known viviparous species posses a relatively complete eggshell, an occurrence that might be expected if thinning of the shell occurs subsequent to evolving viviparity (Blackburn, 1995). In every viviparous species examined to date, the shell is considerably reduced or absent (Blackburn, 1993). Nonetheless, these observations do not negate the possibility that the shell reduction occurs after viviparity is in place. A stronger supporting evidence for this idea is provided by the existence of oviparous species of lizards that retain their eggs to relatively advanced stages of embryonic development and produce shells that are thinner than those of conspecifics that retain eggs to less advanced stages of development (Qualls, 1996; Smith *et al.*, 1997). The idea that thinning of the shell occurs concurrently with longer periods of retention has gained a widespread acceptance only relatively recently, although the initial formulations of this idea can be traced back at least as far as Weekes (1935). However, an alternate view, and one that was more prevalent in the historical discussions on this topic, is that the shell reduction and simple placentation occur subsequent to viviparity (Guillette *et al.*, 1993; Blackburn *et al.*, 1993, 1995, 2002). The rationale behind this model is that a progressive thinning of the shell prior to viviparity would result in eggs that become more prone to dehydration. Thus, if longer periods of egg retention and thinning of the egg-shell occur concurrently, then any selective benefit of the shell reduction in utero would tend to be offset by the selection against oviposition such an egg in a nest, particularly in arid environments (Weekes, 1935).

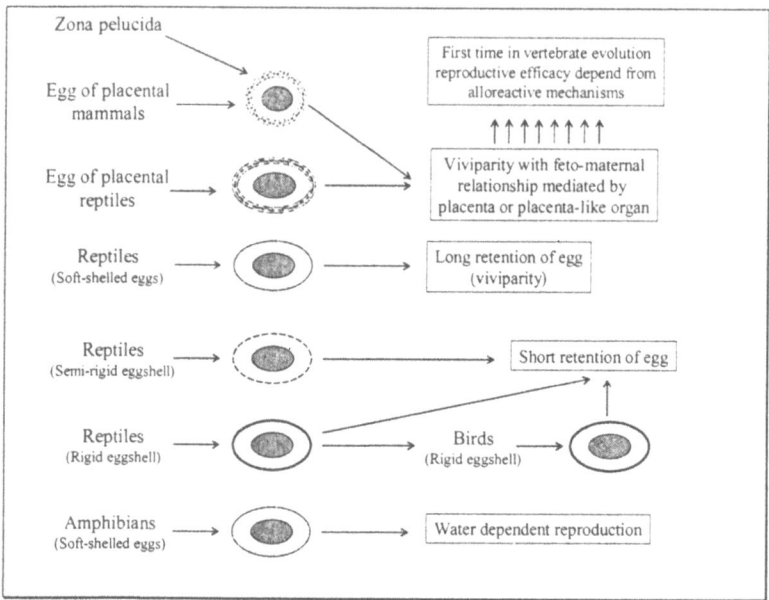

Figure 2.2. Different models of eggshell and non-shelled eggs.

The most widely accepted model for the transition from oviparity to viviparity in reptiles posits that viviparity evolves through the selection for the increasingly longer periods of egg retention which results in an increase in the amount of the embryonic development that takes place in the oviducts prior to oviposition. This model also posits that a "thinning" of the eggshell occurs concurrently with the longer periods of retention (Blackburn, 1982; Guillette 1993). Heulin *et al.* (2002) investigated that the eggs of two distinct oviparous clades of the lizard *Lacerta vivipara* were compared. For example, the eggs laid by females from Slovenian and Italian populations have thicker eggshells, contain embryos on average less developed at the time of oviposition, and require a longer incubation period before hatching than the eggs laid by the females from French oviparous populations. Also, the studies on *sceloporine* lizards demonstrate that the embryonic responses to egg retention that is extended beyond the time of normal oviposition range from the developmental arrest to the normal development (Andrews, 2002).

The thinning of the eggshell would eventually culminate in what is termed a simple placenta (Weekes, 1935); a close apposition of the chorioallantois of the embryo (and the shell membrane or the remnant thereof) and the uterine tissues. Such an arrangement is thought to facilitate the respiratory exchange between the maternal and fetal tissues (Blackburn, 1993). New World skinks of the genus *Mabuya* exhibit a unique form of viviparity that involves the ovulation of tiny, about 1 mm, eggs and the provision of virtually all of the nutrients for the embryonic development by the placental means. In the Brazilian species, *Mabuya heathi* the uterine lining is intimately apposed to the chorioallantois, with no trace of an intervening shell membrane or of epithelial erosion; thus, the placenta is epitheliochorial. The apposed chorionic epithelium is absorptive in morphology and contains giant binucleated cells that bear microvilli. Several specializations of the placental membranes of *Mabuya heathi* are found among eutherian mammals, signifying the evolutionary convergence that extends to histological and cytological levels (Blackburn *et al.*, 2002).

2.2.1.3 Viviparity in Mammals

The placenta is the hallmark of the eutherian mammals, yet, rather than being the most anatomically conserved mammalian organ, it is possibly the most diverse. The placentation ranges from the invasive hemochorial type, as in the human, where the trophoblast surface is in a direct contact with the maternal blood, to the epitheliochorial (e.g., pig) where the uterine epithelium is not eroded (Padykula *et al.*, 1982; Kaufman *et al.*, 1994).

The number of tissue layers separating fetal blood from maternal in the chorioallantoic placenta of the eutherian mammals is a striking indicator of the great structural diversity of placenta. Ungulates such as cattle, sheep, horses and pigs show the basic (epitheliochorial) pattern, in which the chorionic epithelium of the conceptus is in contact with the uterine epithelium of its mother, so that two layers of endothelium, maternal and fetal, and the two epithelial layers separate the two circulations. The placentation in ruminants, such as cattle and sheep, is superficial, relatively non-invasive, and known as a synepitheliochorial cotyledonary. "Synepitheliochorial" describes the fetal-maternal syncytium formed by the fusion of trophoblast binucleated cells and uterine epithelial cells, whereas "cotyledonary" describes the gross structure of the placenta and specifically the tufts of villous trophoblast (cotyledons) that insinuate themselves into the crypts of the maternal caruncles. These regions of interdigitated and partially fused fetal cotyledonary and maternal caruncles are the placentomes and are the main sites for the nutrient and gas exchange in the placenta. The binucleated cells, which compose about 20% of the surface epithelium (trophectoderm), migrate and fuse with the maternal uterine epithelial cells and deliver their secretory products directly to the maternal system (Blackburn *et al.*, 1988; Kaufman *et al.*, 1994).

In carnivores such as dogs, cats and seals the chorionic epithelium becomes invasive and burrows through the uterine epithelium and connective tissue to establish the contact with the epithelium of the uterine capillaries. This endotheliochorial type of placenta is associated with the fusion of chorionic epithelial cells to form an invasive syncytiotrophoblast layer (Shmidt, 1979; Kaufman *et al.*, 1994).

Table 2.1. This table indicates, incompletely though, the amazing variety of placental forms and types.

Order	Species	Placental form	Maternal-fetal Interdigitation	Maternal-fetal separating membrane
Insectivora	*Europ. Mole*	Discoid	Labyrinth	Hemochorial
	Scalopus	Diffuse	Labyrinth	Hemochorial
Chiroptera (Bats)		Discoid	Labyrinth	Endotheliochorial Occas. Hemochorial
Primates	*Tupaia*	Bidiscoid	Labyrinth	Endotheliochorial
	Galago	Diffuse	Folded	Epitheliochorial
	Rhesus	Bidiscoid	Villi	Hemochorial
	Ape, Human	Discoid	Villi	Hemochorial
Lagomorpha	*Rabbit, hare*	Discoid	Labyrinth	Hemochorial
Rodentia	*Guinea pig*	Discoid	Labyrinth	Hemomonochorial
	Rat, mouse	Discoid	Labyrinth	Hemotrichorial
	Beaver	Discoid	Labyrinth	Hemodichorial
Carnivora	*Dog, cat*	Zonary	Labyrinth	Endotheliochorial
Sirenia	*Manatee*	Zonary	Labyrinth	Hemochorial
Artiodactyla	*Pig*	Diffuse	Folded	Epitheliochorial
	Sheep	Cotyledonary	Villi	Epithelio-Syndesmochorial
	Cow, ?goat	Cotyledonary	Villi	Epitheliochorial
Perissodactyla	*Horse*	Diffuse, spec.	Villi, "cups"	Epitheliochorial
Cetacea	*Whale, dolphin*	Diffuse	Villi	Epitheliochorial

Primates and rodents also show the syncytiotrophoblast formation, which in these species leads to an even deeper invasion of the uterine wall; the walls of the small maternal vessels are destroyed and eroded away so that the syncytial surface is washed with the maternal arterial blood unconfined by vessels. In this hemochorial type of placenta, there is no maternal component to the barrier between the blood circulations (Novacek, 1992).

It is clear that different groups of mammals have exploited the basic placental concept in various ways, all of which seem to work. The multilayered epitheliochorial placenta of the pig is no better or worse a placenta than the hemochorial human placenta in which all the maternal layers have been destroyed. However, the cellular mechanisms by which the effective function is achieved differ considerably between different species. For example, the human placenta has receptor-mediated transport mechanisms both in the syncytiotrophoblast and the fetal endothelium which are responsible for the transport of the IgG class of antibodies from the maternal to fetal blood; this confers passive immunity until the immune system of the newborn gets going. In the guinea pig, which has a structurally very similar hemochorial placental barrier, there is no evidence for IgG transport at this site but there is a very busy receptor-mediated antibody-transporting mechanism in the yolk sac (which never acquires a placental function in the human) (Kaufman *et al.*, 1994; Malek *et al.*, 1996).

In cattle and pigs, the passive immunity is conferred on the newborn by the absorption of an antibody through the gut wall from the globulin-rich first milk. The evolution of multiple strategies to achieve a similar effect is characteristic of the placental transport in different mammalian species. The pattern of blood circulation can have profound effects on the placental transfer, particularly on the highly diffusible molecules such as oxygen whose transfer is limited by blood flow rather than by barrier properties. For example, in the human placenta the fetal microcirculation runs through chorionic villi which extract oxygen from a slow-flowing pool of the maternal blood, whereas in the guinea pig maternal blood flows in narrow trophoblast-lined channels running counter to the fetal placental capillaries. This provides a highly efficient counter current exchanger which goes a long way to explaining the success of guinea pigs in carrying litters of large fetuses to term in the low oxygen partial pressures of their high-mountain natural environment (Leiser *et al.*, 1994).

Although cytokine production seems to be a specific phenomenon in the mammalian reproduction, the specific roles of these substances in different species are still not clear. However, a balance of different cytokine activities appears to be crucial for the regulation of the establishment and survival of the semi-allogeneic embryo in maternal tissues. The apparent immunological role of placental cytokines in the mechanisms of implantation and embryonic development in mammals has raised the question of whether cytokines are also involved in the reproduction of non-mammalian vertebrates. Some studies have shown that the production of cytokines fetal-placental-maternal unit is not limited to mammals, but that IL-1α, IL-1β, and TGF-β are secreted by the placenta of a viviparous squamate reptile, *Chalcides Chalcides*. This finding of parallelism between the reptilian and mammalian reproduction suggests that immunological mechanisms, possibly mediated by the secretion of cytokines, played an important role in the ontogenic and phylogenic evolution of placenta and viviparity (Paulesu, 1997).

2.2.1.4 Expression of MHC molecules on Different Types of Placenta in Mammals

In many species of eutherian mammals, the mechanisms of pregnancy survive due to the reducing of placental expression of the MHC genes.

Unexpectedly, in some species the MHC expression is often re-established in the most invasive trophoblast cells. It is not known why the transplantation antigen expression in the fetal cells most exposed to the maternal immune system is advantageous. It is possible that such an expression aids the process of invasion or exerts an immuno-protective effect on the fetus. It may prove possible to identify the essential steps that all eutherian fetuses take to ensure their survival in the face of the potential maternal immune attack by studying the common features of the placental immunology of different species.

There is a large body of data that in most mammals, the fetus limits its presentation of the paternal MHC molecules to the mother immune system. In the horse, however, functional, polymorphic MHC class I antigens are expressed at high levels on the invasive trophoblast cells of the chorionic girdle between days 32 and 36 of pregnancy, although not on the adjacent non-invasive trophoblast of the chorion and allantochorion membranes. Bacon *et al.* (2002) found that 33-34 days old conceptus tissue revealed both transcriptional and posttranscriptional regulation of cell surface class I expression in horse trophoblast. The invasive class I positive trophoblast showed levels of steady-state mRNA nearly as high as those in lymphoid tissues from adult horses, whereas non-invasive class I negative trophoblast also contained transcripts for class I, but at lower levels similar to those present in adult horse non-lymphoid tissue. Also, the source of fetal MHC antigens in the pregnant mare appears to be the specialized trophoblast cells of the chorionic girdle region of the developing placenta. These cells invade the endometrium between days 36 and 38 after the ovulation to form the endometrial cups. The progenitor girdle cells express the high levels of paternal MHC antigens, while the non-invasive trophoblast cells of the allantochorion and the differentiated trophoblast cells in the mature endometrial cups do not. This expression of MHC antigens by the chorionic girdle cells is unusual for a trophoblast tissue, and differs from most forms of trophoblast studied in other species (Antzak *et al.,* 1989).

In pig, trophoblast becomes attach to the endometrial epithelium of the uterus between days 14 and 22. Around this time, the outer endodermal surface of the developing allantois begins to fuse with the inner endodermal layer of the chorion, starting at the embryo and progressing to the both ends of the elongated blastocysts. The mesoderm then develops between the two endodermal layers to innervate the allanto-chorionic sac. Practically the entire surface of the allanto-chorion forms the placenta, hence the name placenta diffusa. The trophoblast remains a non-invasive single layer in the pig. Beginning around midgestation, the capillary plexuses at the tips of the chorionic villi penetrate between the trophoblast cells to about 2 μm from maternal epithelium at term (Ramsoondar *et al.,* 1999).

The lack of the polymorphic MHC molecules on pig trophoblast follows that of both the sheep (Gogolin-Ewens *et al.*, 1989) and the horse, in which, except for the expression of class I MHC on the invasive trophoblast of the transient chorionic girdle cells, the non-invasive trophoblasts of the chorioallantoic membranes are class I negative (Donaldson *et al.*, 1990; Maher *et al.*, 1996). In contrast, in the cow placenta, which is structurally similar to that of the sheep (syndesmochorial), the non-invasive trophoblast of the interplacentomal allanto-chorion has been found to express class I in some instances (Low *et al.*, 1990). This is especially perplexing since the same monoclonal anti-sheep class I antibody (SBU-1) was used in both the cow and sheep studies (Gogolin-Ewens *et al.*, 1989; Low *et al.*, 1990). Since the monoclonal antibodies (mAb) is directed against pig thymocytes, cells that very likely do not express monomorphic, pregnancy-associated class I MHC antigens, it may not recognize these unique forms; hence, the possibility that monomorphic forms are expressed on pig trophoblast cannot be formally excluded. Human extravillous trophoblast subpopulations express the HLA-G molecules (Ellis *et al.*, 1986; Mc Master *et al.*, 1995), while the basal trophoblast of the rat expresses the so-called the PA molecules (Macpherson *et al.*, 1986; Kanbour *et al.*, 1987) both are unique monomorphic class Ib molecules. Recently, HLA-G was shown to be restricted to a differentiated cytotrophoblast (Mc Master *et al.*, 1995), to be co-dominantly expressed in first-trimester trophoblast cells (Hviid *et al.*, 1998), and to present peptides in a manner similar to that of polymorphic class I HLA molecules (Diehl *et al.*, 1996; Gobin *et al.*, 1997). The mAb directed against a monomorphic determinant of class I MHC detects both polymorphic and monomorphic MHC antigens in humans (Barnstable *et al.*, 1978). Therefore, it is possible that the mAb would also detect putative monomorphic forms of class I MHC molecules in the pig.

In the developing human embryo, trophoblasts directly contact the maternal tissues and could be the targets for the maternal immune cells. Extravillous interstitial and endovascular trophoblasts that invade the uterus and uterine blood vessels in early pregnancy also express the non-classical MHC class I molecules HLA-G and classical HLA-C. The mRNA for another non-classical MHC class I molecule, HLA-E, also is expressed on the placental tissues. The recent demonstration with *in vitro* experiments that HLA-G can activate the KIR expressed on the cells of lymphoid and myelomonocytic origin is consistent with the hypothesis that HLA-G plays an important role in establishing the maternal-fetal tolerance (Gobin *et al.*, 1997; Hviid *et al.*, 1998).

However, there is a paucity of *in vivo* evidence supporting the functional significance of trophoblast-immune cell interactions in human pregnancy. A more complete understanding of the *in vivo* roles of placental MHC class I molecules and HLA-G in particular, would be significantly advanced by the use of appropriate animal models. There is a close homology of the implantation and placentation among the Old World primates. The rhesus monkey MHC contains orthologs of the human classical MHC class I loci Mamu-A and -B, and the non-classical loci Mamu-E, -F, and -G. The identification of a novel MHC class I locus expressed in the rhesus monkey placenta that shares unique molecular characteristics of HLA-G and has a tissue distribution closely matching that seen for HLA-G provides a unique opportunity for the study of non-classical MHC class I molecules in the primate placenta (Boyson *et al.,* 1999).

2.2.2   Evolutionary Pressure of Alloimmunity

It is difficult to fathom the nature of the evolutionary pressure that led to the emergence of alloimmunity in vertebrates. Therefore, the likeliest seems the assumption that alloimmunity did not have a direct evolution, but was an accidental consequence or a by-product of the mechanism of the adaptive immunity. The strength and efficacy of the destructive mechanisms of alloimmune reaction increased from cartilaginous fish to mammals, due to the evolutionary advancement of the adaptive immunity, the mechanisms of recognition and killing, as well as the enhancement of the MHC molecules variability. However, although one of the most effective and vigorous models of immune reaction, alloimmunity still remained dormant as the factor of an evolutionary pressure, probably until the emergence of the first viviparous species. The reasons for this are simple - the transplantation of tissues does not exist as a natural phenomenon among vertebrates.

With the appearance of viviparity and placentation as the possible models of natural allotransplantation, the phenomenon of alloimmunity acquired the role of factor of selection pressure. Thus, a by-product of evolution of the adaptive immunity which emerged under the selection pressure of microbes and "latently" accumulated enormous strength eventually became the factor of selection pressure. With the emergence of viviparous mammals, the further evolution of the adaptive immunity and reproduction became closely associated with alloimmunity - a new powerful factor of selection pressure in a direct connection with the reproductive efficacy (*Figure* 2.3).

The selection pressure of alloimmunity, particularly in mammals, could have significantly influenced the evolution of the adaptive immunity, possibly by decreasing the MHC genes variability and reorganizing the mechanisms of "self-non-self" recognition. This, however, was not likely to have happened. In relation to all other vertebrates, the MHC genes variability has reached the culmination point in mammals. As in the case of auto-immunity, the selection pressure of alloimmunity caused the emergence of fine mechanisms for the control if the immune reaction on the one hand and the absence of the expression of the MHC molecules on placental tissues, on the other - except for monomorphic or low-polymorphic MHC class Ib molecules. The controlling mechanisms for the prevention of auto-immunity are very similar to the mechanisms of immunotolerance in pregnancy, and are based on the activity of sex steroids, immunomodulatory/suppressive cytokines, domination of Th2 type immune response, and the like. Therefore, the controlling mechanisms of immunotolerance in pregnancy could represent a more effective and sophisticated advancement of the controlling mechanisms for the prevention of auto-immunity (*Figure* 2.3).

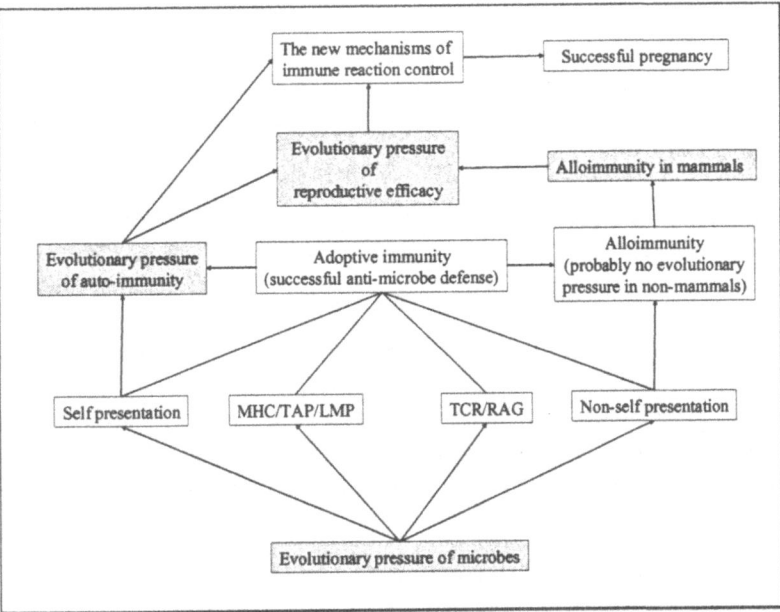

Figure 2.3. "Quiet" or indirect evolution of alloimmunity and its sudden expression as a factor of selection pressure in viviparous mammals.

The connection between alloimmunity and the reproductive efficacy in non-mammalian viviparous species is largely unexplored. The number of viviparous species of non-mammalian vertebrates, such as *squamate* and sharks, is relatively small (about 10%), so that within their class, viviparous non-mammalians are more an exception than a rule. In addition, the phenomenon of viviparity in different classes of vertebrates almost certainly has no common ancestor, i.e. it developed independently as an adaptation mechanism in response to diverse selection pressures.

By contrast to mammals, reptiles and especially sharks do not display a high MHC genes variability, nor is their immune system on the same organisational level as mammalian, so one should not expect significant immune conflicts to occur between the maternal immune system and placenta-like organs of non-mammalian species. We could, therefore, conclude that alloimmunity most probably holds very little or no significance as a factor of selection pressure in these species.

The question of how the influence of selection pressure of alloimmunity influenced the evolution of the adaptive immunity and controlling mechanisms of the immune reaction is a highly complex one. This phenomenon has been poorly explored even from the theoretic aspect, and the experimental data enable only a partial insight into the evolutionary events that possibly occurred. The answers to the following questions could serve to bring into a closer view the evolution of vertebrate immune system and the influence of the evolutionary pressure of alloimmunity on the developing of the adaptive immunity and powerful controlling mechanisms in mammals.

1. Is the ultimate grouping and clustering of MHC genes class I and II in mammals related to a better control of the MHC molecules expression, alloimmunity and reproductive efficacy? It is worth noting that equine are the only mammals which do not have class I and II genes located on the same chromosome, and also that equine have the epitheliochorial placenta.

2. How strong is the influence of the link between alloimmunity and pregnancy with "sudden escape" of TAP/LMP genes into the region of class II genes in mammals? In all other vertebrates, TAP/LMP genes are associated with class I genes.

3. Could alloimmunity in mammals be connected with the phenomenon of high class I genes variability? In some non-mammalian classes of vertebrates, class II genes are more variable than class I genes, whereas in mammals class I genes probably are the most variable genes in the genome. The significance of the innate immunity and class II molecules in initiating of alloimmune reaction is almost crucial. This is conclusive from the fact that the transplant having been purified from APCs, will not be rejected, but will almost certainly induce the tolerance of the recipient's immune cells.

4. What is the influence of alloimmunity in mammals on the evolution of class Ib genes, their low polymorphism or monomorphism?

5. Could the alloimmunity in mammals have been the factor of the evolutionary pressure, responsible for the emergence of controlling mechanisms mediated by Th1, Th2 and Th3 cells?

6. Is there an evolutionary connection between alloimmunity, pregnancy and position of TAP/LMP genes within class II region on one side, with the function of Th cells on the other?
7. What is the influence of alloimmunity in mammals on the diversification of cytokines and the emergence of specialised cytokines, functioning as suppressors/modulators of the immune reaction such as IL-4 and IL-10?
8. Are the adaptation mechanisms, emerging in mammals as a response to alloimmunity, in fact the "advanced" adaptation mechanisms that developed against auto-immunity - one of the forms of selection pressure closely associated with adaptive immunity?

The answers to these and other questions may be helpful in clearing the mystery of the evolution of the adaptive immunity as a powerful anti-microbial system. This system might have produced "along its way" at least two or even more side-effects, which were expressed as the factors of new forms of selection pressure, and took a major part in the evolutionary remodelling of the immune system itself.

REFERENCES

Ackerman R.A., Dmi'el R. (1984). Energy and water vapor exchange by parchment-shelled reptile eggs. *Physiol Zool.* 58:129-137.

Andrews R.M. (2002). Low oxygen: a constraint on the evolution of viviparity in reptiles. Physiol *Biochem Zool.* 75:145-154.

Antczak D.F., Allen W.R. (1989). Maternal immunological recognition of pregnancy in equids. *J Reprod Fertil Suppl.* 37:69-78.

Arnall J.C., Horton J.D. (1987). In vivo studies on allotolerance perimetamorphically induced in control and thymectomized *Xenopus. Immunol.* 62:315-319.

Auchincloss, H.J., Lee, R., Shea, S., *et al.* (1993). The role of "indirect" recognition in initiating rejection of skin grafts from MHC class II-deficient mice. *Proc Natl Acad Sci USA.* 90:3373-3377.

Bacon S.J., Ellis S.A., Antczak D.F. (2002). Control of expression of MHC genes in horse trophoblast. *Biol Reprod.* 66:1612-1620.

Bainbridge D.R. (2000). Evolution of mammalian pregnancy in the presence of the maternal immune system. *Rev Reprod.* 5:67-74.

Barnstable C.J., Bodmer W.F., Brown G. *et al.* (1978). Production of monoclonal antibodies to group A erythrocytes, HLA and other human cell surface antigens-new tools for genetic analysis. *Cell.* 14:9-22.

Blackburn D.G. (1982). Evolutionary origins of viviparity in the Reptilia. I. Sauria. *Amphib-Reptilia.* 3:185-205.

Blackburn D.G., Taylor J.M., Padykula H.A. (1988). Trophoblast concept as applied to eutherian mammals. *J Morphol.* 196:127-136.

Blackburn D.G. (1993). Chorioallantoic placentation in squamate reptiles: structure, function, development, and evolution. *J Exp Zool.* 266:414-430.

Blackburn D.G. (1995). Saltationist and punctuated equilibrium models for the evolution of viviparity and placentation. *J Theor Biol.* 174:199-216.

Blackburn D.G., Vitt L.J. (2002). Specializations of the chorioallantoic placenta in the Brazilian scincid lizard *(Mabuya heathi)*: a new placental morphotype for reptiles. *J Morphol.* 254:121-131.

Benichou G., Takizawa P.A. Ho P.T. *et al.* (1990). Immunogenicity and tolerogenicity of "self"-MHC peptides. *J Exp Med.* 172:1341-1346.

Benichou G., Takizawa A.P., Olson A.C. *et al.* (1992). Donor MHC (MHC) peptides are presented by recipient MHC molecules during graft rejection. *J Exp Med.* 175:305-308.

Benichou G., Fedoseyeva E., Lehmann P.V. *et al.* (1994). Limited T cell response to donor MHC peptides during allograft rejection. Implications for selective immune therapy in transplantation. *J Immunol.* 153:938- 945.

Benichou G., Fedoseyeva E., Tam R.C. *et al.* (1998). The presentation of "self" and allogeneic MHC peptides to T cells: Influence on transplant immunity. *Human Immunol.* 59:540-548.

Boyson J.E., Iwanaga K.K., Urvater J.A. *et al.* (1999). Evolution of a new nonclassical MHC class I locus in two Old World primate species. *Immunogen.* 49:86-98.

Botazzo G.F., Pujol-Borrell R., Hanafusa T. (1983) Role of aberrant HLA-DR expression and antigen presentation in induction of endocrine autoimmunity. *Lancet.* 2:1115-1119.

Cateni C., Paulesu L., Bigliardi E. et al (2003). The IL-1 system in the uteroplacental complex of a cartilaginous fish, the smoothhound shark, Mustelus canis. *Reprod Biol Endocrinol.* 1:25-30.

Ciubotariu R., Liu Z., Colovai A.I. *et al.* (1998). Persistent allopeptide reactivity and epitope spreading in chronic rejection of organ allografts. *J Clin Invest.* 101:398-405.

Cooper D.K.C. (1998). Xenoantigens and xenoantibodies. *Xenotransplantation.* 5:6-17.

Dausset J. (1981). The MHC in man. *Science.* 213;1469-1474.

Di Marzo S.J., Cohen N. (1982). An in vivo study of the ontogeny of alloreactivity in the frog *(Xenopus laevis).* *Immunol.* 451:39-48.

Diehl M., Munz C., Keilholz W. *et al.* (1996). Non-classical HLA-G molecules are classical peptide presenters. *Curr Biol.* 6:305-314.

Donaldson W.L., Zhang C.H., Oriol J.G. *et al.* (1990). Invasive equine trophoblast expresses conventional class I MHC antigens. *Development.* 110:63-71.

Du Pasquer L., Flajnik M. (1999). Origin and evolution of the vertebrate immune system. In Fundamental Immunology, 4th (ed W.E. Paul) Lippincot-Raven Publishers. Philadelphia. pp. 605-650.

Ellis S.A., Sargent L., Redman C.W.G. *et al.* (1986). Evidence for a novel HLA antigen found on human extravillous trophoblast and a choriocarcinoma cell line. *Imunol.* 59:595–601.

Fedoseyeva E.V., Tam R.C., Popov I.A. *et al.* (1996). Induction of T cell responses to a "self"-antigen following allotransplantation. *Transplant.* 61:679-683.

Fedoseyeva E., Zhang F., Orr P.L. *et al.* (1999). *De novo* autoimmunity to cardiac myosin after heart transplantation and its contribution to the rejection process. *J Immunol.* 162:6836-6842.

Feder M.E., Satel S.L., Gibbs A.G. (1982). Resistance of the shell membrane and mineral layer to diffusion of oxygen and water in flexible-shelled eggs of the snapping turtle *(Chelydra serpentina).* *Respir Physiol.* 49:279-291.

Forsdyke D.R. (1991). Early Evolution of MHC Polymorphism. *J Theor Biol.* 150:451-456.

Galili U., Mandrell R.E., Hamadeh R.M. *et al.* (1988). The interaction between the human natural anti-a galactosyl IgG (anti-Gal) and bacteria of the human flora. *Infect Immun.* 57:1730-1737.

Galili U. (1993). Interaction of the natural anti-Gal antibody with a galactosyl epitopes: a major obstacle for xenotransplantation in humans. *Immunol Today.* 14:480-482.

Gaspari A.A., Jenkins M.K., Katz S.I. (1988). Class II MHC-bearing keratinocytes induce antigen-specific unresponsiveness in hapten-specific TH1 clones. *J Immunol.* 141:2216-2220.

Gobin S.J.P., Wilson L., Keijsers V. *et al.* (1997). Antigen processing and presentation by human trophoblast-derived cell lines. *J Immunol.* 158:3587-3592.

Gogolin-Ewens K.J., Lee C.S., Mercer W.R. *et al.* (1989). Site-directed differences in the immune response to the fetus. *Imunol.* 66:312-317.

Guillette L.J.Jr. (1993). The evolution of viviparity in lizards. *Bioscience.* 43:742-751.

Hamlett W.C., Wourms J.P., Hudson J.S. (1985). Ultrastructure of the full term shark yolk sac placenta. I. Morphology and cellular transport at the fetal attachment site. *J Ultrastructure Res.* 91:192-206.

Hamlett W.C., Miglino M.A., DiDio L.J.A. (1993). Fine structure of the term umbilical cord in the Atlantic sharpnose shark *(Rhizoprionodon terraenovae). J Submicrosc Cytol Pathol.* 25:547-557.

Hashimoto K., Nakanishi T., Kurosawa Y. (1992). Identification of a shark sequence resembling the major histocompatibility complex class I α 3 domain. *Proc Natl Acad Sci USA.* 89:2209-2212.

Hayry P., VonWillebrand E. Partehnais E. *et al.* (1984). The inflammatory mechanisms of allograft rejection. *Immunol Rev.* 77:85-142.

Heulin B., Ghielmi S., Vogrin N., *et al.* (2002). Variation in eggshell characteristics and in intrauterine egg retention between two oviparous clades of the lizard Lacerta vivipara: insight into the oviparity-viviparity continuum in squamates. *J Morphol.* 252:255-262.

Hviid T.V., Moller C., Sorensen S. *et al.* (1998). Co-dominant expression of the HLA-G gene and various forms of alternatively spliced HLA-G mRNA in human first trimester trophoblast. *Hum Immunol.* 59:87–98.

Hughes A.L., Nei M. (1993). Evolutionary relationships of the classes of MHC genes. *Immunogen.* 37:337-341.

Itescu S., Kwiatkowski P., Atrip J.H. *et al.* (1998). Role of natural killer cells, macrophages and accessory molecule interactions in the rejection of pig-to-primate xenografts beyond the hyperacute period. *Human Immunol.* 59:275-286.

Kanbour A., Ho H-N., Misra D.N. *et al.* (1987). Differential expression of MHC class I antigens on the placenta of the rat. *J Exp Med.* 166:1861–1882.

Kaufmann P., Burton G. (1994). Anatomy and genesis of the placenta. In The Physiology of Reproduction 2nd Ed. (eds E. Knobil and J.D. Neill) Ch 8, pp 441-484, Raven Press, Ne w York.

Kaufman J., Salomonsen J. (1997). The "minimal essential MHC" revisited: both peptide-binding and cell surface expression level of MHC molecules are polymorphisms selected by pathogens in chickens. *Hereditas.* 127:67-73.

Lanza R.P., Cooper D.K.C., Chick W.L. (1997). Xenotransplantation. *Sc. Am.* 277:54-59.

Lehtonen L., Vainio O., Toivanen P. (1989). Ontogeny of alloreactivity in the chicken as measured by mixed lymphocyte reaction. *Dev Comp Immunol.* 13:187-195.

Leiser R., Kaufmann P. (1994). Placental structure: in a comparative aspect. *Exp Clin Endocrinol.* 102:122-134.

Low B.G., Hansen P.J., Drost M. *et al.* (1990). Expression of MHC antigens on the bovine placenta. *J Reprod Fertil.* 90:235-243.

Liu Z., Braunstein N.S., Suciu-Foca N. (1992). T cell recognition of allopeptides in context of syngeneic MHC. *J Immunol.* 148:35-40.

Lo D., Burkly L.C., Flavell R.A. *et al.* (1989). Brinster: Tolerance in transgenic mice expressing class II MHC on pancreatic acinar cells. *J Exp Med.* 170:87-104.

Macpherson T.A., Ho H-N., Kuntz H.W. *et al.* (1986). Localization of the Pa antigen on the placenta of the rat. *Transplant.* 41:392-394.

Malek A., Sager R., Kuhn P., *et al.* (1996). Evolution of maternofetal transport of immunoglobulins during human pregnancy. *Am J Reprod Immunol.* 36:248-255.

Maeno M., Nakamura T., Tochinai S., *et al.* (1987). Analysis of allotolerance in thymectomized *Xenopus* restored with semiallogeneic thymus grafts. *Transplant.* 44:308-314.

Maher J.K., Tresnan D.B., Deacon S. *et al.* (1996). Analysis of MHC class I expression in equine trophoblast cells using in situ hybridization. *Placenta.* 17:351-359.

Matzinger P., Bevan M.J. (1977). Hypothesis: Why do so many lymphocytes respond to major histocompatibility antigens. *Cell Immunol.* 29:1-7.

Minanov O., Itescu S., Neethling F. *et al.* (1997). Anti-Gal IgG antibodies in sera of newborn humans and baboons and its significance in xenotransplantation. *Transplant.* 63:182-186.

Mc Kinney E.C., Schmale M.C. (1992). Primed allogeneic reactions to leukocytes and epithelial cells in the bicolor damselfish. *Transplantat.* 54:313-317.

Mc Master M.T., Librach C.L., Zhou Y. *et al.* (1995). Human placental HLA-G expression is restricted to differentiated cytotrophoblasts. *J Immunol.* 154:3771-3778.

Nakanishi T., Aoyagi K., Xia C. *et al.* (1999). Specific cell-mediated immunity in fish. *Vet Immunol Immunopathol.* 72:101-109.

Nelson R.J., Demas G.E. (1996). Seasonal changes in immune function. *Q Rev Biol.* 71:511-548.

Novacek M.J. (1992). Mammalian phylogeny : Shaking the tree. *Nature.* 356:121-125

Padykula H.A., Taylor J.M. (1982). Marsupial placentation and its evolutionary significance. *J Reprod Fertil Suppl.* 31:95-104.

Paulesu L. (1997). Cytokines in mammalian reproduction and speculation about their possible involvement in nonmammalian viviparity. *Microsc Res Tech.* 38:188-194.

Ramsoondar J.J., Christopherson R.J., Guilbert L.J. *et al.* (1999). Lack of Class I MHC Antigens on Trophoblast of Periimplantation Blastocysts and Term Placenta in the Pig. *Biol Reprod.* 60:387-397.

Rothchild I. (2003). The yolkless egg and the evolution of eutherian viviparity. *Biol Reprod.* 68:337-357.

Rose A.G., Cooper D.K.C. (1996). A histopathologic grading system of hyperacute (humoral, antibody-mediated) cardiac xenograft and allograft rejection. *J Heart Lung Transplant.* 15:804-817.

Saad A.H., El Ridi R. (1984). Mixed leukocyte reaction, graft-versus-host reaction, and skin allograft rejection in the lizard, Chalcides ocellatus. *Immunobiol.* 166:484-93.

Sandrin M.S., Vaughan H.A., Dabkowski P.L. *et al.* (1993). Anti-pig IgG antibodies in human serum react predominantly with Gal(a 1-3)Gal epitopes. *Proc Natl Acad Sci USA.* 90:11391-11395.

Snell G.D. (1981). Studies in histocompatibility. *Science.* 213:172-178.

Smith S.A., Shine R. (1997). Intraspecific variation in reproductive mode within the scincid lizard Saiphos equalis. *Aust J Zool.* 45:435-445.

Shoskes D.A., Wood K.J. (1994). Indirect presentation of MHC antigens in transplantation. *Immunol Today.* 15:32-38.

Shmidt G.A. (1979). Evolution of the allantoid placenta in placental mammals. *Arkh Anat Gistol Embriol.* 77:5-16.

Tsuji S., Char D., Bucy R.P. *et al.* (1996). $\gamma\delta$ T cells are secondary participants in acute graft-versus-host reactions initiated by CD4+ $\alpha\beta$ T cells. *Eur J Immunol.* 26:420-427.

Qualls C.P. (1996). Influence of the evolution of viviparity on eggshell morphology in the lizard *(Lerista bougainvillii). J Morphol.* 228:119-125.

Weekes H.C. (1927). Placentation and other phenomena in the scincid lizard *Lygosoma (Hinulia) quoyi. Proc Zool Soc.* 52:499-554.

Weekes H.C. (1935). A review of placentation among reptiles, with particular regard to the function and evolution of the placenta. *Proc Zool Soc.* 2:625-645.

Chapter 3

Anti-tumor Immunity Failure in Mammals

3.1   MECHANISMS OF ANTI-TUMOR IMMUNITY

It must be assumed that all tumor cells produce proteins which do not belong to a normal cell. These are called the tumor-associated or tumor-specific antigens. In the classic immune surveillance theory it is believed that the cellular immune defence continuously discovers and eliminates newly arisen tumor cells which express such tumor-specific antigens. Since then it has been shown that one of the preconditions for the T cell system to be able to recognize antigens is that they are presented by MHC class I molecules. There is a continual processing and presentation of all intracellular proteins in a cell. Thus, a tumor cell which produces an abnormal protein will also present this and thereby expose itself to being killed by cytotoxic T cells. The antigens are presented in the form of short peptides (8-9 aminoacids), which arise as a result of controlled degradation of the original proteins. The peptides thus formed are transported by specialised molecules in the so-called endogenous antigen processing and presentation pathway, and are eventually bound to and presented by MHC class I molecules. It has been shown that many tumors express less MHC class I on their surface compared to the normal tissue from which they have arisen, and also that patients with a reduced immune function have an increased incidence of certain forms of cancer. It is therefore widely believed that a low MHC class I level and weak immune response contributes to the ability of tumor cells to avoid the T-cell-mediated immune defence (Garrido et al., 1997). Except this mechanism, tumor cells generally do not activate antigen-specific T cells. Many reasons have been put forward to explain the poor immunogenicity of most tumors, e.g., the production of

85

inhibitory factors that suppress an effective immune response, lack of co-stimulatory molecules and poor maturation of tumor-surrounding/infiltrating dendritic cells, anti-inflammatory cytokine network, elevated Th2 activity and antigen escape etc. (Palmer *et al.*, 1997; Kufer *et al.*, 2001).

3.1.1   Antigen Processing/Presenting Machinery

There are two major pathways of antigen processing within the APCs and target cell: the endogenous and exogenous. The endogenous pathway processes proteins that have been synthesised within the APCs. The tumor associated antigens undergo the same or similar processing, as all other intracellular and extracellular molecules which are presented to the effectory immune cells through APCs. The processing of antigens is a multi-step process which involves: antigen uptake, the degradation of the molecules, binding of fragments (peptide) to the newly synthesized MHC molecules, transport and expression of the MHC molecule/peptide complex on the cell surface (Germain *et al.*, 1994; Cella *et al.*, 1996, 1997).

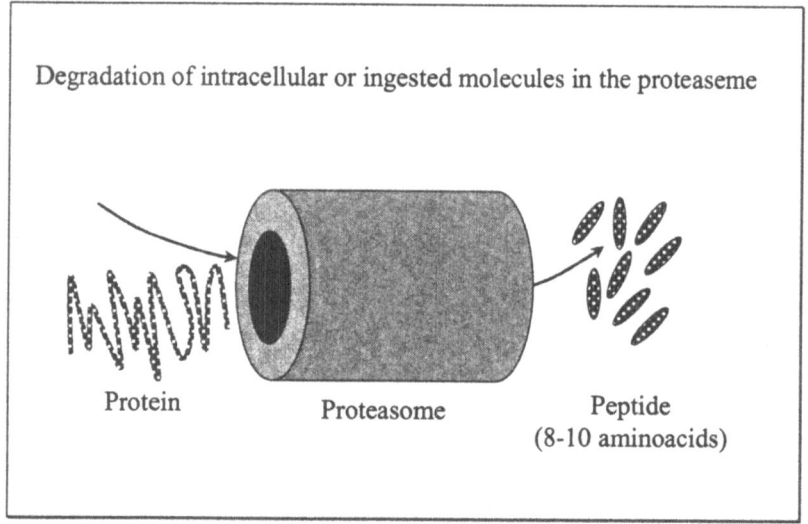

Degradation of intracellular or ingested molecules in the proteaseme

Protein   Proteasome   Peptide
(8-10 aminoacids)

Figure 3.1. Ingested or synthesised molecules in the cytoplasm undergo limited
proteolytic degradation by proteasome.

As with all cytoplasmic proteins, the "non-self" molecules are continuously degraded via the 26S proteasome complex into 8-10 aminoacids long fragments. Proteasomal components also include molecules such as LMP2 and LMP7. Proteasome complex activity varies from minimal to very high. One of the most important activators of proteasome, related to the immune reaction and final presentation of antigens, is IFN-γ. In mammals, LMP2 and LMP7 genes are associated with class II genes on the same chromosomal sequence, which makes the mechanism of antigen processing and class II molecules expression a joined, well-coordinated action (Zhang *et al.*, 2001; Guo *et al.*, 2002).

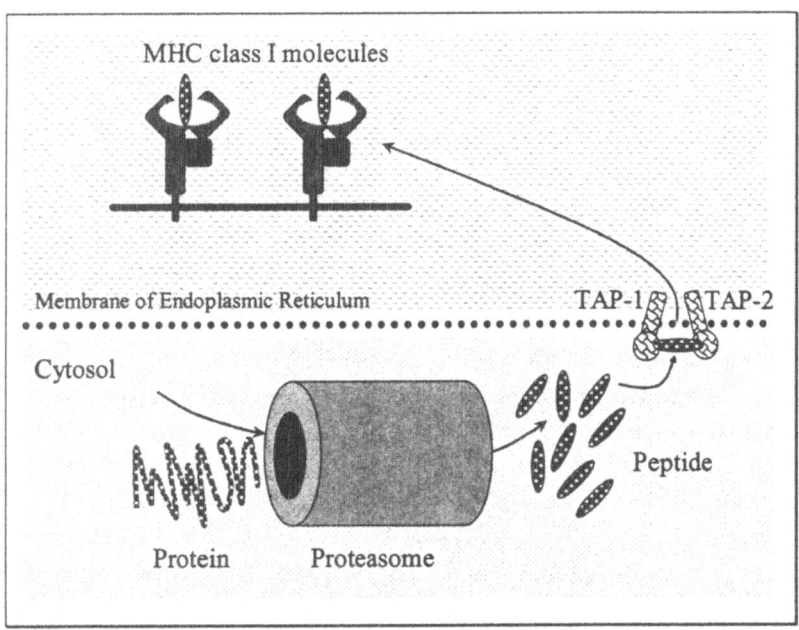

Figure 3.2. Transporters associated with antigen processing and accessing of peptides to the ER in order to be loaded onto MHC class I molecules.

The expression of cell surface MHC class I/peptide complex requires a coordinated transcription of multiple genes such as MHC class I heavy chain, β2m, TAP1, TAP2, LMP2 and LMP7. All of these genes are expressed and defined at distinct levels in normal tissues, and are inducible by IFN-γ. There are two independent elements that are sufficient to activate the transcription of a reporter gene. One (hereby called TAP2 P1) is located 5' to the TAP2 exon 1, while the other (hereby called TAP2 P2) is a transcription initiator residing in intron 1. The analysis of the 5' sequence of TAP2 mRNA indicates that both promoters are active. Moreover, while the TAP2 promoter region contains *cis* elements that can mediate TAP2 induction by IFN-γ, such as γ-activation site and IFN Response Factor Binding Element (IRFE). Only the IRFE is required for IFN-γ induction of

TAP2 promoter *in vitro*. The IRFE appears to work as an enhancer for the initiator (P2). Together with another promoter recently identified by others, TAP2 therefore has three independent promoters that can be differentially regulated (Lobigs *et al.*, 1999; Knittler *et al.*, 1999; Guo *et al.*, 2002).

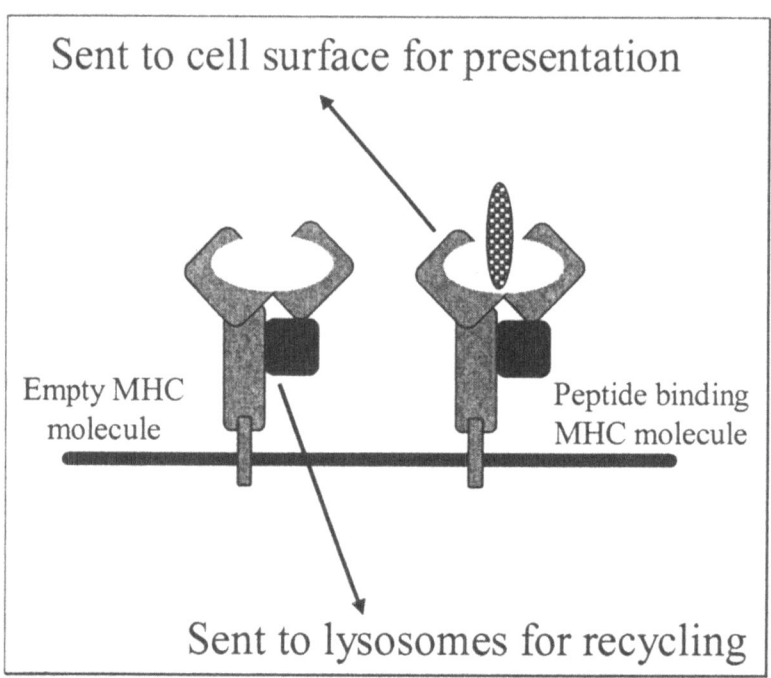

Figure 3.3. Fate of MHC class I molecules depends on peptide binding.

During proteasome mediated degradation of antigens, the MHC molecules are synthesized in ER, but before the products of proteasomal activity bind the MHC molecule, they first need to be transported from cytosol to ER. TAP1 and TAP2 are known to mediate in the control of immune reaction, and function as the transporters of peptides. The importance of their function has been demonstrated on the cells with mutated TAP genes (Lobigs *et al.*, 1999; Knittler *et al.*, 1999). This mutation results in the lessening of the number of expressed MHC molecules, i.e., an aberrant intracellular expression of the MHC/peptide complex. The transfection of the normal TAP genes into mutant cells restores the stable surface MHC expression. The MHC molecules which did not bind the peptide cannot be expressed on the cell surface. Moreover, they show a high instability, so they are sent to undergo a proteolytic degradation and recycling (*Figure* 3.3.). TAP activity can be changed under the influence of various factors, such as:

cytokines, viruses, hormones, prostaglandine etc. The dependence of the MHC expression on the activity of TAP molecules can be demonstrated on the cell infected with the Herpes Simplex Virus (HSV) and adenoviruses.

HSV inhibits TAP1 and TAP2 proteins, preventing the processing of the peptides towards MHC molecules. Most of the MHC molecules which have not bound the peptide are being recycled, while only a small number of those binding the viral or some other peptide will be expressed on the cell surface. The HSV-infected cell expresses a low number of the MHC molecules, and will most probably be eliminated by NK cells, not by CTL. Adenoviruses "block" the already expressed MHC/peptide complex, thus inhibiting its recycling and increasing its half-time. The result of adenovirus activity is the increase of the number of MHC/peptide complexes (mostly non-viral) on the cell surface (Lobigs *et al.*, 1999; Knittler *et al.*, 1999; Zhang *et al.*, 2001; Guo *et al.*, 2002).

It is necessary for the activation of effectory mechanisms of anti-tumor response that the antigen, i.e. its peptide, is first presented to the T cells on APCs. Although macrophages and partly B lymphocytes play the role of APCs, dendritic cells have shown the highest potential for stimulating the anti-tumor immune response (Cella *et al.*, 1997). Non-activated, or commonly defined as "immature", dendritic cells (DCs), whether they are Langerhans or the so-called intestinal DCs, have a high potential for incorporating the antigens from the external environment. The mechanism of antigen internalisation by DCs works mainly in two ways. According to the first model, antigen internalisation is carried out via mechanism of macropinocytosis, while the second model precludes the interaction of membranous receptors (mannose or Fc-receptor). From the aspect of stimulation of anti-tumor effectory mechanisms, it is important whether the antigen will be internalised via pinocytosis or the mannose-receptor. The effectory mechanisms developing after "sugar-dependent way" of antigen internalisation are 100-10,000 times more intensive than those developing after the internalisation of antigens via pinocytosis. The mechanism of antigen molecule processing after the internalisation via Fc-receptor depends on the proteolytic activity of proteasome, whereas the expression of MHC/peptide complex depends on the transporting mechanisms including TAP1 and TAP2 molecules (Cella *et al.*, 1997; Guo *et al.*, 2002).

There is yet another mechanism of antigen internalisation and its further processing. This mechanism could be described as an apoptotic cell dependent way. It has been proved that the DCs containing vesicles with the parts of apoptotic cells undergo a direct processing towards the class I. The ability of DCs to undergo phagocytosis of the apoptotic cells depends on the expression of membranous receptors, such as: $\alpha\beta$-5-integrin and CD36. The expression of these two molecules is down-regulated by the maturation of

DCs, so that phagocytic ability of activated and matured DCs decreases (Konig *et al.* 1992; Cammarota, 1992; Flores-Romo *et al.*, 2001; Zou *et al.*, 2002).

In the exogenous pathway, soluble proteins are taken up from the extracellular environment, generally by specialised or "professional" APCs. The antigens are then processed in a series of intracellular acidic vesicles, called endosomes. During this process, the endosomes intersect with vesicles that are transporting MHC class II molecules to the cell surface. CD4$^+$ T cells recognise antigens that are presented by MHC class II molecules. As with CD8, the CD4 molecule functions as a co-receptor, increasing the strength of the interaction between the T cell and the APCs (Konig *et al.* 1992; Cammarota, 1992). For both systems of antigen presentation, the recognition of the antigen by the T cells is described as being MHC restricted; that is, the T cells recognise only antigen presented by "self" MHC molecules. The nature of the process of antigen recognition by T cells has profound implications on the design of assays to measure T cell function (Konig *et al.*, 1992).

3.1.2   Antigen Recognition by TCR

T cells recognise protein antigens in the form of antigenic peptide fragments that are presented at the cell surface by MHC class I or MHC class II molecules. When the antigen-specific TCR on the T cell surface (specifically the *zeta* chains of the CD3 complex) interacts with the appropriate MHC/peptide complex, it triggers phosphorylation of the intracellular domains of the CD3 *zeta* chains. Subsequently, the Zeta Associated Protein-70 (ZAP-70) binds to the phosphorylated zeta chains, and is activated. The simultaneous co-ligation of the cell marker CD4 (or CD8) with the MHC class II (or class I) molecule results in the phosphorylation of the *lck* kinases. These events stimulate the activation of at least three intracellular signalling cascades. T cell activation also requires a second co-stimulatory signal such as the interaction between the cell markers CD28 on the T cell, and CD80 on the antigen-presenting cell. This interaction also triggers several intracellular signalling pathways. The activation of T cells can lead to the cell division, lymphokine secretion by the T cell and expression by the T cell of antigens associated with the activated state. Alternatively, in the case of CTLs, interaction with antigen via the specific TCR leads to the destruction of target cells (Elder, 1998; Vu, 2000).

3.1.3 DSs Maturation/Activation

DCs play a critical role in the activation of naive T lymphocytes and in the generation of primary T cell responses (Cella *et al.*, 1997; Steinman *et al.*, 1998). These cells reside in a resting or immature state in non-lymphoid tissues, where they efficiently capture and process antigens. Upon the activation they initiate a differentiation process that results in the decreased antigen-processing capacities, enhanced expression of MHC and co-stimulatory molecules, and migration into the secondary lymphoid organs, where they trigger naive T cells. This *in vivo* the maturation process is efficiently regulated and controlled by a complex array of signals present in the DCs microenvironment. A number of cytokines and other factors have been proposed to promote DCs growth and differentiation from myeloid progenitor cells, including GM-CSF, TNF, IL-4, and possibly also stem cell factor, IFN-γ, TGF-β, PGE2 and others (Cella *et al.*, 1997; Steinman *et al.*, 1998). *In vitro*, and possibly *in vivo* as well, inflammatory stimuli such as (IL-1, IL-6, and TNF-α) (Jonuleit *et al.*, 1997) and contact with T cells (via CD40/CD40L interaction) (Cella *et al.*, 1996) further activate DCs, resulting in mature DCs with strong T cell stimulatory potential.

Although DCs make up a system which spreads diffusely across all tissues and organs, their function is often dependent on microenvironmental events. For these reasons, the experimental models of DCs examining are primarily based on *in vitro* generated cultures. CD34$^+$ bone marrow stem cells or circulating blood monocytes are most often used for the generation of DCs cultures. Based on the findings obtained from *in vitro* research, DCs in non-activated state show a high affinity for the ingestion of exogenously derived antigens. After maturation, DCs down-regulate the ingestion affinity and up-regulate the antigen presenting mechanisms. The process of maturation of DCs is followed by an increased expression of the whole set of membranous molecules, such as CD1, CD83, OX40L, CD80, CD86, CD40, class I and II MHC molecules. After the expression of these and other molecules, the activated DCs assume the character clearly discernible from other immune cells and microenvironment. Type, number and the relationship between the molecules expressed on DCs will essentially define the character of the immune reaction (Cella *et al.*, 1997; Flores-Romo *et al.*, 2001; Zou *et al.*, 2002).

The nature of antigens and type of the pre-existing cytokine network in DCs microenvironment play a critical role in the process of maturation/activation of these cells. Monocytes incubated in the presence of cytokines, such as GM-CSF and IL-13 soon gain the receptor repertoire of the activated DCs, including the up-regulation of CD1 and MHC molecules. The non-activated DCs very quickly recycle class II MHC molecules, with a half-time of about 10 hours; however, in the presence of pro-inflammatory cytokines, the up-regulation of synthesis and expression of the MHC molecules also occurs, though with a slowdown in the recycling of these molecules, whose half-time is now about 100 hours. The increased synthesis and cumulative effect of the slower recycling of class II molecules results in the significant increase of the number of these molecules on DCs, as well as their better communication with the Th cells (Flores-Romo *et al.*, 2001; Zou *et al.*, 2002).

DCs can be activated by the apoptotic cells or particles as a result of the apoptotic cell degradation. However, only a small number of apoptotic particles will not lead to DCs activation, which favours the hypothesis that DCs will not activate in the cases of physiological death of one cell or a small group of cells. Massive cell apoptosis (viral infections, hypoxic necrosis) results in the high number of apoptotic particles, which may represent a strong signal for the activation of DCs, leading consequently to the activation of the cellular immunity. DCs activation induced by apoptotic signals may lead to the presentation of "self" molecules to effectory immune cells; however, in the absence of inflammatory stimuli or "danger signals", the presentation of "self" molecules probably will not cause the activation and clonal expansion of "self-reactive" CTL. Nevertheless, in the presence of pro-inflammatory cytokines and "danger signals" auto-immunity is the possible consequence of the presentation of "self" peptides by the activated DCs (Cella *et al.*, 1997; Steinman *et al.*, 1998; Buchler *et al.*, 2003).

3.1.3.1   Heat Shock Proteins (HSP)

According to the theory of the "danger signal", the first step in the activation of effectory mechanism is the releasing of "danger signals" from the cells with disturbed function. This signal comes from the virus-infected cells, tumor cells, mutated cells, cells in hypoxia or cells which may in any way jeopardize the environment. This signal is the first warning of the cell to its microenvironment and to immune cells that "something has happened" (Matzinger, 1994).

The expression of MHC/peptide complex on the endangered cell could be the second step in the activation of effectory mechanisms and, in the light of the theory of "dangerous signal", one of the mechanisms of precise identification and fine selection of the endangered cell. "Dangerous signal" is presumed to have a non-specific nature, while the specific character of the immune reaction is obtained through the expression of MHC/antigen peptide complex, as well as its recognition by Th and CTL. Although being non-specific, the "dangerous signal" is necessary for the developing of the immune reaction, since a full antigen expression in the absence of the "dangerous signal" may produce a weak immune reaction, tolerance, or even the antigen protective manner of immune reaction (Singh-Jasuja *et al.,* 2000, 2001).

LPS, viruses, cytokines, necrotic death of the cell or its peptide, all represent different forms of the "dangerous signal". However, the research carried out on dying cells showed that the family of HSP molecules play a key role in the molecular basis of "dangerous signals". HSPs are protein molecules which are involved in the cell metabolism on various levels. The control mechanisms of HSPs synthesis and secretion have not been clearly identified, probably because a great number of poorly defined factors, e.g. "cell distress", are involved in these mechanisms. The increase of "intracellular stress", such as occurs during viral infection or hypoxia, results in the up-regulation of HSPs gene expression and accumulation of the proteins. Surprisingly, HSPs can bind peptides and take part in their transport directly to MHC class I molecules (Singh-Jasuja *et al.,* 2000, 2001; Milani *et al.,* 2002).

HSPs work as activating signals for macrophages, DCs and T cells. The tissue containing cells "under stress" is infiltrated by lymphocytes, macrophages and immature DCs. The immature DCs can directly present antigens to the T cells by HSP/antigen peptide complex through the cross priming mechanism. However, the release of HSPs does not occur when the cell dies under physiological conditions, which makes apoptotic particles non-immunogenic or poorly immunogenic. In the case of microenvironmental "cell distress", a huge amount of HSPs is released, thus making the apoptotic particles highly immunogenic for DCs. Massive cell apoptosis in the absence of HSPs may also induce the activation and maturation of DCs (Singh-Jasuja *et al.,* 2000, 20001; Milani *et al.,* 2002).

Sauter *et al.* (2000) propose that HSPs from necrotic, but not apoptotic cells, can induce DCs maturation. The maturation was characterized as the up-regulation of costimulatory molecules like HLA-DR, CD40, CD86, as well as the development of DCs maturation restricted markers such as CD83 and DCs lysosome-associated membrane glycoprotein, and enhanced stimulatory capacity of both $CD4^+$ and $CD8^+$ T cells. Only factors released by necrotic transformed cells rather than primary cells such as monocytes or T cells were effective in this regard. Notably, neither apoptotic cells nor their culture supernatants induced maturation. Similar distinctions between necrotic and apoptotic transformed cell lines have been made with murine DCs (Gallucci *et al.,* 1999). However, the factor(s) responsible for this effect have yet to be fully characterized. Many authors have shown that HSPs, conserved molecular chaperones within all cells, are released by necrotic but not apoptotic cells (Basu *et al.,* 2000; Berwin *et al.,* 2001). For example, necrotic cell lysates induce the activation of murine DCs and the partial up-regulation of MHC class II and costimulatory molecules. Purified HSPs (gp96 and hsp70) stimulate peritoneal macrophages to produce IL-1, TNF-α, and IL-12 and induce the maturation of murine DCs (Basu *et al.*, 2000). The injection of gp96 into mice leads to the migration of DCs in draining lymph nodes (Binder *et al.*, 2000), as well as the maturation of human DCs (Singh-Jasuja *et al.*, 2000). These results suggest that the release of HSPs during cell fragmentation may be a critical determinant of APCs activation.

3.1.3.2   Role and Expression of Co-Stimulatory Molecules

The maturation of DCs is not a simple transition of the cell from the inactive to active state. Besides, these are not the only states in which DCs can be found. Between them, there is an entire scale of interstates which, depending on the microenvironmental conditions, may become the definite state or the final stadium of DCs maturation. Because of the important role it plays in the regulation of the immune reaction, the degree of DCs maturation is determined from the number, type and ratios of various, mostly antigen-presenting and co-stimulatory molecules. The absence of co-stimulatory molecules on the relatively immature DCs which presents MHC/peptide complex causes the tolerance or even extrathymic, microenvironmental clonal deletion of lymphocytes. The mature and activated DCs express a high number of different co-stimulatory molecules, which in the presence of pro-inflammatory cytokines strongly activate all killer mechanisms of the immune system. The most important co-stimulatory mechanisms that can be verified on a mature DCs are: CD40, CD80, CD86, CD554 and OX40 (Chaux *et al.*, 1996; Enk *et al.*, 1997; Pasquini *et al.*, 1997; Kufer *et al.*, 2001).

The degree of T cell activation is usually dependent on at least two signals provided by an APC. The first signal is antigen-specific and is mediated through the T cell receptor via a MHC/peptide complex on the APCs. A second signal (as well as any other additional signals) is required for T cell cytokine production and proliferation. This second signal is mediated by the interaction of one or more co-stimulatory molecules expressed on the surface of APCs with its ligand on the T cell surface. The degree of efficacy of DCs in activating T cells seems to be related, at least in part, to the degree and the level of expression of certain costimulatory molecules on the DCs (Garrido *et al.*, 1997).

One of the factors of the inefficacy of anti-tumor immune response may be contained within the incomplete maturation of DCs and inadequate expression of co-stimulatory molecules. There is a lot of evidence suggesting that tumor-infiltrating DCs of the majority of malignant tumors poorly express CD80 and CD86, whereas the level of class II expression is not affected. The DCs isolated from the progressing melanoma express a low number of CD86 molecules, secrete significantly high amount of IL-10 and induce anergy of CD4⁺ cells, whereas the DCs from regressing melanoma express CD80 and CD86 molecules, secrete IFN-γ and IL-12, and do not induce the anergy of CD4⁺ cells (Chaux *et al.*, 1996; Enk *et al.*, 1997; Pasquini *et al.*, 1997).

Although CTL are main effectory cells, the activation and help by Th cells are necessary for achieving the full effects of anti-tumor immune reaction. The cooperation between Th and CTL is mainly reached indirectly, the mediation by DCs. There are a number of presumed models of the interaction between Th, CTL and DCs. One of the possibilities is that Th and CTL communicate through the same DCs, but more likely is the assumption that the activated DCs comprise the dominant microenvironmental subpopulation of APCs, which by the developing of the specific cytokine network determine the behaviour and communication of Th and CTL. Regardless of the model of intercellular communication, the expression of co-stimulatory molecules, such as CD40, CD80 and CD86 on DCs represents an important factor in the further course of the immune reaction (Kuchroo *et al.*, 1995; Chaux *et al.*, 1996; Pasquini *et al.*, 1997; Kufer *et al.*, 2001).

After the expression of MHC/peptide complex on DCs and the activation of Th cell, the number of CD40 ligands (CD40L) on Th cells begins to increase, which results in the strong interaction of DCs and Th cells through CD40-CD40L complex. This signal initiates the secretion of Th1 cytokines, which are of critical importance for the clonal expansion of CTL and maintenance of DCs in the state of "good condition". Moreover, the activation of CD40 molecule results in the increased secretion of cytokines by the same DCs, so that microenvironment of the activated DCs after the interaction with Th lymphocytes abounds in the cytokines, such as INF-α, IL-1β, IL-2, IFN-γ and IL-12. Of all these, IL-12 produces the strongest effects on the activity and maturation of DCs (Enk *et al.*, 1997; Pasquini *et al.*, 1997; Kobata *et al.*, 2000).

A member of TNF family, labelled as OX40 molecule, represents a significant co-stimulatory molecule in the activation of T lymphocytes. The interaction between OX40 and OX40 ligand (OX40L) results in the significant increase of the secretion of cytokines like IL-12, IL-1β and IL-6. A direct consequence of the secretion, along with the activation of OX40L⁺ Th lymphocytes, causes prolonged T cell proliferation and long-term Th memory. Besides, the interaction OX40-OX40L is involved in the mechanisms of B cell activation and the phenomenon of T cell homing. In the experiments with OX40L⁻ᐟ⁻ deficient mice, OX40 molecule was proved to function as a co-stimulatory receptor which correctly stimulates DCs. DCs obtained from these mice are not able to correctly stimulate and coordinate T cell response, as well as the production of cytokines (Weinberg *et al.*, 1998a, 1998b).

3.1.3.3   DCs and Th1/Th2 Dichotomy

Although the type of cytokine network in the anti-tumor immune response is not predetermined, it is of vital importance for the final effects of the anti-tumor immunity. The type of dominant cytokines depends largely on the degree of DCs maturation and the type of pre-existing cytokine network. Besides, the secretion of certain cytokines in the tumor environment depends on various microenvironmental and current circumstances, like the pre-existing viral infection, the activity of resident viruses, pregnancy, immunosuppressive therapy, auto-immune focuses, the cytokine activity of tumor cells, etc. (Mantovani *et al.*, 1998; Tao *et al.*, 1997; Kufer *et al.*, 2001; Bubanovic, 2003d). Basically, the cytokine network profile is polarized between Th1 type of cytokines (IL-2, IL-12, IFN-γ and TNF-α) and Th2 type (IL-4, IL-6 and IL-10). The Th1 and Th2 types are only two extreme states of the phenomenon of Th1/Th2 dichotomy. In reality, the type of cytokine network resides in one of the many interstates, still not defined enough, between Th1 and Th2 extremes (Parra *et al.*, 1993; Kuchroo *et al.* 1995; Mantovani *et al.*, 1998; Kufer *et al.*, 2001).

The cytokines from Th1 group stimulate the activity of CTL, NK cells and APCs, acting as strong generators of the anti-tumor immunity. Unlike these, Th2 cytokines inhibit the secretion of Th1 cytokines, as well as the functions of CTL, NK, APCs and Th cells. Th2 cytokines stimulate the production of the class IgG$_1$ antibodies, apart from the activation of CTL (O' Garra *et al.*, 1998; Mantovani *et al.*, 1998; Kufer *et al.*, 2001). The immune cells isolated from a large number of different types of human tumors secrete high amounts of Th2 cytokines, whereas the secretion of Th1 cytokines is on a low level. The concentration of Th2 cytokines in the tumor tissue is significantly higher than in healthy tissues, while the concentration of Th1 cytokines is extremely low. Regressing tumors reveal a quite inverse picture of the cytokine network, with the domination of Th1 cytokines and strong cellular anti-tumor immune response (Bianchi *et al.*, 1996; Kufer *et al.*, 2001).

In the case of CD4$^+$ T lymphocytes, the typical of Th1 cells secrete IFN-γ, while the Th2 cells secrete cytokines like IL-4 and IL-5, but none of which are produced by naive CD4$^+$ cells (Palmer et. al., 1997). Interestingly, the costimulation through CD80 led to strong secretion of IFN-γ, while no IL-4 and IL-5 was detectable. Thus, naive CD4$^+$ T cells primed in the presence of the exclusive costimulatory signal CD80 could be shown to predominantly differentiate into Th1 cells. Virtually identical results were obtained in culture of naive CD4$^+$ T lymphocytes essentially free of contaminating MHC

class II positive cells. In contrast to CD80 and CD86 mediated costimulation, targeted LFA-3, although inducing a strong T cell proliferation,
did not provide naive CD4$^+$ T lymphocytes with a second signal sufficient
for T cell priming as indicated by CD45 isotype expression and cytokine
secretion (Kuchroo *et al.*, 1995; Palmer et. al., 1997).

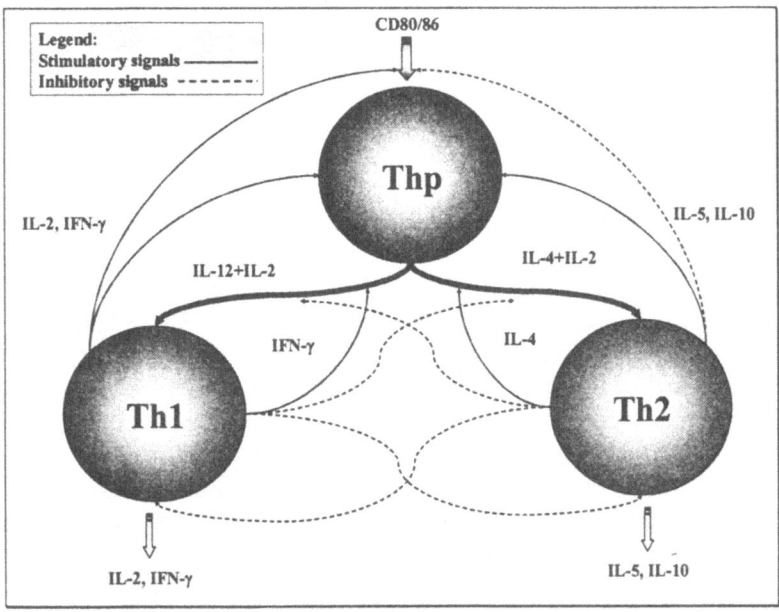

Figure 3.4. Crossroads of Th cells differentiation and maturation.

Several factors have been implicated in the regulation of naive Th cell
differentiation into Th1 or Th2 effectors: (i) the intensity of the primary
signal, (ii) the nature of the second signal, i.e. CD80 or CD86-mediated
costimulation, (iii) the expression of further accessory molecules such as
ICAM-1 or the OX40L by APCs, and (iv) soluble differentiation factors
such as IL-12 and IL-4. It has been reported that a weak primary signal
together with CD28-dependent costimulation induces Th2 differentiation
(Kuchroo *et al.*, 1995; Palmer et. al., 1997; Tao *et al.*, 1997; Kufer *et al.*,
2001). With murine T cells CD80 and CD86 co-stimulatory molecules have
previously been found to distinctly trigger Th1 and Th2 differentiation
respectively (Kuchroo *et al.*, 1995), while in some experiments CD86
costimulatory signal clearly induced a Th1 cytokine pattern (Kufer *et al.*,
2001).

Besides the primary and the secondary T cell signals, a couple of "third signal", either mediated by further accessory molecules or by soluble differentiation factors, have been implicated in Th1/Th2-polarization. Salomon *et al.* (1998) and Luksch *et al.* (1999) reported that the expression of ICAM-1 on antigen presenting cells inhibits the Th2 developmental pathway in the murine system. Accordingly, it has been speculated that ICAM-1 may favour Th1 differentiation of human CD4$^+$ T cells as well. However, some authors clearly demonstrate the development of Th1 cells although the tumor cell line used for the stimulation of naive CD4$^+$ T lymphocytes does not provide any autochthonous ICAM-like signals (Parra *et al.*, 1993). Moreover, Kufer *et al.* (2001) have not found any changes in Th1 polarization in the presence of targeted ICAM-1. Thus, results of these authors do not support the view that ICAM-1 exerts a major influence on Th1/Th2 differentiation.

According to the minimal signalling conditions that were applied, Th1 polarization does not seem to require additional signals. The absence of Th2 differentiation under minimal priming conditions indicates that human CD4$^+$ lymphocytes absolutely require a "third signal" in order to follow the Th2 developmental pathway. The "third signal" may be mediated by antigen-related factors or by ligation of the OX40 molecule through APCs carrying the OX40L, which has recently been shown to promote high levels of IL-4 production in naive human CD4$^+$ cells (Ohshima *et al.*, 1998). IL-4 is known to be required as a soluble differentiation factor for the development of Th2 cells, which themselves are a major source of IL-4 in a kind of positive feedback mechanism (O' Garra *et al.*, 1998). Accordingly, addition of IL-4 induced naive CD4$^+$ T cells to enter the Th2 developmental pathway under the described priming conditions. Another important soluble factor implicated in Th1 polarization is IL-12, which is produced by APCs (Trinchieri *et al.*, 1995). However, Kufer *et al.* (2001) found that IL-12 proved to be indispensable for the development of Th1 cells, confirming the notion that the described minimal priming conditions (i.e. first and second signal only) were sufficient for Th1 differentiation.

3.1.4   Th1 Type and Th2 Type of Anti-tumor Immune Response

It has been widely accepted that the immune response to any antigen, including tumor antigens, is generally divided into two basic types, called simply type-1 and type-2. This classification has been made mainly according to the type of cytokine network and the type of immune cells that are involved in the immune reaction (Salgame *et al.*, 1991; Croft *et al.*, 1994, Mosmann *et al.*, 1996). Type-1 of immune response is characterized by the domination of the cytokines like IL-2, IFN-γ and TNF-α, as well as by the participation of CTL and NK cells. Type-2 comprises cytokines like IL-4, IL-6, IL-9 and IL-13, as well as effectory mechanisms largely associated with the production of antibodies and Th2 cell activity. Although cytokines like IL-5 and IL-10 favour type-2 of immune reaction, they are secreted by cells like APCs, Th1 and Th2 lymphocytes (Croft *et al.*, 1994; Mosmann *et al.*, 1996).

The type of immune reaction against tumor antigens is not predetermined, but depends on the pre-existing cytokine network, cytokine activity of tumor cells, and expression of MHC and co-stimulatory molecules by APCs, etc. Generally, a certain type of cytokine network favours the same type of the immune reaction. For example, IFN-γ selectively stimulates the activity of type-1, while strongly inhibiting type-2 of the immune response. On the other hand, IL-4 selectively inhibits type-1, while stimulating the activity of type-2 (Croft *et al.*, 1994; Mosmann *et al.*, 1996). It is, therefore, logical to assume that the pre-existing conditions of microenvironment are of vital importance for the determination of the type of the immune reaction.

If tumor cells or tumor-infiltrating DCs and NK cells, secrete the cytokines such as IL-10, IL-6 and IL-4, the anti-tumor immune response will necessarily develop according to type-2 model of immune reaction. The secretion of IFN-γ by NK cells, IL-2 by tumor infiltrating APCs and/or Th1 cells directs the immune reaction towards the type-1 model, while strongly inhibiting type-2 (Bendelac *et al.*, 1997). In addition, the presentation of antigens in the absence of co-stimulatory molecules results in an inadequate communication between APCs and Th cells, and also can inhibit the feedback activation of APCs by Th cells. Finally, this aberrant communication results in the failure of APCs to release IL-12 (Croft *et al.*, 1994; Nastala *et al.*, 1994; Bianchi *et al.*, 1996; Stolina *et al.*, 2000). In addition, the concentration of antigens is another factor determining the type of the immune reaction. Thus, the "optimal" concentrations of antigens will direct the immune reaction towards type-1, whereas the low, as well as high concentrations direct it towards type-2 (Hosken *et al.*, 1995; Constant *et al.*, 1997).

Aruga *et al.* (1997) provided a clear demonstration of the effects of cytokines on determining the type of anti-tumor immune response, by using the model of already enacted type-1 of anti-tumor immune response. In experimental animals, type-1 of anti-tumor immune response was inhibited by anti-IFN-γ monoclonal antibodies, and finally translated into type-2 by treating these animals with IL-10. Similar effects were obtained by the treatment with IL-4, whereas treating the animals with IL-12 resulted in high concentrations of IFN-γ in the tumor lesion and low concentrations of IL-10 and IL-4 (Aruga *et al.*, 1997; Tsung *et al.*, 1997).

Le *et al.* (1997) used the inoculation of B-cell lymphoma to demonstrate the difference between the types of the immune response in lymphoma-resistant and non-resistant experimental animals. These authors found that the essential difference between these two groups of animals was the type of activated Th cells. The animals resistant to the inoculation of lymphoma cells mostly activated Th1 lymphocytes, developing type-1 cytokine network, whereas non-resistant animals which activated Th2 type cells and developed type-2 cytokine network eventually became affected with lymphoma (Lee *et al.,* 1997).

The obtained findings were supported by several clinical studies showing type-1 of immunity to be associated with the regression of malignant tumor. One of these studies is related to the ability for the secretion of IFN-γ from TIL, where TIL from regressing tumors was shown to secrete significantly higher amounts of this cytokine, unlike TIL from tumors in progression (Kawakami *et al.,* 1994). Another study related to the levels of mRNA of Th1 cytokines in CTL from melanoma in regression, and melanoma in progression. The level of mRNA of Th1 cytokines in CTL from progressing melanoma was significantly lower compared to the level of mRNA of the cytokines in CTL from melanoma in regression (Lowes *et al.,* 1997). These data corroborate the assumption that type-1 of immune response is essential for effective anti-tumor immunity. Contrary to this, type-2 of the immune response is ineffective as an anti-tumor response, so its mechanisms could be some of the sources of failure of the anti-tumor response in mammals.

Experiments using series of anti-tumor vaccines regarding different tumors unambiguously showed that the usual anti-immune response was type-2, which could be used to account for the inefficacy of the tested anti-tumor vaccines. However, anti-tumor vaccination accompanied by IL-12 leads to the polarization of the immune response towards type-1, a more effective anti-tumor response and consequently, in many cases, the suppression of tumor process (Hu *et al.,* 1998; Hung *et al.,* 1998).

3.2   ARE TUMORS IMMUNOGENIC?

There is no exact answer as to whether tumors are immunogenic. The fact itself that tumors are a new creation, points to the possibility of their being associated with the newly arisen "non-self" molecules which work as potential targets for the immune system. On the other hand, tumor cells are altered cells of the host, so they have the same or almost the same genome as other, normal cells. However, the dilemma over tumor immunogenicity does not originate from our conceiving them as "self" or "non-self" structures, but from the experimental and clinical findings regarding the activation of tumor-specific lymphocytes whose activity fails in the effective anti-tumor action. A number of researches indicate a high disproportion between the immunogenicity of tumor cells and the intensity and efficacy of the anti-tumor immune reaction. Although the nature of these paradoxical relationships has not been precisely defined yet, the mechanisms of failure of the anti-tumor response are most probably not contained within only one factor, molecule or event. A large body of data suggest that the failure of anti-tumor immunity is a multileveled controlled phenomenon involving mechanisms of immune tolerance, intercellular communication, immune recognition, disturbed balance between Th1 and Th2 type of immune response, etc. (Dunn *et al.*, 2002; Zhang *et al.*, 2003).

Numerous infiltrating T lymphocytes are often found in tumors, but this by itself does not prove that an immune response is slowing the progression of the tumor. Zhang *et al.* (2003) provide a clear demonstration that patients with ovarian carcinoma can expect to have much longer overall and progression free survival if the tumor is infiltrated by T cells than if it lacks infiltrating T cells. Much longer survival was also observed among the patients in whom a complete response to chemotherapy occurred if the tumor was infiltrated by lymphocytes.

3.2.1 Tumor-Associated/Specific Antigens

The proteins which represent specific molecular markers of tumor cells are products of gene activity of the tumor cell. There is a group of tumor antigens which are the products of gene activity during the embryonic development and/or cell differentiation. The activation of these genes is the integral part of carcinogenesis and is most probably the result of ontogenic regression of the altered cell, and the failure of control mechanisms of the cell cycle. As most of these proteins express early during the embryonic development, according to Burnett's theory of clonal selection, clonal repertoire of T lymphocytes in adults should not contain the clone specific for oncofetal and differentiation antigens. Therefore, we could define these antigens as "self" molecules with small or without immunogenicity. There are also the so-called cancer-related antigens, as well as antigens which are the products of over-expressed genes of tumor cells. These molecules can be identified on normal tissue with a 100 times weaker expression, compared to the tumor cells. The structure of these molecules in tumor cells remains unaltered, so that cancer-related antigens and the products of over-expressed genes might also belong in the category of "self" molecules (Van den Eynde *et al.*, 1997; Coulie *et al.*, 2001; Dunn *et al.*, 2002).

Proteins which are the products of mutated genes or genes altered through translocation comprise the second group of tumor antigens. There are also the so-called fusions or chimeric proteins, which are the products of "new" genes arisen at the point of re-coupling of chromosome parts. These can be defined as "non-self" molecules and, in most cases, may be good antigen targets for the immune system. Finally, the third group of tumor antigens are virus-associated tumor antigens. These are also "non-self" molecules which show a high level of immunogenicity (Van den Eynde *et al.*, 1997; Dunn *et al.*, 2002).

In mammals, tumor antigens associated/specific antigens, whether encoded by normal or mutant genes, may be lost by more malignant variants that arise during tumor progression (Chiari *et al.*, 2000). Indeed, this might be expected since the loss of antigens may give a selective survival advantage to a loss variant. It is somewhat surprising then that certain tumor-specific antigens are retained during tumor progression. For example, it was found many years ago that certain tumor-specific antigens recognized by CD4$^+$ cells can be retained (Van Waes *et al.*, 1986), even though CD4$^+$ cells, upon adoptive transfer, can be very effective in eliminating tumor cells (Monach *et al.*, 1995). Mumberg *et al.*, (1999) and others (Qin *et al.*, 1997) have shown that this killing occurs by an indirect mechanism requiring IFN-γ. Since these CD4$^+$ cells can eliminate MHC class II negative tumor cells even in the absence of CD8$^+$ cells, such an attack may be effective even when MHC antigens are lost by the tumor cells during tumor progression. For example, Mumberg *et al.* (1999) have shown that CD4$^+$ cells specific for the mutated form of so-called L9 molecules can eliminate *in vivo* MHC class II negative tumor cells expressing this antigen, without the selection for antigen loss variants. Tumor-specific antigens recognized by CD8$^+$ cells can also be retained by variants during progression even though these antigens are direct targets for T cell immunity (Ward *et al.*, 1990), and in one recent study, extensive efforts to select for antigen loss variants of a methylcholanthrene-induced sarcoma in vitro failed using CD8$^+$ cells specific for a tumor antigen (Ikeda *et al.*, 1997).

As tumor cells grow and multiply, some of tumor-specific/associated antigens can increase and leak into the bloodstream or other fluids. Depending upon the tumor antigens, it can be measured in blood, urine, stool or tissue. Some tumor antigens are associated with many types of cancer; others, with as few as one. Also, some tumor markers are always elevated in specific cancers; most are less predictable. However, no tumor marker is specific for cancer and most are found in low levels in healthy persons, or can be associated with non-neoplastic diseases (inflammatory and auto-immune diseases) as well as cancer (Dunn *et al.*, 2002).

3.2.1.1 Carcinoembryonic Antigen (CEA) and Alpha-fetoprotein (AFP)

CEA (CD66 family) is a whole family of glycoprotein molecules which are the members of the Ig supergene family. The tissue distribution and level of expression of these molecules significantly change during ontogenesis. The highest level of CEA expression was recorded on embryonic intestines and some other embryonic tissues, whereas it can be found in low concentrations on healthy adult tissues like mucosa of the colon. Inflammatory, auto-immune and benign lesions of the colon have also been associated to enhance CEA expression. Nevertheless, the highest CEA expression, comparable with the level of expression on fetal tissues, was recorded only on the colon, lung, pancreatic, breast and ovarian cancer cells or a few other malignant tumors (Tsang *et al.,* 1995; Pinto *et al.,* 1997).

The studies on transgenic mice expressing human CEA (CEA-Tg) with the similar levels of expression and tissue distribution as in humans have shown a complete immune areactivity against basal CEA expression, and also against immunization by this molecule. However, the vaccination of CEA-Tg mice with CEA molecule and different adjuvant or with recombinant *vaccinia* virus expressing CEA resulted in the activation of lymphoproliferative response and cytotoxic T lymphocyte reaction. Similar results were obtained by testing the CEA immunogenicity in humans (Hasegawa *et al.,* 1992).

Elevated AFP typically indicates a primary liver tumor or a germ cell tumor of the ovary or testicle. AFP is a glycoprotein produced in high amounts by fetal tissue and is elevated during pregnancy. It is most widely used as a marker for hepatocellular carcinoma and testicular cancer but is also associated with ovarian cancer. Seventy percent of people with liver cancer have increased AFP levels. In China, where liver cancer rates are high, AFP is used as a screening test for that disease. AFP levels indicate the extent of cancer, but non-cancerous liver conditions such as cirrhosis and viral hepatitis have moderately increased levels of AFP (Tsang *et al.,* 1995; Pinto *et al.* 1997).

3.2.1.2 Differentiation Antigens (DA)

Unlike CEA which normally appear only during embryogenesis, DA elicits its effects in almost all stages of the cell and individual development, especially in proliferative tissues or tissues which permanently regenerate, such as bone marrow or mucous membranes. The level of expression and type of tissue distribution of DA under normal circumstances depend on the phase of ontogenic development and current microenvironmental conditions, such as the need for the tissue regeneration after a trauma. Concerning the role of DA in the life cycle of the cell, it is obvious that they belong to "self" molecules, so there is a high level of immune tolerance to these molecules. However, a number of malignant tumors also express DA, which may seem unusual both in the type and intensity of the expression. Although in cancer bearers the mechanisms of immune tolerance normally suppress the development of the immune reaction on DA, some researches have shown that within the TCR repertoire of every individual there are DA specific lymphocyte clones, which could possibly indicate the potential immunogenicity of these molecules. The immunogenicity of DA molecules is generally very low both in normal circumstances and during the expression of these by tumor cells, which is primarily due to the absence of T lymphocytes with high TCR affinity for DA. The rare cases of strong and effective anti-tumor reaction directed towards DA molecules are probably the consequence of the activation of the mechanisms which are similar to the mechanisms for the destruction of "self" molecules in auto-immune diseases (Brichard *et al.*, 1993; Boon *et al.*, 1994; Paul *et al.*, 1994; Bronte *et al.*, 2000).

3.2.1.3   Viral Antigens (VA)
Approximately 15 percent of human cancers are associated with the presence of viral infection. This percentage is certainly not definite as the introduction of more sensitive methods of virus detection that revealed the significant increase of the percent linkage between carcinoma and viral infection. Tumor associated VA fall in the group of "non-self" molecules, so they are normally expected to be rapidly recognized by the immune system and eliminated thereafter, possibly leading to the elimination of tumor cells. However, that does not happen very often in reality. The true reasons for a relatively poor immunogenicity of VA are largely unexplored yet. It is assumed that the mechanisms of anti-VA immunity failure may be connected with a rather poor expression of TAP and MHC molecules in the viruses infected cells, and also with the control mechanisms of immune reaction that are normally involved in preventing the development of too strong, "self" destructive, auto-immune or auto-immune-like reaction (Boon *et al.*, 1994; Liu, 1997).

3.2.1.4   Oncogenic Fusion Proteins (OFP)
Chromosomal translocations belong in the group of somatic adopted cytogenetic mutations. Basically, this disorder occurs due to disruption of the chromosome within the exon of two different genes, the translocation of its whole part and the re-coupling on a different location of the same chromosome, or even of some other chromosome. As a consequence of such chromosomal aberration, a completely new, the so-called fusion gene is formed whose products are fusion or chimeric proteins, unknown to the

immune system. OFP have been identified in a number of human tumors including various forms of leukaemia, lymphoma and cancers. Although being products of own genome, OFP are still "non-self" molecules, and theoretically could be the representatives of tumor-specific antigens which are a clear target for the mechanisms of the immune recognition and the effectory anti-tumor cells (Drexler *et al.,* 1995; Zhu *et al.,* 1999; Chansky *et al.,* 2001).

3.3 BASIC MECHANISMS OF ANTI-TUMOR IMMUNITY FAILURE IN MAMMALS

Although there is a lot of evidence that tumors can be immunogenic, and as such recognized and eliminated by the immune system, a relatively poor success of a large number of anti-tumor immunotherapeutic procedures revealed the possibility that the mechanisms of anti-tumor immunity failure might not have been explored enough. There is an opinion, that although most of the mechanisms of anti-tumor immunity failure have been identified and relatively well explored, immunotherapeutic procedures focused on the partial phenomena are responsible for the failure of anti-tumor immunotherapy.

The mechanisms of anti-tumor immunity failure are probably deeply ingrained in the varying mechanisms of immune reaction, and closely connected with the mechanisms for the prevention of auto-immunity and, in mammals, with the mechanisms of immunotolerance in pregnancy. Therefore, the immunotherapy of malignant diseases should be based on a more complex and overall approach. It would also include the correction of all the regulatory mechanisms that are responsible for the phenomenon of anti-tumor immunity failure, including the antigen processing/presenting machinery of tumor cells and the effectory immune mechanisms.

3.3.1 Altered Expression of MHC Molecules and Tumor Antigens by Tumor Cells

The loss or reduced MHC class I surface expression due to structural alterations and/or dysregulation of the MHC class I molecules has frequently been identified in malignant cells and was associated with the resistance of tumor cells to CTL mediated lysis. In some cases, the reversion of the malignant phenotype was obtained by the introduction of MHC class I genes (Garrido *et al.,* 1993, 1997). However, the transfection of cell lines with MHC class I genes have not always resulted in the increased MHC class I surface expression although enhanced MHC class I mRNA levels were obtained (Seliger *et al.,* 1998).

The poor prognosis of the malignant diseases is associated with the loss of class I molecules on tumor cells. The selective or complete loss of class I expression on tumor cells is not a rare phenomenon; for example, a complete loss of class I expression on breast cancer cells has been identified in more than 50% patients, whereas a selective loss of class I molecules has been identified in approximately 35% patients. The full expression of class I molecules on breast cancer cells has been proved in only 12% of patients (Cabrera *et al.*, 1996). The same authors propose that the degree of class I expression on breast cancer cells can be associated with the reduced anti-tumor activity of CTL (Garrido *et al.*, 1993, 1997; Cabrera *et al.*, 1996).

From studies using cultured tumor cell lines, antigen processing mutants and surgically removed malignant lesions it has been postulated that the loss or dysfunction of the LMP and/or TAP genes may account for abnormalities of MHC class I surface expression. The expression of MHC class I molecules could be independently controlled at the level of transcription, synthesis, assembly and transport to the cell surface. The recent identification and characterization of different components of the antigen processing machinery, the TAP1 and TAP2 and the LMP2 and LMP7, provided new tools to elucidate how tumors evade the immune recognition (Seliger *et al.*, 1998).

A large number of authors agree that one of the mechanisms responsible for the altered expression of class I molecules on tumor cells could be the loss of TAP genes expression. In the absence of TAP molecules, the MHC molecules are not capable of binding antigenic peptides, leading to the loss of their expression on cell surface. It was proved earlier that the peptide-bound MHC molecules cannot be expressed on cell membrane, but instead undergo recycling. The inhibition of the TAP genes transcription or inhibition of the function of the already synthesized TAP molecules has been proved on several malignant tumors. The functional deficiency of TAP genes can be restored using the treatment of tumor cells with IFN-γ or by the transfection of IFN-γ genes into class I deficient tumor cells. These procedures result in the activation of TAP genes and enhanced expression of class I molecules on tumor cells (Seliger *et al.*, 1998). The importance of TAP molecules in treatments can be demonstrated by the fact that TAP positive cells are not capable of producing tumor in normal mice, while they are capable of producing it in nude, athymic mice (Johnsen *et al.* 1999).

The defects in the ability to process and present antigenic peptides to CD8[+] CTL were first described in small cell lung carcinoma cell lines, which failed to transcribe LMP and TAP mRNA. This resulted in impaired MHC class I surface expression and resistance to CTL-mediated lysis. Both deficiencies could be restored by IFN-γ treatment (Restifio *et al.*, 1993).

The expressions of MHC class I molecules and the TAP genes were investigated in some hepatocellular carcinoma cell lines. Two cell lines, Hep-3B and HuH-7, showed a reduced level of TAP, which might cause the low surface expression of HLA class I. On the same tumor sample it was proved that IFN-γ can restore functions of the genes (Matsui *et al,* 2002). Murray *et al.* (2000) found that tumor infiltrating lymphocytes from a patient with melanoma, expanded *in vitro* in the presence of IL-2, effectively kill the tumor cells expressing class I molecules (64%), but not the class I negative cells (18%). In a number of tumor specimens, including cervical cancer, colorectal cancer and melanoma, the lack of or reduced TAP and/or LMP expression has been observed (Cromme *et al.,* 1994; Kaklamanis *et al.,* 1995). Also, in Burkitt's lymphoma, prostate carcinoma, melanoma and renal cell carcinoma cell lines showed reduced MHC class I expression which appeared to be due to down-regulation of LMP2, LMP7, TAP1 and/or TAP2 genes (Restifio *et al.,* 1993).

In the last decade, tumors induced by the highly oncogenic adenovirus (Ad)-12 have been employed as models to demonstrate a direct correlation between tumorigenicity and the lack of host immune response due to suppression of MHC class I expression. The Ad-12 mediated down-regulation of MHC class I surface expression was associated with the reduced TAP1, TAP2, LMP2, and LMP7 expression. The re-expression of the transporters genes and proteasome subunits by gene transfer was able to at least partially reconstitute MHC class I surface expression (Seliger *et al.,* 1998).

3.3.1.1 Expression of HLA-G on Tumor Cells as a Mechanism of Immunosurveillance

It is well known that tumors frequently are infiltrated by NK cells. The effectory functions of these cells depend on the level of MHC class I molecules expressed at the surfaces of tumor cells. Although cytotoxic CTL recognize tumor associated antigenic peptides presented by MHC class I molecules, NK cells are cytotoxic for tumor cells that have lost MHC class I expression (Cabestre *et al.,* 1999). This mechanism of anti-tumor immunity could be effective if tumor cells did not express the so-called non-classical MHC class I or Ib molecules, such as: HLA-G and HLA-E molecules. Class Ib molecules activate the KIR on NK cells, which is a very strong inhibitory signal for NK cells. Eventually, the result of the interaction between class Ib molecules and KIR receptors is a strong inhibition of the NK cell anti-tumor activity (Paul *et al.,* 1998; Cabestre *et al.,* 1999).

One of the most significant molecules within the class Ib group, HLA-G have a highly selective and restrictive type of tissue distribution. Basically, HLA-G are only expressed on thymic epithelial cells, trophoblast cells and cells of a large number of malignant tumors. HLA-G genes also belong to the least polymorphic MHC genes with a low number of alleles. According to some authors, polymorphism of HLA-G is so low that they can be considered as monomorphic (Rouas-Freiss *et al.*, 2000; Davies *et al.*, 2001). In contrast to classical HLA class I genes, the primary transcript of the HLA-G gene generates at least five mRNAs resulting from alternative splicing that potentially encode five protein isotype: HLA-G1, HLA-G2, HLA-G3, and HLA-G4, which have the capacity anchor to the cell membrane, and the soluble HLA-G5 isotype. HLA-G is expressed on cytotrophoblasts at the feto-maternal interface, where it plays an important role in the maternal tolerance of the allogeneic placenta. Some authors recently showed that, in addition to HLA-G1, the HLA-G2 isotype strongly inhibits NK cell lysis *in vitro* and that HLA-G molecules protect cytotrophoblasts from the lytic activity of maternal uterine NK cells *ex vivo* (Pascale *et al.*, 1998; Cabestre *et al.*, 1999). The fact that HLA-G is expressed selectively on trophoblast cells is of particular interest. Indeed, although the trophoblast is a normal tissue, it shares several common features with neoplastic cells, such as highly mitotic activity, invasive nature, secretion of growth factors and hormones, as well as expression of growth factor receptors and proto-oncogenes, which have led to defining the trophoblast as a "pseudo-malignant" type of tissue.

The physiological expression of MAGE-3 and MAGE-4 genes, expressed in metastatic human melanoma, is restricted to placenta and testis. Melanoma cell adhesion molecules are expressed in both melanoma and trophoblast cells. Taken together, these features suggest that genes preferentially expressed in trophoblast cells, such as HLA-G, also might be expressed preferentially in neoplastic cells (Bubanovic, 2003a; 2003c; Bubanovic *et al.*, 2003b).

3.3.2   Altered Expression of Co-stimulatory and Adhesion Molecules by Tumor Cells and APCs

The effective anti-tumor response is in most cases closely connected with the presentation of tumor antigen peptides and co-stimulatory molecules by DCs. Contrary to this, the aberrant function of DCs such as the low expression of co-stimulatory molecules or secretion of anti-inflammatory cytokines and prostaglandines results in the failure of the anti-tumor immune response. DCs isolated from the infiltrate of colorectal carcinoma show normal expression of class II molecules, but generally do not express co-stimulatory molecules like CD80 and CD86. Contrary to this, DCs isolated from the inflammatory infiltrate of Crohn's disease suffering patients show a quite normal expression of class II molecules as well as CD80 and CD86 molecules (Chaux *et al.,* 1996). Hepatocellular carcinoma infiltrating cells secrete significant quantities of TNF-α into the tumor microenvironment, but the stimulation of the surrounding lymphocytes is weaker in relation to the stimulation of lymphocytes by DCs from cyrotic lesions or from normal liver (Ninomiya *et al.,* 1999). As little as 5-10% of the DCs isolated from certain skin tumors, and less than 5% of those isolated from benign tumors, express co-stimulatory molecules. A lot of researches have shown that the expression of peptides in the presence of class II molecules and absence of co-stimulatory signals may lead to the anergy of specific CD4$^+$ cells. DCs from progressing tumors had depressed CD86 expression and secreted IL-10, inducing the anergy of autologous CD4$^+$ cells. DCs from regressing tumors, however, secrete IL-12 and IFN-γ, and show a good expression of co-stimulatory molecules (Enk *et al.,* 1997).

IL-10 exerts inhibitory effects related to the expression of co-stimulatory molecules on DCs. In addition, IL-10 pre-treated DCs secrete high amounts of IL-4, important for the forwarding of Th0/Thp maturation towards Th2 lymphocytes. The presentation of peptides on DCs in the presence of IL-10 results in the inhibition of CTL activity and anergy, or even deletion of the specific clone of effectory lymphocytes (Moore *et al.,* 1993; Qin *et al.,* 1997).

Tumor cells may also express CD80 and CD86 molecules. The cells of the primary gastric carcinoma expressed CD80 and CD86 molecules; however, metastases rarely expressed CD80, while showing a better expression of CD86 than on cells of primary lesion. Some researches have shown that enhanced CD86 expression may favour the selective stimulation of Th2 cells and up-regulate IL-4 production (Liu *et al.,* 1998).

The cells of melanoma may also express other co-stimulatory molecules, highly relevant for anti-tumor immunity, such as CD40. von Leoprechting *et al.* (1999) propose that cells of certain forms of primary melanoma lesions express CD40 molecules, whereas metastatic cells are basically CD40 negative. The same authors claim that CD40$^+$ melanoma cells are sensitive to the specific CTL activity, while CD40 negative cells cannot be recognised and eliminated by the same CTL clone.

A successful anti-tumor immune reaction depends, among others, on the interaction of activated effectory cells and non-specific adhesion molecules on the every target cell. The tumor cells of different histological origin display relatively lowered level of expression of important adhesion molecules, such as ICAM-1. There is some evidence indicating that CTL can quite easily kill autologous melanoma cells if there is a significant expression of adhesion molecules on tumor cells, while the loss or reduced expression of these result in the resistance of melanoma cells to CTL activity. There is also certain evidence that the degree of the expression of adhesion molecules on tumor cells depends on the cytokine activity of TIL. On the other hand, the inhibition of the cytokine secretion from TIL induced by soluble tumor factors, like IL-10, closes the cycle of control mechanisms for the expression of adhesion molecules, resulting in the aberrant expression of these molecules on tumor cells (Moore *et al.*, 1993; Yue *et al.*, 1997; Salomon *et al.*, 1998).

3.3.3   Downregulation of Anti-tumor Immune Response by Cytokines

A large number of experiments and studies have clearly shown that cytokines secreted from the tumor cells or immune cells infiltrating and surrounding the tumor tissue affect the final outcome of the anti-tumor immune reaction by exerting immunomodulatory/suppressive effects. While it is true that cytokine network of tumor microenvironment contains the pro-inflammatory and immunostimulatory cytokines, it seems that their effects are largely subject to the effects exerted by modulatory and suppressive cytokines, such as IL-4, IL-6, IL-10 and TGF-β (Krasagakis *et al.* 1998; Pawelec, 1999). The serum concentration of IL-6 is quite often enhanced in patients suffering from malignant diseases. In patients with malignant lung carcinoma, the serum concentration of this cytokine is several times higher than in patients with an obstructive disease or acute infection (Dowlati *et al.*, 1999). In melanoma suffering patients who show a good response to treatment, the concentration of this cytokine is only two times higher in relation to the healthy control. Also, in terminal patients with disseminated melanoma the serum concentration of IL-6 is even 11 times higher in relation to the healthy control (Lowes *et al.*, 1997).

Higher values of IL-10 in sera have also been frequently reported in various tumor sufferers. At the same time, a significantly lower concentration of IL-2 in the tumor tissue and sera of these patients has been reported. The intralesional treatment with IFN-γ was associated with tumor regression and down-regulation of IL-10 mRNA. The importance of IL-10 secretion by tumor cells is evidenced by the fact that five-year survival of the patients suffering from IL-10 non-secreting tumors is approximately 90%, while as little as 15% of five-year survivors was reported in patients with IL-10 secreting tumors. As well as secreting IL-10, malignant tumors also may induce IL-10 and immunosuppressive prostaglandin E2 (PGE2) production by monocytes. The reduced activity of CD8$^+$ and CD4$^+$ cells in TIL population can be enhanced by IFN-γ and TNF-α, which strongly inhibit the synthesis and the secretion of anti-inflammatory cytokines, such as IL-10 (Moore *et al.*, 1993; Fortis *et al.*, 1996; Pawelec, 1999).

IL-10 is one of most important cytokines that prevents the class I molecules expression at the cell surface. Terrazzano *et al.* (2000) investigated the IL-10 effects in a human lymphoblastoid cell defective for TAP1 and TAP2 genes after TAP1 and TAP2 genes transfection. In this experimental system, the down-regulation of antigen presenting/processing machinery was observed in TAP transfected cells in the presence of IL-10, whereas the processing and presenting of peptides in IL-10 non-treated transfected cells was almost normal.

IL-10 and TGF-ß, is a cytokine that exerts multiple effects on antigen presentation, B and T cell proliferation, cytokine production, and monocyte/macrophage function. IL-10 prevents up-regulation of CD80/86 expression during macrophage activation and down-modulates the expression of a broad range of cytokines in peripheral blood mononuclear cells (PBMC), including IFN-γ, IL-2 and TNF-α (Moore *et al.,* 1993).

Tumor cells incubated or transfected with IL-10 had decreased but peptide-inducible expression of class I, decreased sensitivity to class I restricted CTL, and increased NK sensitivity. IL-10 signal inhibit the TAP dependent translocation of peptides to the ER, resulting in the accumulation of immature class I molecules in the endoplasmic reticulum and subsequently a low expression of cell surface class I molecules. This finding is explained by a down-regulation of expression of TAP1 and TAP2, observed in IL-10 transfected tumor cells. IL-10 is the first example to demonstrate that a cytokine can decrease the expression and function of the TAP1 and TAP2 molecular complex and, in more general terms, the first example of a cytokine with an inhibitory effect on class I-mediated peptide presentation (Salazar-Onfray *et al.,* 1997; Petersson *et al.,* 1998).

Virally infected cells degrade intracellular viral proteins proteolytically and present the resulting peptides in the association with MHC class I molecules to CTL. These cells are normally prone to CTL-mediated elimination. However, several viruses have evolved strategies to avoid detection by the immune system that interfere with the pathway of antigen presentation. Epstein-Barr virus expresses a predominantly late protein, the BCRF1 gene product EBV-IL-10 that is similar in sequence to the human IL-10. The EBV-IL-10 affects the expression of one of the TAP1 and TAP2 to the same degree as the host's IL-10. The expression of the LMP2 and TAP1 genes but not the expression of TAP2 or LMP7 is efficiently down-regulated, indicating a specific IL-10 effect on the two divergently transcribed TAP1 and LMP2 genes. The downregulation of TAP1 by IL-10 hampers the transport of antigenic peptides into the ER, as shown in the TAP-specific peptide transporters assay, their loading onto empty class I molecules, and the subsequent translocation to the cell surface. As a consequence, IL-10 causes a general reduction of surface class I molecules on B lymphocytes that might also affect the recognition of EBV-infected cells by cytotoxic T cells (Zeidler *et al.*, 1997).

A high level of TGF-β was also found in the sera of patients suffering from progressing tumors, with a much lower level of this cytokine secreted by regressing tumors. Disseminated processes as well as processes resistant to conventional therapy are most frequently associated with a high TGF-β concentration in the sera. Consistent with these data, the addition of anti-TGF-β antibodies to autologous lymphocyte/tumor cells co-cultures increased the frequency of cultures showing effective anti-tumor response immune (Krasagakis *et al.*, 1998).

3.3.4   Th1/Th2 Immunity and Tumor Escape

Th1 and Th2 types of immune response are identified in the immune reactions against most antigens. The mutual interaction of these two types of the immune reaction is a variable category depending on many factors and current circumstances, like the concentration and nature of antigens, genetic factors and the nature of other antigens simultaneously engaging the immune system. Given the fact that the activation of Th cells necessitates the interaction of CD80/86 molecules of APCs with CD28 molecules of the lymphocytes, it is highly important whether APCs will express predominantly CD80 or CD86. If mostly CD86 molecules are expressed, the activation will affect primarily Th2 lymphocytes - leading to the developing of the cascade of the humoral immune response. Naturally, the dominant expression of CD80 molecules on APCs would be expected to result mostly in the activation of Th1 lymphocytes, but this inference has not been proved yet (Reiner *et al.*, 1995; Kuchroo *et al.*, 1995; Desmedt *et al.*, 1998; Kufer *et al.*, 2001).

CD4$^+$αβTCR$^+$ (Th1 and Th2) cells trace their origin from the common precursor - Thp lymphocytes. Most authors agree that Th1 and Th2 cells represent the mature form of Th cells, whether appearing as their effectory or memory iso-forms. Cytokines are one of the most significant factors responsible for the promotion of Th cell differentiation towards either Th1 or Th2 lymphocytes. IL-2, IL-15, IFN-γ, TNF-α, and especially IL-12, promote the maturation of Th0 cells toward Th1 lymphocytes. Th2 lymphocytes are favoured largely through the domination of the cytokines such as IL-6, IL-10, TGF-β2, and particularly IL-4 (*Figure* 3.4.) (Kuchiro *et al.*, 1995, 1998; Mosmann *et al.*, 1996; Constant *et al.*, 1997).

Among the cytokines capable of promoting the differentiation of non-differentiated Th precursors the most important are IL-12 and IL-4. There is evidence that the following stimulation by antigens, Thp cells enhance the equal numbers of receptors for IL-4 and IL-12, while the number of receptors for IFN-γ remains at the initial value. This stadium of Th lymphocytes is nominated as Th0 stadium. Depending on the dominant signal, which can appear as IL-12 or IL-4, a further differentiation will be promoted either towards Th1 or Th2 lymphocytes (Mosmann *et al.*, 1996; Seder, 1997; Constant *et al.*, 1997; Kuchiro *et al.*, 1998). The domination of Th2 cytokines in the anti-tumor immune response could be associated with the down-regulation of Th1 cytokines, so that it could be the one of critical event for tumor escape. The dominant Th2 type of the cytokine network inhibits the secretion of IFN-γ by the mechanisms of down-regulation of CD40-ligand-induced IL-2 production. In addition, the CTL isolated from tumor tissue very often display the features of Th2 cells like secretion of Th2, but not Th1 cytokine (Cella *et al.*, 1996; von Leoprechting *et al.* 1999).

There is a large body of evidence why Th2 type of the immune response is thought not to be an efficient anti-cancer effector. In a mouse lymphoma model, the presence of tumor-specific CTL was clearly evidenced in both susceptible and resistant hosts. These evidences clearly demonstrate that in susceptible hosts Th2 type was dominant, whereas in resistant hosts type of dominant immune response was Th1. These corroborates the earlier notion that tumors are indeed immunogenic and could initiate tumor-specific CTL, but this CTL-mediated activity largely fails due to the presence of the dominant Th2 type of the immune response (Lee *et al.*, 1997). A large number of studies have shown that IL-10 and TGF-β play a key role in developing of Th2 type of the immune response in the anti-tumor immune response. In addition to inhibiting the activity of NK cells and DCs, these cytokines were also found to inhibit the synthesis and secretion of pro-inflammatory cytokines like IFN-γ and TNF-α, reduce MHC expression and possibly inhibit the whole peptide processing/expressing machinery (Alleva *et al.*, 1994; Aruga *et al.*, 1997; Petersson *et al.*, 1998).

In some models, the successful immunotherapy of established tumor is associated with a change in the balance of T cells subsets from Th2 to Th1 phenotype. Tumor derived TGF-β may induce an over-production of IL-10, stimulating suppressors mechanisms, and finally inhibiting the effectory anti-tumor reaction. Contrary to this, anti-TGF-β antibodies effectively inhibit the development of Th2 type immunity, preventing the establishment of suppressor mechanisms, and eventually leads to the activation of Th1 type, as well as a more effective anti-tumor immune reaction (Alleva *et al.*, 1994; Krasagakis *et al.*, 1998).

There is a large body of data that Th1 immunity is effective in the anti-tumor immune response, and also that the dominant Th2 immunity actually represents some kind of tumor protective immunity. This dichotomy, however, is not universal. Joined with CD8[+] cells, Th2 cells may mediate a successful rejection of tumor tissue. Although IL-4 is known as a strong promoter of Th2 immunity, forwarding the maturation of Thp cells towards Th2 cells, this classical Th2 cytokine can enhance rather than inhibit anti-tumor activity of CD8$^+$ and NK cells. In some therapeutic protocols, IL-4 was used as adjuvant in anti-tumor vaccination therapy (Golumbek *et al.,* 1991).

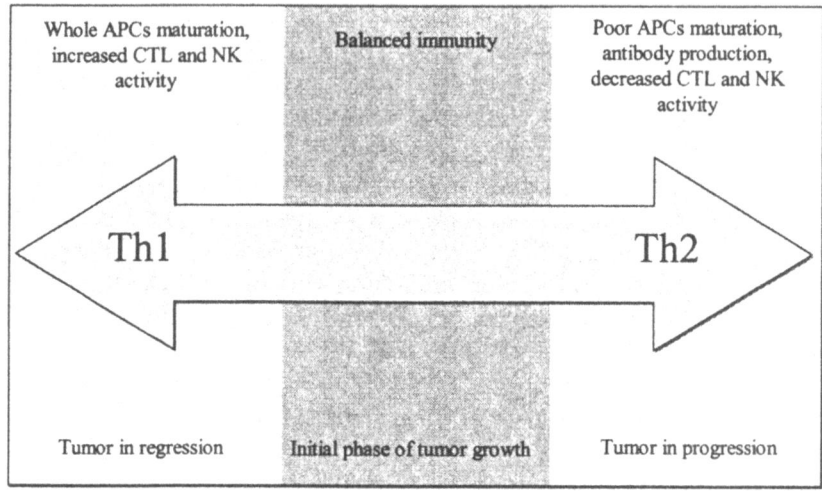

Figure 3.5. Different possibility in development of anti-tumor immunity.

Only a few experimental models indicate that IL-10 may enhance anti-tumor activity. The explanation for the anti-tumor activity of IL-10 has been found in the anti-apoptotic activity of this cytokine against the activated tumor-specific CD4$^+$ and CD8$^+$ lymphocytes and NK cells. There is also evidence indicating the anti-angiogenetic and anti-metastatic effects of IL-10 in some models of murine tumors, so that it remains difficult to dissect out the contradictory activities of this cytokine in tumor immunology (Kirkin *et al.,* 1996; Huang *et al.,* 1999).

3.3.5   Downregulation of Immune Response by Molecules Other than Cytokines

The sera of cancer patients very often contain immunosuppressive proteins other than cytokines. These proteins elicit non-specific immunosuppressive activity, and are capable of inhibiting various cells of the immune system as well as participate in the discontinuing of the immune reaction on different levels. The activity of this and other factors not only decreases the effectiveness of anti-tumor immune response but may also inhibit the immune response against microbes. For example, terminal patients with malignant melanoma show poor reaction or none to influenza matrix peptide; whereas this is a normal reaction to this peptide in healthy people (Scheibenbogen *et al.,* 1997). Tumor cells can produce different proteins, which may under certain conditions exhibit immunomodulatory/ suppressive effects. The increased concentration of soluble CD85 molecule in the sera of tumor bearers may inhibit the lysis of tumor cells mediated by the CD58-CD2 interaction (Altomonte *et al.,* 1993). The soluble factors such as CD54 have been also evidenced to suppress the interaction of tumor-specific TIL and tumor cells (Becker *et al.,* 1993). The concentration of soluble CD54 very often correlates with the disease progression. In addition, there is a whole range of other soluble factors like isoferritin-associated p43 molecule and MUC-1, which inhibit the proliferation and cytotoxicity of tumor-specific lymphocytes. Soluble *FasL* certainly belong to immunomodulatory factors, acting as the suppressors of apoptosis of tumor and/or immune cells mediated by *FasL-FasR* interaction (Rosen *et al.,* 1994; Paul *et al.,* 1999).

3.3.6 Anti-tumor Immunity Failure by Apoptosis and/or Anergy Induction

One of the most radical mechanisms for inducing the failure of anti-tumor immune response is the ability of tumor cells to initiate the apoptosis of peripheral the anti-tumor specific T cells, as well as TIL. In addition, some results indicate that the circulating myeloma protein is processed and presented by thymic APCs, and induces the deletion of antigene-specific thymocytes. The deletion of tumor specific thymocytes may represent a tumor escape mechanism in patients with tumors that secrete or shed tumor antigens (Lauritzsen *et al.,* 1998; Bubanovic, 2003c).

Peripheral clonal deletion is most often mediated by *FasL* expressed on tumor cells. A large number of human tumors, such as malignant lung tumors, malignant melanoma, hepatocellular carcinoma, express *FasL* capable of inducing apoptosis of *FasR*[+] T cells. Furthermore, lung carcinoma cells are capable of killing a *Fas*-sensitive human T cell line in co-culture experiments; this killing was inhibited by a recombinant form of the soluble portion of the *FasR* (Niehans *et al.* 1997). In melanoma metastatic lesions, *FasR*[+] T cell infiltrates were proximal to *FasL*[+] tumor cells. *In vitro,* the apoptosis of *Fas*-sensitive target cells occurred upon the incubation with melanoma tumor cells; and *in vivo,* injection of *FasL*[+] mouse melanoma cells in mice led to rapid tumor formation. In contrast, tumorigenesis was delayed in *Fas*-deficient mutant mice in which immune effector cells cannot be killed by *FasL* (Hahne *et al.,* 1996).

The quantity of the expression of *FasL* on tumor cells depends on the invasiveness of the tumor, and whether it is the primary lesion or metastasis. For instance, the tumor cells of primary lesion of colorectal carcinoma more rarely express *FasL* than the metastatic lesions of the same tumor. The findings of Shiraki *et al.* (1997) and other data suggest the mechanism that enables tumors to evade the immune destruction by inducing apoptosis in activated T lymphocytes. Furthermore, the constitutive expression of *FasL* in hepatic metastatic tumors suggests that *FasL* may also be important in their colonization in the liver through the induction of apoptosis in the surrounding *Fas* expressing hepatocytes.

The expression of functional *FasL* on tumor cells capable of inducing apoptosis of tumor-specific TIL has been found in astrocytoma, ovarian carcinoma, pancreatic adenocarcinoma, carcinoma of oesophagus and many other malignant tumors. There is another mechanism completely opposite to this, but it is largely unexplored yet. Namely, these are co-stimulatory effects exerted by *FasL* of tumor cells on TIL. In this case, CD8$^+$ and CD4$^+$ do not respond by apoptosis to interaction *FasL-Fas*R. Probably depending on the microenvironmental conditions, *Fas*-mediated signals may stimulate TIL anti-tumor response (Rivoltini *et al.*, 1998). In addition, TIL expressing the *FasL* can interact with *FasL* of tumor cells and may induce the apoptotic death of tumor cells themselves. In some experimental models, regressing tumor masses after IL-12 treatment exhibited a massive lymphoid cell infiltration and expressed significant levels of *FasL* mRNA, suggesting the infiltration of *FasL* expressing cells to tumor sites. These results indicate that IL-12 induces the expansion of lymphoid cells that exhibit *FasL*-mediated cytolytic activity and accumulate into regressing tumor masses (Nastala *et al.*, 1994; Tsung *et al.*, 1997; Stolina *et al.*, 2000).

*FasL-Fas*R is not an only form of molecular interaction between TIL and tumor cells with apoptotic consequences. MUC-1 protein often expressing different forms of human carcinoma may also induce apoptosis of TIL (Paul *et al.*, 1999). Agrawal *et al.* (1998a) did not prove MUC-1 mediated apoptosis of TIL; however, they showed that soluble MUC-1 protein may inhibit the proliferation of T lymphocytes and induce their anergy on various stimulatory signals. MUC-1 mediated inhibition of TIL is reversible after the treatment with IL-12. Regardless of whether MUC-1 protein induces apoptosis or temporary inhibition of the TIL proliferation, its role seems to be vital in inducing the failure of the anti-tumor immunity (Agrawal *et al.*, 1998b).

The anergy of T lymphocytes is most often defined as a state of lymphocyte activity where, regardless of the antigen signals (activation of TCR), T lymphocytes are not able to proliferate and perform their cytotoxic or helper function. The induction of T lymphocytes has been evidenced in most diverse experimental conditions as well as models, so that this phenomenon is regarded as one of the physiological mechanisms for the control of the immune reaction and prevention of auto-immunity. Moreover, the anergy of TIL is considered to be the one of the mechanisms working to induce the phenomenon of the failure of the anti-tumor immune reaction.

There are several pathways for the induction of TIL anergy, such as the loss or poor expression of MHC and co-stimulatory molecules on DCs, loss or reduced expression of MHC molecules on tumor cells and microenvironmental cytokine network of the tumor tissue (domination of Th2 cytokines, particularly IL-10), extrathymic maturation of lymphocytes etc. The research carried out on different peptide sequences of tumor-specific/associated antigens that could be used as an antigenic base for the anti-tumor vaccines, have shown that vaccination with peptide sequences in the presence of IL-2 and/or IL-12 could prevent the development of TIL anergy and contribute to the enacting of a strong anti-tumor immune response. These data in part indicate the possibility that the anergy of TIL is largely accompanied by the deficiency of the helper function which could bear certain connection with the deficiency of presenting/stimulatory function of DCs. The basic research of the mechanisms of $CD4^+$ anergy showed that antigen stimulation of $CD8^+$ cells without helping function of $CD4^+$ or its products will lead to the loss of proliferation and cytotoxic reaction of $CD8^+$ cells, and in some cases to their apoptotic death (Hosken *et al.*, 1995; Constant *et al.*, 1997; Hung *et al.*, 1998).

3.3.6.1 Extrathymic Lymphocyte Maturation and Selection

The majority of T cells located in peripheral lymphoid organs are dependent on the thymus for regular differentiation and function. Only a minority of T lymphocytes is thymus independent. These cells pass by extrathymic maturation processes and become mature T lymphocytes. Some data suggest that the mechanism of extrathymic lymphocytes maturation (eTLM) includes the migration, proliferation, differentiation and selection of lymphocytes, like in the thymus (Antonia *et al.*, 1995; Abo, 2001). With ageing and the progression of thymic involution or in accidental thymic involution, the pathways of eTLM gains emphasis (Ohteki *et al.*, 1992). Some authors propose that T cells from extrathymic pathways probably can polarize the action of thymic dependent T cells or participate in the immune reaction in antigen-destructive or antigen-protective manners. Consequently, extrathymic pathways can be the source of the "self" reactive T cells or cells which participate in the mechanisms of tumor escape.

The results of eTLM probably are not preset, but depend on many factors and microenvironmental "snapshots". The factors like cytokines, prostaglandine, microbes, MHC molecules, hormones, *FasL*, HSPs, phenotypes of DCs and APCs, probably can polarize the course of eTLM pathways. Finally, the definitive course of extrathymic derived cells action presumably is resultant of microenvironmental relations and the interactions of foregoing factors. The hypothesis that microbes, especially viruses, can be the promoters of extrathymic (self)antigen reactive lymphocytes maturation is real as well as the hypothesis that the extrathymic lymphocytes selection and products of selected lymphocytes can be included in the mechanisms of tumor, trophoblast and transplant rejection or escape (Bubanovic, 2003d).

Except the major sites of extrathymic T cells maturation like the intestine and liver, it is well known that pregnant uterus, decidua, microenvironments of malignant tissues, allogeneic transplants, auto-immune and infective focuses also can be the real sites of eTLM. Since the sites of eTLM in more cases are temporary and poorly anatomical defined, any similarity and/or differences between the thymic and eTLM pathways are difficult to demonstrate. However, the extrathymic pathways of lymphocyte maturation probably play important role in the mechanisms of auto-immune diseases, malignancies, pregnancy and response to microbes. Finally, the mechanism of eTLM can be the one of universal mechanisms of the immune reactions control (Tamauchi *et al.,* 1988; Abo, 2001; Bubanovic, 2003d).

Lee *et al.* (1996) have detected that the population of tumor infiltrating lymphocytes contains about 15% of double positive (DP) immature lymphocytes. These data maintained opinion about tumor's tissues like the microenvironment of the intensive eTLM. Although DP T cells are rarely present in human peripheral blood the relative percentage of this lymphocyte population can increase spontaneously in healthy individuals and in persons suffering from certain disease conditions. These cells can also be found among those T cells infiltrating arthritic joints, rejected kidney grafts and certain tumors (Zuckermann, 1999). The experiments with neonatally thymectomized mice showed that they fall victim to such auto-immune diseases as gastritis and pancreatitis with ageing. Self-reactive T cell clones can be consistently generated in the absence of the thymus, and these clones via the extrathymic pathways as an alternative pathway, probably are responsible for the auto-immune disease induction (Ikebe *et al.,* 2001).

Some authors propose the possibility that switching of the immune system from the thymus to the alternative eTLM sites might be regulated by the autonomic nervous system as well as by cytokines. The extrathymic T cells are very few in number at any extrathymic sites in adolescence, but they increase in number as a function of ageing (Abo, 2001). This process probably is parallel with the progression of thymic involution. Moreover, acute thymic atrophy always accompanies the activation of eTLM. Therefore, reciprocal regulation between extrathymic derived T cells and thymus-derived T cells might be present (Ohteki *et al.*, 1992; Bubanovic, 2003d).

Cytokines like IL-7 have a very important role in the extrathymic T cell development. Under the influence of this cytokine, human haematopoietic stem cells can develop into mature αβTCR lymphocytes and immature progenitors in the bone marrow of athymic mice (Maerurer *et al.*, 1998; Tsark *et al.*, 2001). Extrathymic αβTCR can be detected, but not γδTCR, in IL-7 gene-deleted animals suggesting that alternative cytokines may be involved in eTLM (Maki *et al.*, 1996; Maerurer *et al.*, 1998). Porter *et al.* (2000) have discovered that eTLM is regulated by cytokines like IL-2, IL-7 and IL-15, but IL-2 is important for thymic pathway in the same way as IL-15 is important for eTLM (Di Santo *et al.*, 1997; Porter *et al.*, 2000).

Extrathymic dendritic cells and APCs can be associated with eTLM and clonal selection as well as thymic dendritic cells and APCs. There is evidence about the antigen-specific induction of apoptosis in $CD4^+CD8^+$ thymocytes cultured in suspension, by thymic and/or splenic APCs. Thus the recognition of antigen by $CD4^+CD8^+$ thymocytes may lead to the deletion, suggesting that this is the central mechanism of tolerance induction, which is not limited by the antigen-presenting ability of the thymic stroma (Swat *et al.*, 1991). Also, steroid hormones probably can be the activators of eTLM; at the same time, extrathymic matured lymphocytes are more resistant to steroids activity in relation to thymic lymphocytes (Shimizu *et al.*, 2000). The data about activation of eTLM by stress in adolescence indicate the relationship between stress, steroids, thymic atrophy and eTLM. In addition, pregnancy and malignant tumors may be also associated with thymic atrophy and eTLM activation.

Microbes and immuno-stimulators such as Th1 cytokines probably might be the factors of eTLM polarization, after the (self)antigen-reactive manner of immune reaction. Allowing for this model, the (self)antigen-reactive manner of eTLM possibly include the positive selection of (self)antigen-reactive thymus independent lymphocytes, activation of thymus dependent lymphocytes like CTL and Th1 cells, NK activity stimulation, suppression and/or negative selection of Th2 and suppressor cells. In the events of microenvironmental domination of Th2 cytokines and other immuno-suppressive/modulatory factors, the mechanism of eTLM probably include the negative selection of antigen-reactive thymus independent lymphocytes, suppression of thymus dependent lymphocytes like CTL and Th1 cells, NK activity suppression, activation and/or positive selection of Th2 antigen-reactive and other suppressor cells (Bubanovic, 2003d).

3.3.6.2   Failure of Blood-thymus Barrier

Central tolerance induction by clonal deletion is achieved in T cells during their maturation in the thymus. There are two critical events in the establishment of a functionally effective yet non-autoreactive T cells pool: the positive selection of T cells with a TCR capable of recognizing complex MHC/peptide and the negative selection of T cells (Nossal, 1994; Chidgey *et al.*, 1998). In the positive selection only those cells which recognize MHC/peptide complexes are selected for survival. However, those cells whose receptors are not MHC restricted, do not interact with thymic epithelial cells and consequently do not receive the protective signal, thus leading to their death within 3-4 days via apoptosis (Page *et al.*, 1993; Nossal, 1994). The negative selection involves the population of MHC-restricted reactive thymocytes that survive the positive selection comprising some cells with the low affinity receptors for "self"-peptide presented within the "self"-peptide-MHC molecules complexes and other cells with the high affinity receptors. During the negative selection, dendritic cells and macrophages bearing class I and class II MHC molecules are thought to interact with thymocytes bearing the high affinity receptors for the "self"-peptide-MHC complex, or self MHC molecules alone. The nature of the interaction is unknown, but cells with receptors for "self"-peptide-MHC complex or MHC molecules alone undergo death by apoptosis (Page *et al.*, 1993; Vasquez *et al.*, 1994).

Another mechanism of immunologic "self"-tolerance is based on the mechanisms called "clonal anergy". The remaining non-deleted, auto-reactive thymocytes are temporarily unable to mediate immune reactions and functionally silent. Some authors have found that a low level antigen intrathymic expression results in thymic lymphocytes anergy, which also shows a reduction of Th1 activity with no decrease in Th2 activity (Antonia *et al.*, 1995). Although thymic macrophages have been implicated in deleting auto-reactive thymocytes, they also present antigens to an IL-4-producing Th2 cells after IFN-γ treatment as evidenced by T cell proliferation and the release of IL-3 and IL-4. However, these thymic macrophages are inefficient at stimulating IL-2 producing Th1 clones (Jayaraman *et al.*, 1992). Antonia *et al.* (1995) propose that the loss of Th1 activity as a consequence of the thymic epithelium being encountered by tissue-specific proteins results in the functional tolerization of CTL *in vivo*, despite the fact that CTL are fully functional *in vitro*. The thymic expression of peripheral proteins may therefore be an additional way in which the tolerance to peripheral proteins can be achieved.

The experiments with the acquired thymic tolerance induction to the foreign MHC molecules in neonatal mice have also shown the presence of Th1 anergy as the possible mechanisms of the thymic tolerance. In MLR, spleen cells from normal mice proliferated in response to the "tolerogen" generated Th1 and CTL producing IL-2 and IFN-γ, but not IL-4 or IL-5. In contrast, although spleen cells from tolerant mice proliferated and produced IL-2, they failed to generate cytotoxic cells or produce IFN-γ, but produced large amounts of IL-4 and IL-5. Overall, the results point to a profound switch in the peripheral tolerogen-specific responses from a Th1 biased response in normal mice to a Th2 biased response in tolerant mice (Wood *et al.*, 1993).

The studies of blood-thymus barrier permeability in late fetal and adult stage showed that cortical thymocytes are protected from circulating macromolecules. The adult blood-thymus barrier contains capillary endothelial cells, basal lamina of endothelium, perivascular space, basal lamina of the epithelial-reticular cells and the epithelial-reticular cells. In contrast to cortical thymocytes, the medullar thymocytes are unprotected from circulating macromolecules. Blood vessels in the thymic cortex were not permeable to macromolecules, but the post-capillary venules of the thymic medulla permitted the macromolecules to leak along the clefts between migrating lymphocytes and endothelial cells. The macromolecules, which crossed medullary venule walls, had a limited distribution in the thymic parenchyma because macrophages in the perivascular space ingest and retain much of the leaked macromolecules (Ranga *et al.*, 1982; von Gaudecker, 1991).

The most important function of blood-thymus barrier in postnatal life is the prevention of foreign antigen penetration into the thymus. The intra-thymic inoculation of the antigen induces acquired and specific immune tolerance to the antigen. The model of the acquired thymic tolerance in adulthood by intrathymic antigen inoculation is utilized for tolerance induction in transplantation immunology (Nieuwenhuis *et al.*, 1988; Najman *et al.*, 1990; Chen *et al.*, 1998).

Chen *et al.* (1997, 1998) demonstrate that antigen intrathymic injection results in apoptotic cell death of both CD4$^+$ and CD8$^+$ thymocytes. Furthermore, thymectomy after intrathymic injection abrogates the effect of the acquired thymic tolerance and restores the antigen-dependent clonal expansion *in vivo*. Moreover, these authors conclude that intrathymic injection of antigen induces Th1 cell irresponsiveness and prevents the peripheral expansion of antigen-specific CD4$^+$ T cells *in vivo*. Many authors used the intrathymic inoculation of antigens for the prevention of some auto-immune disease. The intrathymic injection of guinea pig myelin basic protein without otherwise compromising the peripheral lymphocyte pool in adult rats dramatically inhibits the onset of experimental allergic encephalomyelitis caused by the usual peripheral inoculation with incomplete Freund's adjuvant (Ellison *et al.*, 1970; Wilson *et al.*, 1998).

Other authors used the inoculation of thymic epithelial cells, MHC molecules, lymphocytes or other MHC expressing cells of a donor, into the thymus of a recipient of the allogeneic graft. Adult animals, the recipients of donor cells or MHC molecules and donor allogeneic graft, reject the graft more slowly in comparison with control animals. These authors instigate that the intrathymic inoculation of MHC expressing cells from a donor to a recipient induce the acquired and specific thymic tolerance to allogeneic graft. In addition, the intrathymic injection of an immunodominant peptide induces the acquired thymic tolerance and suggests an indirect pathway of allorecognition in the thymus (Najman *et al.*, 1990; Garrovillo *et al.*, 1999).

There is now evidence that some factors can disturb the permeability of blood-thymus barrier and contribute to the intrathymic penetration of foreign antigens. Sex steroids like estradiol, progesterone or glucocorticoids can impair blood-thymus barrier. For now, we know for the thymic involutive factors like stress, ACTH and steroid hormones as factors, which can partly unclose blood-thymus barrier for circulating molecules and possible foreign antigens (Martin *et al.*, 1995; Bubanovic, 2003c).

Tumors are associated with the involution of the thymus, accompanied by a massive depletion of the cortical region and the alteration in the distribution of thymocytes, with a decrease in $CD4^+CD8^+$, $CD4^+CD8^-$ and $CD4^-CD8^+$ thymocytes. However, $CD4^-CD8^-$ population shows an increase, suggesting impairment in the thymocytes differentiation at an early maturation stage (Shanker *et al.*, 2000). Adkins *et al.* (2000) investigated three possible mechanisms leading to this thymic atrophy: (i) increased apoptosis, (ii) decreased proliferation and (iii) the disruption of normal thymic maturation. Their findings suggest that the thymic hypocellularity seen in breast tumor bearers is not due to a decreased level of proliferation but, rather, to an arrest at an early stage of the thymic differentiation along with a moderate increase in apoptosis.

The enhanced levels of glucocorticoids are known to produce similar effects on the thymus, but the adrenalectomy of mice followed by tumor implantation did not result in reversal of the thymic atrophy. Furthermore, a study of serum corticosterone levels in tumor bearers indicated no significant changes during tumor development. Because no major changes were observed in tumor bearers, either at their capacity to repopulate the thymus or at the patterns of subsequent redistribution of thymocytes, it was postulated that the thymic atrophy might be caused by a direct or indirect effect of the tumor or tumor-associated factor(s). The intrathymic injections of tumor cells into young normal recipient mice resulted in a significant reduction of the thymus weight and cellularity (Fu *et al.*, 1998). Lauritzsen *et al.* (1998) have found that circulating myeloma antigens are processed and presented by thymic APCs, and induces the deletion of the antigen-specific thymocytes. The deletion of tumor-specific thymocytes may represent a tumor escape mechanism in patients with cancers that secrete or shed tumor antigens. However, there is no much evidence about the influence of cytokines and other factors of tumor escape on the blood-thymus barrier permeability. For now, we know that some of pro-inflammatory cytokines induce thymic involution. At the same time, some of anti-inflammatory cytokines and growth factors are strong anti-involutive agents. Cytokines like IFN-β are associated with the massive apoptotic death of thymocytes and thymic involution. In this process, $CD4^+CD8^+$ T cells decreased significantly, whereas there were relative increases in $CD4^-CD8^-$, $CD4^+CD8^-$ and $CD4^-CD8^+$ T cells pool. The high doses of IL-2 also cause the thymic involution (Cesario *et al.*, 1991). In contrast to TNF-β and IL-2, cytokines like IL-10 and IGF-I are strong anti-apoptotic and anti-involutive factors of the thymus (Yamaguchi *et al.*, 1998).

The current cognition maintained the hypothesis that involutive factors increase the blood-thymus barrier permeability as well as the hypothesis that the increased permeability of the barrier might be a factor of central and peripheral thymic tolerance. Many cytokines, oncofetal antigens, soluble receptors, prostaglandine, progesterone, estrogens, trophoblast factors and other factors, which characterized pregnancy and tumor's microenvironment, probably can impair the permeability of blood-thymus barrier. As results of these events, circulating tumor-specific, trophoblast-specific or paternal antigens can permeate the barrier and induces the acquired thymic tolerance. In view of the acquired thymic tolerance as a phenomenon of selection and anergy of thymocytes, the failure of blood-thymus barrier as a reason of the thymic tolerance in the mechanisms of tumor and trophoblast escape is particularly possible. The mechanism of specific thymic selection in tumor bearers is already described, but similar mechanisms in pregnancy are yet unexplored. The mechanism of Th1 lymphocyte anergy in pregnancy and in tumor bearers as a result of blood-thymus barrier failure also is unexplored (Lauritzsen *et al.*, 1998; Bubanovic, 2003c).

3.3.7   Immunomodulatory Role of Prostaglandine

The elevated levels of PGE2 have been detected in a variety of common human tumors. PGE2 is known to possess the properties that promote malignant growth. For example, PGE2 stimulates angiogenesis and inhibits immune reactivity (Mc Lemore *et al.*, 1988; Le Fever *et al.*, 1990; Rigas *et al.*, 1993; Yoshimatsu *et al.*, 2001).Various types of human cancer cell derived PGE2 can orchestrate an imbalance in the production of IL-10 and IL-12 by lymphocytes and macrophages (Huang *et al.*, 1998). IL-10 and IL-12 probably are critical regulatory elements of cell mediated anti-tumor immunity. Although IL-10 inhibits the important aspects of cell-mediated immunity, IL-12 induces type 1 cytokine production and effective anti-tumor cell mediated responses (Nastala *et al.*, 1994; Beissert *et al.*, 1995). Although IL-10 overproduction at the tumor site has been implicated in tumor mediated immune suppression (Qin *et al.*, 1997; Kim *et al.*, 1995) IL-12 is critical for effective anti-tumor immunity (Bianchi *et al.*, 1996; Colombo *et al.*, 1996). In both *in vitro* experimental models and patients, the tumor bearing state induces lymphocyte and macrophage IL-10 but inhibits macrophage IL-12 production (Huang *et al.*, 1998; Kobayashi *et al.*, 1998). Because PGE2 appears to be pivotal in the reciprocal regulation of IL-10 and IL-12 (Strassmann *et al.*, 1994; Van der Pouw Kraan *et al.*, 1995), the pathways responsible for its high level production at the tumor site can be mediated by tumor cyclooxygenase-2 (COX2), that the elevated activity is a pivotal determinant of the expression of these cytokines in the tumor-bearing host (Alleva *et al.*, 1994; Yoshimatsu *et al.*, 2001).

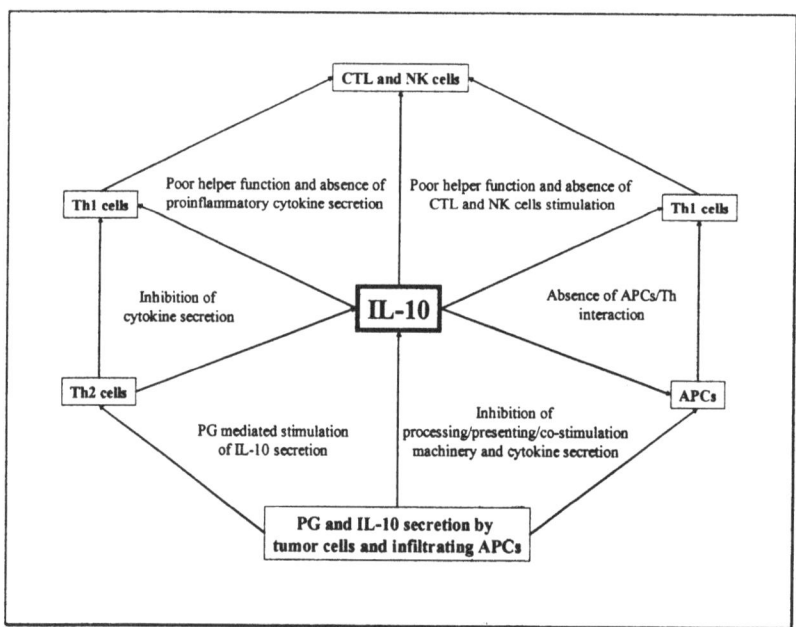

Figure 3.6. Central role of IL-10 in mechanisms of anti-tumor immunity failure.

Two isoenzymes have been identified: a constitutive form COX1 and an inducible isoenzyme COX2. COX2 is up-regulated in response to a variety of stimuli, including growth factors and cytokines (Herschman, 1996). Because it can lead to an enhanced PGE2 production and subsequent cytokine imbalance *in vivo*, tumor expression of COX2 may be instrumental in the generation of tumor-induced abrogation of T cell-mediated anti-tumor responses (Yoshimatsu *et al.*, 2001). COX2 has been implicated in the development of colon cancer and may play a role in promoting invasion, metastasis, and angiogenesis in established tumors (Tsujii *et al.*, 1998). In addition to lung and colon carcinomas, COX2 has recently been reported to be expressed in a variety of human malignancies (Tucker *et al.*, 1999; Chan *et al.*, 1999).

Multiple cytokines have been demonstrated to alter COX2 expression *in vitro*. Pro-inflammatory cytokines including TNF-α (Minghetti *et al.*, 1999), IL-1 (Vlahos *et al.*, 1999), and IFN-γ (Matsuura *et al.*, 1999) have been demonstrated to induce COX2 expression, whereas anti-inflammatory cytokines such as IL-4 (Niiro *et al.*, 1997), IL-13 (Diaz-Cazorla *et al.*, 1999), and IL-10 (Niiro *et al.*, 1995) can inhibit COX2 induction. The pleiotropic cytokine TGF-β can enhance or inhibit COX2 expression, depending upon the cell type tested. The functional significance of these *in vitro* observations has not been established *in vivo* (Reddy *et al.*, 1994; Saha *et al.*, 1999).

IL-10 may regulate COX2 mRNA expression through either indirect or direct means. A direct regulation of COX2 by IL-10 may occur at the transcriptional level, as well as the inhibition of inflammatory cytokine production (Lentsch *et al.*, 1997). Also, IL-10 may regulate COX2 gene expression at the posttranscriptional level, and it can be the one of the mechanisms by which IL-10 suppresses inflammation is via the induction of instability in the mRNAs of inflammatory cytokines, resulting in their degradation (Brown *et al.*, 1996). Indeed, LPS and IL-1β enhance COX2 mRNA stability (Barrios-Rodiles *et al.* 1999), whereas exogenous IL-10 *in vitro* accelerates the degradation of COX2 mRNA in human monocytes (Niiro *et al.*, 1995). Therefore, the absence of IL-10 may have resulted in the increased COX2 mRNA through an increase in the mRNA stability.

Another of the mechanisms by which PG (especially PGE2) supports tumor growth is by inducing the angiogenesis necessary to supply oxygen and nutrients to tumors >2 mm in diameter (Hanahan *et al.*, 1996). Masferrer *et al.* (1999) found that the COX2 expression in newly formed blood vessels within tumors grown in animals, whereas under normal physiological conditions the quiescent vasculature expresses only the constitutive COX1 enzyme. This observation shown the expression of the enzyme in cancer cells suggests a predominant role for COX2 in tumor regulation and angiogenesis.

There is a large body of data that the inhibition of COX2 led to a marked lymphocytic infiltration of the tumor and reduced tumor growth. The treatment of mice with anti-PGE2 mAb replicated the growth reduction seen in tumor-bearing mice treated with COX2 inhibitors. The COX2 inhibition was accompanied by a significant decrement in IL-10 and a concomitant restoration of IL-12 production by APCs. Because the COX2 metabolite PGE2 is a potent inducer of IL-10 secretion, many authors hypothesized that the COX2 inhibition led to anti-tumor responses by down-regulating production of this potent immunosuppressive cytokine (Stolina *et al.*, 2000). Although PGE2 mediated IL-10 activity has been shown to have the capacity to directly suppress T cell activities, the inhibitory effects of IL-10 have most often been found to be mediated indirectly through the modulation of APCs function. IL-10 has the capacity to decrease APCs MHC and TAP expression and to down-regulate critical costimulatory molecules including CD54, CD80, and CD86 (Sharma *et al.*, 1999).

REFERENCES

Abo T. (2001). Extrathymic pathways of T-cell differentiation and immunomodulation. *Int Immunopharmacol.* 7:1261-1273.

Adkins B., Charyulu V., Sun Q.L. (2000). Early block in maturation is associated with thymic involution in mammary tumor-bearing mice. *J Immunol.* 164:5635-5640.

Agrawal B., Krantz M.J., Reddish M.A. *et al.* (1998a). Cancer-associated MUC1 mucin inhibits human T cell proliferation, which reversible by IL-2. *Nature Med.* 4:43-49.

Agrawal B., Krantz M.J., Parker J. *et al.* (1998b). Expression of MUC1 mucin on activated human T cells: Implication for a role of MUC1 in normal immune regulation. *Cancer Res.* 58:4079-4081.

Alleva D.G., Burger C.J., Elgert K.D. (1994). Tumor-induced regulation of suppressor macrophage nitric oxide and TNF-α production: role of tumor-derived IL-10, TGF-ß and prostaglandin E2. *J Immunol.* 153:1674-1679.

Altomonte M., Gloghini A., Bertola G. *et al.* (1993). Differential expression of cell adhesion molecules CD54/CD11a and CD58/CD2 by human melanoma cells and functional role in their interaction with cytotoxic cells. *Cancer Res.* 53:3343-3348.

Antonia S.J., Geiger T., Miller J. *et al.* (1995). Mechanisms of immune tolerance induction through the thymic expression of a peripheral tissue-specific protein. *Int Immunol.* 7:715-725.

Aruga A., Aruga E., Tanigawa K. *et al.* (1997). *Type 1* versus *type 2* cytokine release by Vβ T cell subpopulations determines in vivo anti-tumor reactivity: IL-10 mediates a suppressive role. *J Immunol.* 159:664-673.

Barrios-Rodiles M., Tiraloche G., Chadee K. (1999). Lipopolysaccharide modulates COX2 transcriptionally and posttranscriptionally in human macrophages independently from endogenous IL-1β and TNF- α. *J Immunol.* 163:963-969.

Basu S., Binder R., Suto R. *et al.* (2000). Necrotic but not apoptotic cell death releases heat shock proteins, which deliver a partial maturation signal to DCs and activate the NF-B pathway. *Int Immunol.* 12:1539-1545.

Becker J.C., Termeer C., Schmidt R.E. *et al.* (1993). Soluble intercellular adhesion molecule-1 inhibits MHC-restricted specific T cell/tumor interaction. *J Immunol.* 151:7224-7232.

Beissert S., Hosoi J., Grabbe S., (1995). IL-10 inhibits tumor antigen presentation by epidermal antigen-presenting cells. *J Immunol.* 154:1280-1284.

Bendelac A., Rivera M.N., Park S.H. *et al.* (1997). Mouse CD1-specific NK1 T cells: development, specificity, and function. *Annu Rev Immunol.* 1997;15:535-562.

Berwin B., Reed R.C., Nicchitta C.V. (2001). Virally induced lytic cell death elicits the release of immunogenic GRP94/gp96. *J Biol Chem.* 276:21083-21088.

Bianchi R., Grohmann U., Belladonna M. *et al.* (1996). IL-12 is both required and sufficient for initiating T cell reactivity to a class I-restricted tumor peptide (P815AB) following transfer of P815AB-pulsed DCs. *J Immunol.* 157:1589-1593.

Binder R.J., Anderson K.M., Basu S. *et al.* (2000). Heat shock protein gp96 induces maturation and migration of CD11c+ cells in vivo. *J Immunol.* 165:6029-6035.

Boon T., Cerottini J.C., Van den Eynde B. *et al.* (1994). Tumor antigens recognized by T lymphocytes. *Annu Rev Immunol.* 12:337-365.

Brichard V., Van Pel A., Wolfel T. *et al.* (1993). The tyrosinase gene codes for an antigen recognized by autologous cytolytic T lypmhocytes on HLA-A2 melanomas. *J Exp Med.* 178:489-495.

Bronte V., Apolloni E., Ronca R. *et al.* (2000). Genetic Vaccination with "self" tyrosinase-related protein 2 causes melanoma eradication but not vitiligo. *Cancer Res.* 60:253-258.

Brown C.Y., Lagnado C.A., Vadas M.A. *et al.* (1996). Differential regulation of the stability of cytokine mRNAs in lipopolysaccharide-activated blood monocytes in response to IL-10. *J Biol Chem.* 271:2010-2018.

Bubanovic, I. (2003a). Origin of anti-tumor immunity failure in mammals and new possibility for immunotherapy. *Med Hypotheses.* 60:152-158.

Bubanovic I., Najman S. (2003b). Failure of Anti-tumor Immunity Mammals - Evolution of the Hypothesis. The 23rd Annual Meeting of the American Society for Reproductive Immunology. New Haven, Connecticut, USA. June 18-22, *Am J Reprod Immunol.* 49:329-372, Abstract N° 49; p.:351.

Bubanovic, I., (2003c). Failure of blood-thymus barrier as a mechanism of tumor and trophoblast escape. *Med Hypotheses.* 60:315-320.

Bubanovic I. (2003d). Crossroads of extrathymic lymphocytes maturation pathways. *Med Hypotheses.* 61:235-239.

Buchler T., Michalek J., Kovarova L., *et al.* (2003). DCs-based immunotherapy for the treatment of hematological malignancies. *Hematol.* 8:97-104.

Cabestre A.E., Moreau P., Riteau B. *et al.* (1999). HLA-G expression in human melanoma cells: protection from NK cytolysis. *J Reprod Immunol.* 43(2):183-193.

Cabrera T., Angustias Fernandez M., Sierra A. *et al.* (1996). High frequency of altered HLA class I in invasive breast carcinomas. *Human Immunol.* 50:127-134.

Cammarota G. (1992). Identification of a CD4 binding site on the β2 domain of HLA-DR molecules. *Nature.* 356;799-801.

Cella M., Scheidegger D., Palmer-Lehmann K. *et al.* (1996). Ligation of CD40 on DCs triggers production of high levels of IL-12 and enhances T cell stimulatory capacity: T-T help via APC activation. *J Exp Med.* 184:747-751.

Cella M., Sallusto F., Lanzavecchia A. (1997). Origin, maturation and antigen presenting function of DCs. *Curr Opin Immunol.* 9:10-18.

Cesario T.C., Vaziri N.D., Ulich T.R. *et al.* (1991). Functional, biochemical, and histopathologic consequences of high-dose IL-2 administration in rats. *J Lab Clin Med.* 118:81-88.

Chaux P., Moutet M., Faivre J. *et al.* (1996). Inflammatory cells infiltrating human colorectal carcinoma express HLA class II but not B7-1 and B7-2 costimulatory molecules of the T-cell activation. *Lab Invest.* 74:975-983.

Chan G., Boyle J.O., Yang E.K. *et al.* (1999). *COX2* expression is up-regulated in squamous cell carcinoma of the head and neck. *Cancer Res.* 59:991-998.

Chansky H.A., Hu M., Hickstein D.D. *et al.* (2001). Oncogenic TLS/ERG and EWS/Fli-1 fusion proteins inhibit RNA splicing mediated by YB-1 protein. *Cancer Res.* 61:3586-3590.

Chen W., Issazadeh S., Sayegh M.H. *et al.* (1997). In vivo mechanisms of acquired thymic tolerance. *Cell Immunol.* 179:165-173.

Chen W., Sayegh M.H., Khoury S.J. (1998). Mechanisms of acquired thymic tolerance in vivo: intrathymic injection of antigen induces apoptosis of thymocytes and peripheral T cell anergy. *J Immunol.* 160:1504-1508.

Chiari R., Hames G., Stroobant V. *et al.* (2000). Identification of a tumor-specific shared antigen derived from an Eph receptor and presented to CD4 T cells on HLA class II molecules. *Cancer Res.* 60:4855-4863.

Chidgey A.P., Boyd R.L. (1998). Positive selection of low responsive potentially autoreactive T cells induced by high avidity, non-deleting interactions. *Int Immunol.* 10:999-1008.

Colombo M., Vagliani M., Spreafico F. *et al.* (1996). Amount of IL 12 available at the tumor site is critical for tumor regression. *Cancer Res.* 56:2531-2535.

Constant S.L., Bottomly K. *et al.* (1997). Induction of Th1 and Th2 CD4+ T cell responses: the alternative approaches. *Annu Rev Immunol.* 15:297-322.

Coulie P.G., Karanikas V., Colau D., *et al.* (2001). A monoclonal cytolytic T-lymphocyte response observed in a melanoma patient vaccinated with a tumor-specific antigenic peptide encoded by gene MAGE-3. *Proc Natl Acad Sci USA.* 98:10290-10295.

Croft M., Carter L., Swain S.L. *et al.* (1994). Generation of polarized antigen-specific CD8 effector populations: reciprocal action of IL (IL)-4 and IL-12 in promoting *type 2* versus *type 1* cytokine profiles. *J Exp Med.* 180:1715-1728.

Cromme F.V., Airey J., Heemels M. *et al.* (1994). Loss of transporter protein, encoded by the TAP1 gene, is highly correlated with loss of HLA expression in cervical carcinomas. *J Exp Med.* 179:335-340.

Davies B., Hiby S., Gardner L. *et al.* (2001). HLA-G expression by tumors. *Am J Reprod Immunol.* 45:103-107.

Desmedt M., Rottiers, P., Dooms, H. *et al.* (1998). Macrophages induce cellular immunity by activating Th1 cell responses and suppressing Th2 cell responses. *J Immunol.* 160:5300-5308.

Di Santo J.P. (1997). Cytokines: shared receptors, distinct functions. *Curr Biol.* 7:424-426.

Diaz-Cazorla M., Perez-Sala D., Ros J. *et al.* (1999). Regulation of *COX2* expression in human mesangial cells: transcriptional inhibition by IL-13. *Eur J Biochem.* 260:268-272.

Dowlati A., Levitan N., Remick S.C. (1999). Evaluation of IL-6 in bronchoalveolar lavage fluid and serum of patients with lung cancer. *J Lab Clin Med.* 134:405-409.

Drexler H.G., Macleod R.A.F., Borkhardt A. *et al.* (1995). Recurrent chromosomal translocation and fusion genes in Leukemia-Lymphoma Cell-Lines. *Leukemia.* 9:303-318.

Dunn G.P., Bruce A.T., Ikeda H., *et al.* (2002). Cancer immunoediting: from immunosurveillance to tumor escape. *Nat Immunol.* 3:991-998.

Enk A.H., Jonuleit H., Saloga J. *et al.* (1997). DCs as mediators of tumor-induced tolerance in metastatic melanoma. *Int J Cancer.* 73:309-316.

Elder M.E. (1998). ZAP-70 and defects of T-cell receptor signaling. *Semin Hematol.* 35:310-320.

Ellison G., Waksman B. (1970). Role of the thymus in tolerance. IX. Inhibition of experimental autoallergic encephalomyelitis by intrathymic injection of encephalitogen. *Immunol.* 105:322-356.

Flores-Romo L. (2001). In vivo maturation and migration of DCs. *Immunol.* 102:255-262.

Fortis C., Foppoli M., Gianotti L. *et al.* (1996). Increased IL-10 serum levels in patients with solid tumours. *Cancer Lett.* 24;104:1-5.

Fu Y., Paul R.D., Wang Y. *et al.* (1998). Thymic involution and thymocyte phenotypic alterations induced by murine mammary adenocarcinomas. *J Immunol.* 143:4300-4307.

Gallucci S., Lolkema M., Matzinger P. (1999). Natural adjuvants: endogenous activators of DCs. *Nat Med.* 5:1249-1253.

Garrovillo M., Ali A., Oluwole S.F. *et al.* (1999). Indirect allorecognition in acquired thymic tolerance: induction of donor-specific tolerance to rat cardiac allografts by allopeptide-pulsed host DCs. *Transplant.* 68:1827-1834.

Garrido F., Cabrera T., Concha,A. *et al.* (1993). Natural history of HLA expression during tumor development. *Immunol Today.* 14:491-499.

Garrido F., Ruiz-Cabello F., Cabrera T. *et al.* (1997). Implications for immunosurveillance of altered HLA class I phenotypes in human tumours. *Immunol Today.* 18:89-95.

Germain, R.N. (1994). MHC-dependent antigen processing and peptide presentation: providing ligands for T lymphocyte activation. *Cell.* 76:287-299.

Golumbek P.T., Lazenby A.J., Levitsky H.I. *et al.* (1991). Treatment of established renal cancer by tumor cells engineered to secrete IL-4. *Science.* 254:713-716.

Guo, Y., Yang, T., Liu, X. *et al.* (2002). Cis elements for transporter associated with antigen-processing-2 transcription: two new promoters and an essential role of the IFN response factor binding element in IFN-γ-mediated activation of the transcription initiator. *Int Immunol.* 14:189-200.

Hahne M., Rimoldi D., Schroter M. *et al.* (1996). Melanoma cell expression of *fas*(Apo-1/CD95) ligand: implications for tumor immune escape. *Science.* 274:1363-1366.

Hanahan D., Folkman J. (1996). Patterns and emerging mechanisms of the angiogenic switch during tumorigenesis. *Cell.* 86:353-364.

Hasegawa T., Isobe K., Nakashima I. *et al.* (1992). Quantitative analisys of antigen for the induction tolerance in carcinoembryionic antigen transgenic mice. *Immunol.* 77:577-581.

Herschman H. (1996). Review: prostaglandin synthase 2. *Biochim Biophys Acta.* 1299:125-135.

Hosken N.A., Shibuya K., Heath A.W. *et al.* (1995). The effect of antigen dose on CD4+ T helper cell phenotype development in a T cell receptor-α β-transgenic model. *J Exp Med.* 182:1579-1584.

Hu H.M., Urba W.J., Fox B.A. (1998). Gene-modified tumor vaccine with therapeutic potential shifts tumor-specific T cell response from a *type 2* to a *type 1* cytokine profile. *J Immunol.* 161:3033-3341.

Huang M., Stolina M., Sharma S. *et al.* (1998). Non-small cell lung cancer COX2-dependent regulation of cytokine balance in lymphocytes and macrophages: up-regulation of IL 10 and down-regulation of IL 12 production. *Cancer Res.* 58:1208-1212.

Huang S., Ullrich S.E., Bar-Eli M. *et al.* (1999). Regulation of tumor growth and metastasis by IL-10: the melanoma experience. *J IFN Cytokine Res.* 19:697-703.

Hung K., Hayashi R., Lafond-Walker A. *et al.* (1998). The central role of CD4(+) T cells in the antitumor immune response. *J Exp Med.* 88:2357-2368.

Ikebe H., Yamada H., Nomoto M., *et al.* (2001). Persistent infection with Listeria monocytogenes in the kidney induces anti-inflammatory invariant fetal-type γδ T cells. *Immunol.* 102:94-102.

Ikeda H., Ohta N., Furukawa K. *et al.* (1997). Mutated mitogen-activated protein kinase: a tumor rejection antigen of mouse sarcoma. *Proc Natl Acad Sci USA.* 94:6375-6379.

Jayaraman S., Luo Y., Dorf M.E. (1992). Tolerance induction in T helper (Th1) cells by thymic macrophages. *J Immunol.* 148:2672-2681.

Johnsen A.K., France J., Sy M.S. *et al.* (1999). Deficiency of TAP in tumor cells allows evasion of immune surveillance and increase tumorogenesis. *J Immunol.* 163:4224-4231.

Jonuleit H., Kuhn U., Müller G. *et al.* (1997). Pro-inflammatory cytokines and prostaglandins induce maturation of potent immunostimulatory DCs under fetal calf serum-free conditions. *Eur J Immunol.* 27:3135-3140.

Kaklamanis L., Leek R., Koukourakis M. *et al.* (1995). Loss of TAP1 protein and MHC class I molecules in primary and metastatic versus primary breast cancer. *Cancer Res.* 55:5191-5196.

Kawakami Y., Eliyahu S., Delgado C.H. *et al.* (1994). Identification of a human-melanoma antigen recognized by TIL associated with in vitro rejection. *Proc Natl Acad Sci USA.* 91:6458-6462.

Kim J., Modlin R.L., Moy R.L. *et al.,* (1995). IL-10 production in cutaneous basal and squamous cell carcinomas: a mechanism for evading the local T cell immune response. *J Immunol.* 155:2240-2245.

Kirkin A.F., thor Straten P., Zeuthen J. *et al.* (1996). Differential modulation by IFN-γ of the sensitivity of human melanoma cells to cytolytic T cell clones that recognize differentiation or progression antigens. *Cancer Immunol Immunother.* 42:203-212.

Knittler M.R., Alberts P., Deverson E.V. *et al.* (1999). Nucleotide binding by TAP mediates association with peptide and release of assembled MHC class I molecules. *Curr Biol.* 9:999-1008.

Kobata T., Azuma M., Yagita H. *et al.* (2000). Role of costimulatory molecules in autoimmunity. *Rev Immunogenet.* 2:74-80.

Kobayashi M., Kobayashi H., Pollard R. *et al.* (1998). A pathogenic role of Th2 cells and their cytokine products on the pulmonary metastasis of murine B16 melanoma. *J Immunol.* 160:5869-5872.

Konig R., Huang L.Y., Germain R.N. (1992). MHC class II interaction with CD4 mediated by a region analogous to the MHC class I binding site for CD8. *Nature.* 356;796-798.

Krasagakis K., Tholke D., Farthmann B. *et al.* (1998). Elevated plasma levels of TGF-β1 and TGF-β2 in patients with disseminated malignant melanoma. *Br J Cancer.* 77:1492-1494.

Kuchroo V.K., Das M.P., Brown J.A. *et al.* (1995). B7-1 and B7-2 costimulatory molecules activate differentially the Th1/Th2 developmental pathways: application to autoimmune disease therapy. *Cell.* 80:707-718.

Kufer P., Zettl F., Borschert K. *et al.* (2001). Minimal costimulatory requirements for T cell priming and TH1 differentiation: Activation of naive human T lymphocytes by tumor cells armed with bifunctional antibody constructs. *Cancer Immunity.* 1:10-17.

Lauritzsen G.F., Hofgaard P.O., Schenck K. *et al.* (1998). Clonal deletion of thymocytes as a tumor escape mechanism. *Int J Cancer.* 78:216-222.

Liu M.A. (1997). The immunologist's grail: Vaccines that generate cellular immunity. *Proc Natl Acad Sci USA.* 94:10496-10498.

Liu L., Rich B.E., Inobe J. *et al.* (1998). Induction of Th2 cell differentiation in the primary immune response: DCs isolated from adherent cell culture treated with IL-10 prime naive CD4+ T cells to secrete IL-4. *Int Immunol.* 10:1017-1026.

Lee P.P., Zeng D., McCaulay A.E. *et al.* (1997). T helper 2-dominant antilymphoma immune response is associated with fatal outcome. *Blood.* 90:1611-1617.

Lee R.S., Schlumberger M., Caillou B. *et al.* (1996). Phenotypic and functional characterisation of tumour-infiltrating lymphocytes derived from thyroid tumours. *Eur J Cancer.* 32:1233-1239.

Le Fever A., Funahashi A. (1990). Elevated prostaglandin E2 levels in bronchoalveolar lavage fluid of patients with bronchogenic carcinoma. *Chest.* 98:1397-1402.

Lentsch A.B., Shanley T.P., Sarma V. *et al.* (1997). In vivo suppression of NF-kapa B and preservation of I-kapaBα by IL-10 and IL-13. *J Clin Invest.* 100:2443-2447.

Lobigs M., Mullbacher A., Blanden R.V. *et al.* (1999). Antigen presentation in syrian hamster cells: substrate selectivity of TAP controlled by polymorphic residues in TAP1 and differential requirements for loading of H2 class I molecules. *Immunogenetics.* 11-12:931-941.

Lowes M.A., Bishop G.A., Crotty K. *et al.* (1997). T helper 1 cytokine mRNA is increased in spontaneously regressing primary melanomas. *J Invest Dermatol.* 108:914-919.

Luksch C.R., Winqvist O., Ozaki M.E. *et al.* (1999). Intercellular adhesion molecule-1 inhibits IL 4 production by naive T cells. *Proc Natl Acad Sci USA.* 96:3023-3028.

Maeurer M.J., Lotze M.T. (1998). IL-7 knockout mice. Implications for lymphopoiesis and organ-specific immunity. Int Rev *Immunol.* 16:309-322.

Mantovani A., Allavena P., Vecchi A., *et al.* (1998). Chemokines and chemokine receptors during activation and deactivation of monocytes and DCs and in amplification of Th1 versus Th2 responses. *Int J Clin Lab Res.* 28:77-82.

Maki K., Sunaga S., Komagata Y. (1996). IL 7 receptor-deficient mice lack γδT cells. *Proc Natl Acad Sci USA.* 93:7172-7177.

Martin A., Casares F., Alonso L. *et al.* (1995). Changes in the blood-thymus barrier of adult rats after estradiol-treatment. *Immunobiol.* 192:231-248.

Masferrer J.L., Koki A. T., Seibert K. (1999). COX2 inhibitors. A new class of antiangiogenic agents. *Ann NY Acad Sci.* 889:84-86.

Matsui, M., Machida, S., Itani-Yohda, T. *et al.* (2002). Downregulation of the proteasome subunits, transporter, and antigen presentation in hepatocellular carcinoma, and their restoration by IFN-γ. *J Gastroenterol Hepatol.* 17:897-907.

Matsuura H., Sakaue M, Subbaramaiah K. *et al.* (1999). Regulation of COX2 by IFN-γ and TGF-α in normal human epidermal keratinocytes and squamous carcinoma cells: role of mitogen-activated protein kinases. *J Biol Chem.* 274:2913-2918.

Matzinger P. (1994). Tolerance, danger, and the extended family. *Annu Rev Immunol.* 12:991-1045.

Mc Lemore T.L., Hubbard W.C., Litterst C.L. *et al.* (1988). Profiles of prostaglandin biosynthesis in normal lung and tumor tissue from lung cancer patients. *Cancer Res.*, 48:3140-3147.

Milani V., Noessner E., Ghose S. *et al.* (2002). Heat shock protein 70: role in antigen presentation and immune stimulation. *Int J Hyperthermia.* 18:563-575.

Minghetti L., Walsh D.T., Levi G. *et al.* (1999). In vivo expression of *COX2* in rat brain following intraparenchymal injection of bacterial endotoxin and inflammatory cytokines. *J Neuropathol Exp Neurol.* 58:1184-1188.

Monach P.A., Meredith S.C., Siegel C.T. *et al.* (1995). A unique tumor antigen produced by a single amino acid substitution. *Immunity.* 2:45-59.

Moore K.W., O'Garra A., de Waal Malefyt R. *et al.* (1993). IL-10. *Annu Rev Immunol.* 11:165-190.

Mosmann T.R., Sad S. (1996). The expanding universe of T-cell subsets: Th1, Th2 and more. *Immunol Today.* 17:138-146.

Mumberg D., Monach P.A., Wanderling S. *et al.* (1999). CD4+ T cells eliminate MHC class II-negative cancer cells in vivo by indirect effects of IFN-γ. *Proc Natl Acad Sci USA.* 96:8633-8638.

Murray, J.L., Hudson, J.M., Ross, M.I. *et al.*, (2000). Reduced recognition of metastatic melanoma cells by autologous MART-1 specific CTL: relationship to TAP expression. *J Immunother.* 23:28-35.

Najman S., Bubanovic I., Stamenkovic S. *et al.* (1990). Effects of inoculating adherent thymic cells of a donor into the thymus of a recipient on the survival of the skin graft. *Acta Med Medianae.* 1:79-88.

Nastala C.L., Edington H.D, McKinney T.G. *et al.* (1994). Recombinant IL-12 administration induces tumor regression in association with IFN- production. *J Immunol.* 153:1697-1701.

Niehans G.A., Brunner T., Frizelle S.P. *et al.* (1997). Human lung carcinomas express *fas* ligand. *Cancer Res.* 57:1007-1012.

Nieuwenhuis P., Stet R.J., Wagenaar J.P. *et al.* (1988). The transcapsular route: a new way for (self-) antigens to by-pass the blood-thymus barrier? *Immunol Today.* 9:372-375.

Niiro H., Otsuka T., Tanabe T. *et al.* (1995). Inhibition by IL-10 of inducible *COX* expression in lipopolysaccharide-stimulated monocytes: its underlying mechanism in comparison with IL-4. *Blood.* 85:3736-3740.

Niiro H., Otsuka T., Izuhara K. *et al.* (1997). Regulation by IL-10 and IL-4 of COX2 expression in human neutrophils. *Blood.* 89:1621-1625.

Ninomiya T., Akbar S.M., Masumoto T. (1999). DCs with immature phenotype and defective function in the peripheral blood from patients with hepatocellular carcinoma. *J Hepatol.* 31:323-331.

Nossal G.J.V. (1994). Negative selection of lymphocytes. *Cell.* 76:229-240.

O' Garra A. (1998). Cytokines induce the development of functionally heterogeneous T helper cell subsets. *Immunity.* 8:275-283.

Ohshima Y., Yang L.P., Uchiyama T. *et al.* (1998). OX40 costimulation enhances IL-4 (IL-4) expression at priming and promotes the differentiation of naive human CD4(+) T cells into high IL-4-producing effectors. *Blood.* 92:3338-3345.

Ohteki T., Okuyama R., Seki S. *et al.* (1992). Age-dependent increase of extrathymic T cells in the liver and their appearance in the periphery of older mice. *J Immunol.* 149:1562-1570.

Page D.M., Kane L.P., Allison J.P. *et al.* (1993). Two signals are required for negative selection of CD4+CD8+ thymocytes. *J Immunol.* 151:1868-1880.

Palmer E.M., van Seventer G.A. (1997). Human T helper cell differentiation is regulated by the combined action of cytokines and accessory cell-dependent costimulatory signals. *J Immunol.* 158:2654-2662.

Parra E., Wingren A.G., Hedlund G. *et al.* (1993). Human naive and memory T-helper cells display distinct adhesion properties to ICAM-1, LFA-3 and B7 molecules. *Scand J Immunol.* 38:508-514.

Pascale P., Rouas-Freiss N., Khalil-Daher I. (1998). HLA-G expression in melanoma: A way for tumor cells to escape from immunosurveillance. *Immunol.* 8:4510-4515.

Pasquini S., Xiang Z., Wang Y. *et al.* (1997). Cytokines and costimulatory molecules as genetic adjuvants. *Immunol Cell Biol.* 75:397-401.

Paul W.E., Fathman C.G. Metzger H. (1994). Tumor Antigens Recognized by T Lymphocytes. *Ann Rev Immunol.* 12:337-366.

Paul P., Rouas-Freiss N., Khalil-Daher I. *et al.* (1998). HLA-G expression in melanoma: A way for tumor cells to escape from immunosurveillance. *Proc Natl Acad Sci USA.* 95:4510-4515.

Paul S., Bizouarne N., Paul A. *et al.* (1999). Lack of evidence for an immunosuppressive role for MUC1. *Cancer Immunol Immunother.* 48:22-28.

Pawelec G. (1999). Tumour escape from the immune response: the last hurdle for successful immunotherapy of cancer? *Cancer Immunol Immunother.* 48:343-345.

Petersson, M., Charo, J., Salazar-Onfray, F. *et al.* (1998). Constitutive IL-10 production accounts for the high NK sensitivity, low MHC class I expression, and poor transporter associated with antigen processing (TAP)-1/2 function in the prototype NK target YAC-1. *J Immunol.* 161:2099-2105.

Pinto M.M., Greenebaum E., Simsir A. *et al.* (1997). CA-125 and carcinoembryonic antigen assay vs. cytodiagnostic experience in the classification of benign ovarian cysts. *Acta Cytol.* 41:1456-1462.

Porter B.O., Malek T.R. (2000). Thymic and intestinal intraepithelial T lymphocyte development are each regulated by the c-dependent cytokines IL-2, IL-7, and IL-15. *Semin Immunol.* 12:465-474.

Ranga V., Ispas A.T. (1982). Chirulescu A.R. Elements of structure and ultrastructure of the blood-thymus barrier in ACTH involuted thymus. *Acta Anat.* 111:177-189.

Reddy S.T., Gilbert R.S., Xie W. *et al.* (1994). TGF-β 1 inhibits both endotoxin-induced prostaglandin synthesis and expression of the TIS10/prostaglandin synthase 2 gene in murine macrophages. *J Leukocyte Biol.* 55:192-198.

Reiner S.L. (1995). T helper cell differentiation in immune response. *Curr Opin Immunol.* 7:360-366.

Restifo N.P., Esquivel F., Kawakami Y. (1993). Identification of human cancers deficient in antigen processing. *J Exp Med.* 177:265–272.

Rigas B., Goldman I.S., Levine L. (1993). Altered eicosanoid levels in human colon cancer. *J Lab Clin Med.* 122:518-523.

Rivoltini L., Radrizzani M., Accornero P. *et al.* (1998). Human melanoma-reactive CD4+ and CD8+ CTL clones resist *fas* ligand-induced apoptosis and use *fas/Fas* ligand-independent mechanisms for tumor killing. *J Immunol.* 161:1220-1230.

Rosen H.R., Ausch C., Reiner G. *et al.* (1994). Downregulation of lymphocyte mitogenesis by breast cancer-associated p43. *Cancer Lett.* 82:105-111.

Rouas-Freiss N., Paul P., Dausset J. *et al.* (2000). HLA-G promotes immune tolerance. *J Biol Regul Homeost Agents.* 14:93-98.

Saha D., Datta P.K., Sheng H. *et al.* (1999). Synergistic induction of COX2 by TGF-β1 and epidermal growth factor inhibits apoptosis in epithelial cells. *Neoplasm.* 1:508-512.

Salazar-Onfray, F., Charo, J., Petersson, M. *et al.* (1997). Down-regulation of the expression and function of the transporter associated with antigen processing in murine tumor cell lines expressing IL-10. *J Immunol.* 159:3195-3202.

Salgame P., Abrams J.S., Clayberger C. *et al.* (1991). Differing lymphokine profiles of functional subsets of human CD4 and CD8 T cell clones. *Science.* 254:279-82.

Salomon B., Bluestone J.A. (1998). LFA-1 interaction with ICAM-1 and ICAM-2 regulates Th2 cytokine production. *J Immunol.* 161:5138-5142.

Sauter B., Albert M.L., Francisco L. *et al.* (2000). Consequences of cell death: exposure to necrotic tumor cells, but not primary tissue cells or apoptotic cells, induces the maturation of immunostimulatory DCs. *J Exp Med.* 191:423-429.

Scheibenbogen C., Lee K.H., Stevanovic S. *et al.* (1997). Analysis of the T cell response to tumor and viral peptide antigens by an IFN-γ-ELISPOT assay. *Int J Cancer.* 71:932-936.

Seder R.A. (1997). Are differentiated human T helper cells reversible? *Int Arch Allergy Immunol.* 113:163-166.

Seliger B., Harders C., Lohmann S. *et al.* (1998). Down-regulation of the MHC class I antigenprocessing machinery after oncogenic transformation of murine fibroblasts. *Eur J Immunol.* 28:122-133.

Shanker A., Singh S.M., Sodhi A. (2000). Ascitic growth of a spontaneous transplantable T cell lymphoma induces thymic involution. 2. Induction of apoptosis in thymocytes. *Tumour Biol.* 21:315-327.

Sharma S., Stolina M., Lin Y. (1999). T Cell-Derived IL-10 Promotes Lung Cancer Growth by Suppressing Both T Cell and APC Function. *J Immunol.* 163:5020-5028.

Shiraki K., Tsuji N., Shioda T. *et al.* (1997). Expression of *fas* ligand in liver metastases of human colonic adenocarcinomas. *Proc Natl Acad Sci USA.* 94:6420-6425.

Shimizu T., Kawamura T., Miyaji C. *et al.* (2000). Resistance of extrathymic T cells to stress and the role of endogenous glucocorticoids in stress associated immunosuppression. *Scand J Immunol.* 51:285-292.

Singh-Jasuja H., Scherer H.U., Hilf N. *et al.* (2000). The heat shock protein gp96 induces maturation of DCs and down-regulation of its receptor. *Eur J Immunol.* 30:2211-2216.

Singh-Jasuja H., Hilf N., Arnold-Schild D. *et al.* (2001). The role of heat shock proteins and their receptors in the activation of the immune system. *Biol Chem.* 382:629-636.

Steinman R.M., Banchereau J. (1998). DCs and the control of immunity. *Nature.* 392:245-247.

Stolina S., Sharma S., Lin Y. *et al.* (2000). Specific Inhibition of COX2 Restores Antitumor Reactivity by Altering the Balance of IL-10 and IL-12 Synthesis. *J Immunol.* 164:361-370.

Strassmann G., Patil-Koota V., Finkelman F. *et al.* (1994). Evidence for the involvement of IL 10 in the differential deactivation of murine peritoneal macrophages by prostaglandin E2. *J Exp Med.* 180:2365-2370.

Swat W., Ignatowicz L., von Boehmer H. *et al.* (1991). Clonal deletion of immature CD4+8+ thymocytes in suspension culture by extrathymic antigen-presenting cells. *Nature.* 351:150-153.

Tamauchi H., Tamaoki N., Habu S. (1988). CD4+CD8+ thymocytes develop into CD4 or CD8 single-positive cells in athymic nude mice. *Eur J Immunol.* 18:1859-1862.

Tao X., Constant S., Jorritsma P. *et al.* (1997). Strength of TCR signal determines the costimulatory requirements for Th1 and Th2 CD4+ T cell differentiation. *J Immunol.* 159:5956-5963.

Terrazzano, G., Romano, M.F., Turco, M.C. *et al.* (2000). HLA class I antigen downregulation by IL (IL)-10 is predominantly governed by NK-kappaB in the short term and by TAP1+2 in the long term. *Tissue Antigens.* 55:326-32.

Trinchieri G. (1995). IL-12: a pro-inflammatory cytokine with immunoregulatory functions that bridge innate resistance and antigen-specific adaptive immunity. *Annu Rev Immunol.* 13:251-276.

Tsang K.Y., Zaremba S., Nieroda C.A. *et al.* (1995). Generation of human cytotoxic T cells specific for human carcinoembryonic antigen epitopes from patients immunized with recombinant vaccinia-CEA vaccine. *J Natl Cancer Inst.* 87:982-990.

Tsark E.C., Dao M.A., Wang X. *et al.* (2001). IL-7 enhances the responsiveness of human T cells that develop in the bone marrow of athymic mice. *J Immunol.* 166:170-181.

Tsujii M., Kawano S. Tsuji S. *et al.* (1998). *COX* regulates angiogenesis induced by colon cancer cells. *Cell.* 93:705-709.

Tsung K., Meko J.B., Peplinski G.R. *et al.* (1997). IL-12 induces T helper 1-directed antitumor response. *J Immunol.* 158:3359-3365.

Tucker O.N., Dannenberg A.J., Yang E.K. *et al.* *et al.* (1999). *COX2* expression is up-regulated in human pancreatic cancer. *Cancer Res.* 59:987-992.

Van der Pouw Kraan T., Boeije L., Smeenk R. *et al.* (1995). Prostaglandin-E2 is a potent inhibitor of human IL 12 production. *J Exp Med.* 181:775-778.

Van den Eynde B.J., van der Bruggen P. (1997). T cell defined tumor antigens. *Curr Opin Immunol.* 9:684-693.

Van Waes C., Urban J.L., Rothstein J.L. *et al.* (1986). Highly malignant tumor variants retain tumor-specific antigens recognized by T helper cells. *J Exp Med.* 164:1547-1565.

Vasquez N.J., Kane L.P., Hedrick S.M. (1994). Intracellular signals that mediate thymic negative selection. *Immunity.* 1:45-56.

Vlahos R., Stewart A.G. (1999). IL-1α and tumour necrosis factor-α modulate airway smooth muscle DNA synthesis by induction of cyclo-oxygenase-2: inhibition by dexamethasone and fluticasone propionate. *Br J Pharmacol.* 126:1315-1320.

von Gaudecker B. (1991). Functional histology of the human thymus. *Anat Embryol.* 183:1-15.

von Leoprechting A., van der Bruggen P., Pahl H.L. (1999). Stimulation of CD40 on immunogenic human malignant melanomas their cytotoxic T lymphocyte-mediated lysis and induces apoptosis. *Cancer Res.* 59:1287-1294.

Vu C.B. (2000). Recent advances in the design and synthesis of SH2 inhibitors of Src, Grb2 and ZAP-70. *Curr Med Chem.* 7:1081-1100.

Ward P.L., Koeppen H.K., Hurteau T. *et al.* (1990). MHC class I and unique antigen expression by murine tumors that escaped from CD8+ T-cell-dependent surveillance. *Cancer Res.* 50:3851-3858.

Weinberg A.D., Vella A.T., Croft M. (1998a). OX-40: life beyond the effector T cell stage. *Semin Immunol.* 10:471-480.

Weinberg AD. (1998b). Antibodies to OX-40 (CD134) can identify and eliminate autoreactive T cells: implications for human autoimmune disease. *Mol Med Today.* 4:76-83.

Wilson D.B., Wilson D.H., Schroder K. (1998). Acquired thymic tolerance and experimental allergic encephalomyelitis in the rat. I. Parameters and analysis of possible mechanisms. *Eur J Immunol.* 28:2770-2779.

Wood P.J., Cossens I.A. (1993). Loss of Th1-associated function in peripheral T cells but not thymocytes in tolerance to MHC alloantigen. *Immunol.* 79:556-561.

Yamaguchi Y., Okabe K., Miyanari N. *et al.* (1998). TNF-β is associated with thymic apoptosis during acute rejection. *Transplant.* 66:894-902.

Yoshimatsu K., Altorki N., Golijanin D. *et al.* (2001). Inducible Prostaglandin E Synthase Is Overexpressed in Non-Small Cell Lung Cancer. *Cancer Res.* 7:2669-2674.

Yue F.Y., Dummer R., Geertsen R. *et al.* (1997). IL-10 is a growth factor for human melanoma cells and down-regulates HLA class I, class II and ICAM-1 molecules. *Int J Cancer.* 71:630-637.

Qin Z., Noffz G., Mohaupt M. *et al.* (1997). IL-10 prevents dendritic cell accumulation and vaccination with granulocyte-macrophage colony-stimulating factor gene-modified tumor cells. *J Immunol.* 159:770-773.

Zeidler, R., Eissner, G., Meissner, P. *et al.* (1997). Downregulation of TAP1 in B lymphocytes by cellular and Epstein-Barr virus-encoded IL-10. *Blood.* 90:2390-2397.

Zhang L., Conejo-Garcia J.R., Katsaros D. *et al.* (2003). Intratumoral T cells, recurrence, and survival in epithelial ovarian cancer. *N Engl J Med.* 348:203-213.

Zhang L., Pagano J.S. (2001). IFN regulatory factor 7: a key cellular mediator of LMP-1 in EBV latency and transformation. *Semin Cancer Biol.* 11:445-453.

Zhu J., Gianni M., Kopf E. *et al.* (1999). Retinoic acid induces proteasome-dependent degradation of retinoic acid receptor α (RARα) and oncogenic RARα fusion proteins. *Proc Natl Acad Sci USA.* 96:14807-14812.

Zou G.M, Tam Y.K. (2002). Cytokines in the generation and maturation of DCs: recent advances. *Eur Cytokine Netw.* 13:186-199.

Zuckermann F.A. (1999). Extrathymic CD4/CD8 double positive T cells. *Vet Immunol Immunopathol.* 72:55-66.

Chapter 4

Immunosurveillance Mechanisms of the Fetoplacental Unit

4.1 "SEMINAL PRIMING" AND SUCCESSFUL PREGNANCY

A serious debate has been held concerning the role of semen in the reproduction of mammals. The classical notion of the role of seminal plasma presented it as a complex medium enabling the survival and transport of spermatozoa through the male and female genital tracts. Present findings have shown that sperm components interact directly with the epithelial cells of the cervical channel and endometrium, as well as leukocytes of endometrium, triggering a cascade of important molecular and cellular events preceding the implantation of conceptus and successful pregnancy (Thaler, 1989; Robertson *et al.*, 2002, 2003). In a response well documented in rodents, semen triggers an influx of antigen-presenting cells into the female reproductive tract which processes and presents paternal ejaculate antigens to elicit the activation of lymphocytes in the adaptive immune compartment. As well as preventing an aberrant immunity to spermatozoa, these events are implicated in priming an appropriate female immune response to embryo implantation, since many seminal antigens are shared by the conceptus (Robertson *et al.*, 2002, 2003). *In vitro*, seminal plasma has been shown to suppress T and B cell proliferation, neutrophils and macrophage phagocytic activity, as well as killer cell activity. Seminal plasma interacts with complement components C1 and C3 and contains factors that specifically bind the Fc region of IgG. These *in vitro* findings suggest possible seminal plasma-suppressive effects on female alloimmune

135

responses after insemination. Seminal plasma also contains allotypic Trophoblast Lymphocyte Cross-reacting (TLX) antigens that could prime mothers prior to fertilization (Thaler, 1989). Contrary to these, there are claims based on the fact that successful insemination can be performed by washed spermatozoa. However, such a procedure only apparently minimizes the role of sperm in fertilization since, according to some findings, implantation and pregnancy may be seriously compromised if fertilization is void of sperm (Pang *et al.*, 1979; Peitz *et al.*, 1986). Embryo transfer probably produces a lower incidence of implantation than the physiological incidence despite all other factors which seem to be similar. The only factor known to be present physiologically and absent in embryo transfer is the presence of sperm in the uterine cavity. Therefore, fetal loss incidence in the case of *in vitro* fertilization (IVF) by washed spermatozoa and embryo transfer might be 50%. By treating female rats with acellular sperm components before fertilization by washed spermatozoa, the fetal loss incidence is restored to a physiological level (Carp *et al.*, 1984).

The significance of "seminal priming" for normal implantation and pregnancy has been confirmed in a large number of studies. The adequate "seminal priming" has been associated with the percent of vital fetuses in the pregnancies of experimental animals, as well as the weight of the placenta and fetus (Murray *et al.*, 1983). O' *et al.* (1988) found that in the golden hamster, the removal of the male accessory sex gland causes a slower cleavage rate in the embryonic development and a significant embryonic loss during pregnancy. Similar results were obtained during the research into the role of "seminal priming" in humans. The higher rates of implantation, successful pregnancies and live-born children were acquired in IVF procedures if women were pre-treated using seminal plasma. Seminal exposition could reduce the incidence of miscarriages in patients from high-risk groups (Coulam *et al.*, 1995a, 1995b). Also indicative of this is the fact that women fertilized after using barrier contraceptive means for a longer period, carry high-risk of implantation error, spontaneous miscarriages or even preeclamptic syndrome (Klonoff-Cohen *et al.*, 1989).

Sperm contains a high number of potentially antigenic components which, in some women may cause strong immune reaction ranging from mild allergic to anaphylactic reactions. Antisperm reaction may be so intensive that a completely sperm-hostile environment is set up in the vagina cervical channel, uterine cavity and tubes, virtually disabling fertilization (Thaler, 1989; Kelly *et al.*, 1997). This happens due to the antigens of blood groups and MHC system can be proved on spermatozoa, and since the passage to peritoneal cavity is accessible, alloimmunization to these antigens is inevitable in every sexually active female individual (Thaler, 1989; Kelly *et al.*, 1997). Consequently, the immunization by paternal MHC molecules before pregnancy occurs via spermatozoa, the epithelial cells of male genital tract, and leucocytes which are usual cellular component of sperm (Martin Villa *et al.,* 1996). As some of these antigens are situated in the placenta, the immune response developing to cellular component of sperm could be as well directed towards trophoblast antigens and could probably influence the growth and survival of the placenta.

Immunohistochemical researches of the endometrial cytokine network in mice have shown that sperm and its cellular components initiate inflammatory cascade, similar to that seen in the reaction to infection or injury (Robertson *et al.,* 1996). This reaction is initiated by the contact between factors such as TGF-β from sperm and the endometrium. Under the impact of TGF-β and other factors, the cells of the endometrium synthesize and secrete GM-CSF and IL-6 which accompanied by other cytokines, mobilize and lure leucocytes into endometrial stroma. Besides leucocytes, activation also includes IL-1, TNF-α secreting APCs and mature DCs expressing a high level of MHC class II molecules. Activated endometrial macrophages take part in the recognizing and phagocytosis of allogeneic spermatozoa, which contributes to their cytokine activity (Robertson *et al.,* 2002, 2003). The phagocytosis of spermatozoa has been detected on 3H tymidine marked spermatozoa, so that the activity of 3H tymidine is found in the uterine lymph vessels and also in the lymph nodes draining the uterus (Reid, 1966).

Despite strong evidence that the maternal immune system is capable of rendering itself sensitive to paternal antigens via spermatozoa and other cell components of the sperm, cell-mediated hypersensitivity to sperm is a rarity, unlike humoral type of anti-spermatozoid activity. This phenomenon can be explained by the peculiarity of mucous immunoreactivity as well as by immunomodulatory effects elicited by the sperm on the immune cells of the endometrium. To that effect, there is evidence that sperm molecular structure contributes to the development of a type of immunotolerance to sperm antigens, or to switch of the immune reaction towards humoral response, meaning the dominance of Th2 type of immune response. This phenomenon was observed in mice in cases of deferred rejection of skin allograft of male rats following the pre-treatment of the recipient female with the donor's seminal plasma. The female rats of Balb/k strain always rejected Balb/c tumor cells but could not reject them after the copulation with male of Balb/c strain, even when sterilized with the ligature of utero-oviduct junction. This indicated that the exposure of female rats to allogeneic sperm produces a specific type of immune tolerance to the MHC molecules of the male even in the absence of pregnancy. The presence of immune tolerance is evidenced by the fact that the copulation of female rats of Balb/k strain with Balb/k male rats or some other mouse strain will not result in any loss of resistance to Balb/c tumor cells (Robertson *et al.*, 1997).

4.1.1   Role of TGF-β in "Seminal Priming"

Presumptions as to which factors may be responsible for immuno-modulatory effects of the sperm are associated with PGE2 and TGF-β, albeit there are more factors which exert immunomodulatory effects of the sperm, probably via their synergistic action. TGF-β is a potent immune deviating agent, driving active forms of immune tolerance in peripheral tissues through effects on the induction and resolution of inflammatory responses and phenotype shifting in antigen-presenting cells and lymphocytes. The TGF-β content of seminal plasma from human, rodent and other mammalian species is the highest regarding biological fluids. The seminal vesicle gland is the principal source of TGF-β in the semen of mice, where its synthesis is regulated by testosterone. TGF-β present in abundance in seminal plasma, initiates this inflammatory response by stimulating the synthesis of pro-inflammatory cytokines and chemokines in uterine tissues. Lymphocyte activation is evident in lymph nodes draining the uterus, nevertheless T cells shows hypo-responsiveness to paternal alloantigens. TGF-β has potent immune-deviating effects and is likely to be the key agent in shifting the immune response against a type 1 bias. The prior exposure to semen in the context of TGF-β can be shown to be associated with the enhanced fetal/placental development late in gestation (Robertson et al, 2002, 2003). In addition to its regulatory role in the female post-mating inflammatory response, TGF-β has pivotal functions in other aspects of male and female reproductive organs. Mice with a null mutation in the TGF-β1 gene are an important research tool in establishing those functions, and in evaluating exogenous TGF-β supplementation. These mice have multiple striking lesions in reproductive function. For instance, mutant males are 100% infertile, with decreased testosterone production, decreased libido and ejaculation failure (Ingman *et al.*, 2002).

4.2   DECIDUA AND TROPHOBLAST

There is growing evidence that viviparous reproduction depends on a complex sequence of two-way interactions between fetal and maternal tissues that involves humoral, cellular, innate and adaptive immune responses (Hill, 1995). There is currently an extensive support for the view that immune tolerance is established through the innate immunological interactions between maternal uterine NK cells that produce pro-inflammatory Th1 cytokines and trophoblast cells that react by triggering a

Th2 anti-inflammatory cytokine response (Wegmann *et al.,* 1993; Clark, 1999). The complement system regulation (Xu *et al.,* 2000) and shift in balance from Th1 to Th2 cytokines are thought to be critical in determining the success or failure of pregnancy (Wegmann *et al.,* 1993; Clark, 1999). Also implicated in the modulation of maternal NK cell mediated attack is HLA-G, a non-classic MHC class I gene of limited polymorphism, which is paternally expressed in trophoblast cells, primarily during early fetal development (Le Bouteiller *et al.,* 2000). In addition, HLA-G is known to protect melanoma cells from NK cytolysis (Paul *et al.,* 1998). At the classic, highly polymorphic MHC loci, the expression is minimal immediately following the implantation but increases steadily through pregnancy in placental and extravillous cytotrophoblast tissues. The restriction of polymorphic MHC gene expression to later stages of embryonic development probably serves to limit the maternal rejection of the fetus.

Decidualization is an endocrinologically, mainly progesterone-dependent, process that transforms uterine cells that can produce hormones, growth factors and matrix components to support the implanting and post implantation conceptus. A conceptus is not essential for decidualization. This process begins late in the luteal phase of every menstrual cycle in humans and other primates. Decidualization can also be modelled in pseudo-pregnant rodents that are endocrinologically primed and mated with vasectomized males or have foreign material inserted into the uterine lumen. During the process of decidualization, with or without a conceptus, a specialized lymphocyte subset becomes activated. The numbers of these cells increase dramatically. The cells proliferate within the uterus, enlarge, acquire numerous cytoplasmic granules and initiate the production of immune cytokines. Since decidua represents only a minor part of the placenta, and is of entirely maternal origin, it therefore has no antigenic potential. As well as epithelial stroma of the thymus or bone marrow, decidua is also the site of selection, immigration, proliferation and maturation of unique subpopulations of lymphocytes and NK cells. The role of these phenomena is probably in selecting and favouring only certain populations of decidual immunocompetent cells. These cells probably play a pivotal role in the development of immunomodulatory effects, stimulation of placental growth, removal of cellular detritus, processing of antigens, as well as establishment of specific cytokine network and regulation of NK cell activity (Haynes *et al.* 1995; King *et al.,* 1996a).

4.2.1   Decidual Immunocompetent Cells

Decidual immunocompetent cells differ in subpopulation composition from other tissues. CD56[+] cells comprise about 80%, CD3[+] T lymphocytes comprise 10%, whereas CD14$^+$ macrophages make another 10%. Decidual cells accumulate in large number and density in the decidua, particularly in basal decidua (placental bed). Such a subpopulation structure of decidual immunocompetent cells develops as a result of ovarian, trophoblast, humoral, cytokine and other bio-humoral signals, via the mechanisms for the stimulation of apoptosis of some and proliferation, differentiation or migration of other immunocompetent cells. The composition of decidual immunocompetent cells set up through the activity of foregoing mechanisms provides the efficient establishment and maintenance of immunoregulation during pregnancy and probably the regulation and stimulation of placental function and growth (Haynes *et al.* 1995; Ho *et al.,* 1996; Chaouat *et al.,* 1997; Loke *et al.,* 2000).

4.2.1.1 Decidual NK Cells

The features of uterine NK cell population are subject to change. The subpopulation arrangement and number of uterine NK cells depends primarily on the phase of the menstrual cycle, and possibility to discontinue the cycle by implanting the blastocyst. Most NK cells were found in late secretory phase of the cycle, at the time of potential implantation. The absence of implantation causes most of these cells to degranulate, undergo apoptosis several days prior to menstruation. However, if implantation does occur, the granulated NK cells will proliferate rapidly and soon become the dominant population of decidual lymphoid cells (~70%), while some of these get in close contacts with the invading trophoblast (Ferry *et al.,* 1990; King *et al.,* 1991; Croy *et al.,* 1997; Loke *et al.,* 2000).

NK cells from the pregnant uterus and from other tissues in pregnant and nonpregnant mammals can be stimulated by IL-2 during culture to become LAK cells. The susceptibility of cultured trophoblast cells to lysis by LAK cells raises the enigma of why uterine NK cells that are characterized by morphology and by surface phenotyping as "activated", and thus potentially damaging to the placenta, become localized to the implantation sites during normal gestation. Croy *et al.* (1979) found that the early post-implantation, mRNA from migrating uterine NK cells contains transcripts for all three isotype of the IL-2R(α, β and γ), but only IL-2Rγ was expressed at day 12 of gestation; the expression of this gene was also lost by day. The loss of IL-2R transcription did not result in the loss of the protein expression; however, it did coincide with the loss of uterine NK cell viability *in vivo*. Apparently normal differentiation of uterine NK cells occurred in IL-2($^{-/-}$) mice as well as in doubly mutant IL-2($^{-/-}$)/β2m($^{-/-}$) mice.

The changes in the number and relationship among NK cell subpopulations do not only occur when the endometrium has undergone a decidual transformation. A subpopulation rearrangement of endometrial NK cells is a continual process which occurs throughout the menstrual cycle regardless of whether the cycle will be discontinued by pregnancy or continued up to menstrual bleeding. CD56$^+$CD16$^-$ NK cells comprise the dominant NK cell subpopulation in proliferative and secretory phases of the cycle. Upon coming of the blastocyst in the uterine cavity, a subpopulation shift of endometrial NK cells probably takes place. The signals arriving from the blastocyst during the apposition probably lead to the apoptosis of CD56$^+$CD16$^+$ and proliferation of CD56$^+$CD16$^-$ cells, first in the periimplantation zone of the endometrium and later in the whole decidua. It is assumed that such shift of NK cells during pregnancy is necessary because of the specific cytokine activity of CD56$^+$CD16$^+$ cells as well as a high risk of pregnancy failure if CD56$^+$CD16$^+$ cells are a dominant subpopulation of decidual NK cells in early pregnancy (Ferry *et al.*, 1990; King *et al.*, 1991; Croy *et al.*, 1995; Ho *et al.*, 1996).

Table 4.1. Percentage of the lymphocytes and NK cells in the decidua and peripheral blood of healthy pregnant women.

Subpopulation of Lymphocytes	Receptor	Decidual lymphocytes (%)	Peripheral blood lymphocytes (%)
CD4	CD69	$58,3 \pm 12,7$	$0,9 \pm 0,5$
	HLA-DR	$67,7 \pm 14,8$	$11,4 \pm 4,6$
	IL-2Rα	$26,9 \pm 5,0$	$20,5 \pm 5,2$
	IL-2Rβ	$8,6 \pm 4,6$	$1,5 \pm 0,5$
CD8	CD69	$73,3 \pm 14,3$	$3,5 \pm 0,7$
	HLA-DR	$82,9 \pm 13,5$	$18,6 \pm 0,8$
	IL-2Rα	$6,2 \pm 2,9$	$2,5 \pm 1,5$
	IL-2Rβ	$17,7 \pm 7,2$	$10,6 \pm 3,4$
CD16$^-$CD56$^{bright+}$	CD69	$69,4 \pm 13,4$	$7,7 \pm 2,2$
	HLA-DR	$13,4 \pm 5,3$	$77,7 \pm 20,2$

The differences between the populations of decidual NK cells and NK cells from other tissues are not only revealed in their composition and cytokine activity, but in antigenic phenotype features as well. Decidual NK cells express the receptor of early activation (CD69), while the expression of this receptor by lymphocytes and NK cells is absent in the cells of peripheral blood. HLA-DR as a receptor of late activation on decidual lymphocytes can be found on most cells of this population, whereas only a low percent of decidual NK cells express this receptor. Only a small number of leucocytes from peripheral blood express HLA-DR, while this molecule is mainly expressed by NK cells of the peripheral blood. Rukavina *et al.* (1995) found that perforin expressing decidual lymphocytes are prevalently non-classical CD56$^+$CD16$^-$ NK cells (42%) with a low cytolytic activity but fully equipped with the potent cytolytic machinery (perforin$^{bright+}$). Also, the number of perforin[bright+] cells is even higher in decidua parietalis in the vicinity of non-invasive trophoblast, than in decidua basalis.

4.2.1.2   Class Ib MHC Molecules and Activity of Decidual Immune Cells

There is evidence now that the immunological recognition of pregnancy is important for the maintenance of gestation, and that an inadequate recognition of fetal antigens might result in failed pregnancy. In contrast to class I genes that are down-regulated in human trophoblast cells, non-polymorphic class Ib molecules like HLA-G are expressed on placental tissues. The trophoblast does not induce alloimmune response and resists NK and CTL mediated lysis *in vitro*. According to present knowledge, HLA-G presents antigens for γδT cells and at the same time defends the trophoblast from cytotoxic effector mechanisms. Since polymorphic MHC is absent from the trophoblast, the presentation of fetal antigens is unlikely to be MHC restricted. Cells like γδT lymphocytes recognize a distinct group of ligands with a smaller receptor repertoire than αβT cells. Most γδT cells recognize unprocessed foreign antigens without MHC. In the decidua, γδT cells significantly increase in number and the majority of decidual γδT cells are in an activated form due to the recognition of conserved mammalian molecules on the trophoblast. Following the recognition of fetal antigens, the immune system reacts with the setting in of a wide range of protective mechanisms (Mincheva-Nilsson *et al.*, 2000; Szekeres-Bartho, 2002; Bubanovic, 2003a). Also, there is now evidence for the expression of HLA-C, HLA-E, and HLA-F on human trophoblast cell lines. HLA-E and HLA-F have been proposed to be of great importance in HLA-G-induced NK cell regulation (Lee *et al.*, 1998; King *et al.*, 1996b).

The basic hallmark of NK cells is to kill proliferating cells not expressing the MHC molecules (Karre, 1995). Although trophoblast is a tissue which proliferates rapidly and does not express classical MHC molecules, no cytotoxic activity towards the cells of the trophoblast is displayed by NK cells in normal pregnancies. In the case of NK cells activation by IL-2, trophoblast becomes their target and undergoes a very rapid and effective cytolysis (Ober *et al.*, 1997). That NK cells may be fatal to the outcome of pregnancy is best evidenced by the fact that decidua basalis and placenta from pregnant abortion prone CBA/JxDBA/2J mice show a strong infiltration by the activated NK cells, whereas intravenous infusion of anti-NK antibodies prevents the resorption of embryos (Croy *et al.*, 1995). It has been proved on other experimental models that the activated NK cells induce abortions in strains of animals characterized by the low percent of fetal losses. These data indicate that NK cells have the potential for recognizing the trophoblast as a foreign tissue and killing it, and also that in normal pregnancies trophoblast induces the suppression of NK cells by means of a specific arrangement of decidual trophoblast-cytokine network, and presents itself to NK cells as "self" tissue probably through the expression of monomorphic or low polymorphic MHC class Ib molecules (Karre, 1995; Ober *et al.*, 1997).

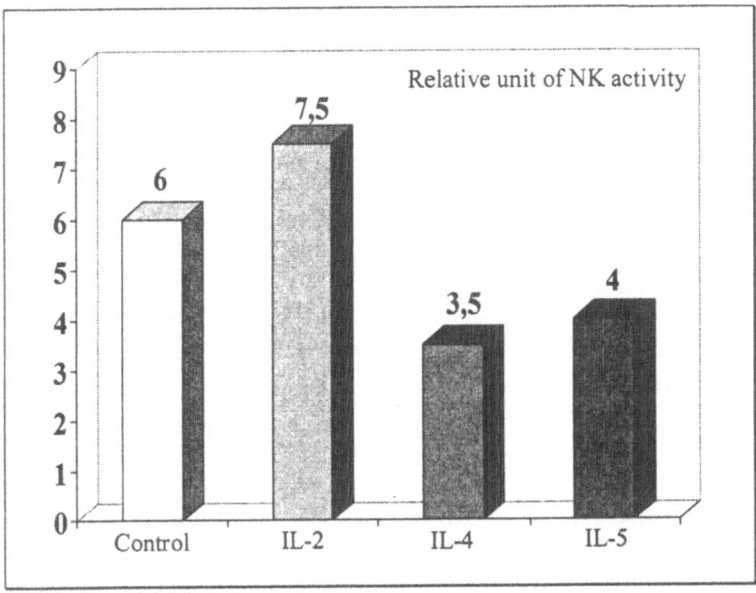

Figure 4.1. Activity (in relative units) of decidual NK cells in the presence of different cytokines.

The notion that the expression of HLA-G may represent a strong inhibitory signal for NK cells has been proved on the phenomenon of high resistance of transgenic lymphoblastic line expressing HLA-G, while lymphoblastic cells not expressing HLA-G are highly sensitive to the NK activity (King *et al.*, 2000). Furthermore, the cytokines of Th2/Th3 group, such as IL-4, IL-5, IL-10 and TGF-β2, elicit inhibitory effects on the activity of NK cells. These cytokines are considered to inhibit NK activity by suppressing the expression of the IL-2 receptor on NK cells (Saito, 2000).

The utero-decidual population of NK cells displays a low cytotoxic potential against trophoblast in spite of the completely developed apparatus for killing target cells. This apparent "friendship" between trophoblast and NK cells is relatively easy to break by treating pregnant animals with IL-2 or *indomethacine*, which results in the activation of NK cells, their attacking the trophoblast and, consequently, resorption of the feto-placental unit. Lala *et al.* (1990) shown that NK lineage cells migrate to the murine decidua but with advancing gestation, they are progressively inactivated in situ by PGE2 secreted by decidual cells and decidual macrophages. They also showed that the same mechanism inactivates all killer lineage cells in the human decidua, and that this inactivation is at least in part due to a down-regulation of IL-2 receptors and an inhibition of IL-2 production in situ. Chronic indomethacin therapy non-selectively blocks the activity of COX-1 and COX-2, or a systemic administration of a high dose of IL-2, or a combination of both agents administered to pregnant mice could activate killer cells in situ and interfere with the progress of pregnancy (Lala *et al.*, 1990). Also, the treatment of pregnant animals with anti-PGE2 antibody revived the NK activity in the decidua, suggesting a PGE2-mediated suppression, while the addition of pure PGE2 but not PGF2α during the NK assay or to the effector cells for a 20-hr period prior to the assay led to an inhibition of NK activity (Scodras *et al.*, 1990).

The limitations of lytic and cytokine activity of decidual NK cells do not mean that these cells have been completely inhibited or inactivated. King *et al.* (1989) found that trophoblast cells are resistant to lysis by NK cells from both peripheral blood and decidua. However, decidual cells do exhibit cytotoxicity against the tumor cell line K562, but unlike the NK cells of peripheral blood, their lytic activity is much weaker. Decidual $CD56^+CD16^+$ NK cells contain mRNA for IFN-γ, TGB-β1 and LIF, but in normal pregnancy, these cells secrete low amounts of the cytokines. By contrast to the decidual $CD56^-$ cells, $CD56^+$ cells could release TGF-β2, which is a very important suppressor of NK cells and CTL cytolytic activity (Clark *et al.*, 1994). In normal pregnancy, the decidual NK cells express a small number of IL-2R, although they contain mRNA for IL-2R. This could further mean that IL-2R (CD 25 and CD 122) could be expressed on decidual NK cells, but only under conditions of the weakened impact of immunoregulatory factors in pregnancy. There are major differences between decidual NK cells and peripheral blood NK cells regarding the effects of IL-2. For peripheral blood NK cells, IL-2 is a factor of proliferation, maturation and activation, while for decidual NK cells IL-2 is only a factor of activation (Lala *et al.*, 1990; Croy *et al.*, 1997).

4.2.1.3 Decidual T Cells

After pregnancy, $CD8^+CD3^+$ lymphocytes were decreased in the decidua. In both endometrium and decidua, more T cells expressed CD69, CD71, HLA-DR, and CD38 molecules than in peripheral blood. After pregnancy, $CD71^+CD3^+$ lymphocytes were further increased. $CD25^+CD3^+$ lymphocytes decreased significantly in the endometrium and decidua of ectopic pregnancies, but not in the decidua of normal pregnancies. These findings indicate that T cells are regionally activated in the first trimester, and it may be the result of the stimulation by fetal (paternal) antigens, as well as hormones and cytokines signals (King *et al.,* 1991, 1996a; 1997; Ho *et al.,* 1996; Loke *et al.,* 2000). Besides, T cells were present in the decidua, as compared to the peripheral blood, CD45RO$^+$, CD29$^+$ and CD45RA$^-$ CD4$^+$ T cells as well as CD45RO$^+$, CD29$^+$ and CD45RA$^-$CD8$^+$ T cells, which are considered to be memory T cells, were in the majority, with only small numbers of CD45RO$^-$, CD29$^-$ and CD45RA$^+$ CD4$^+$ and CD8$^+$ cells, which are naive T cells, present (Saito *et al.,* 1994; King *et al.,* 1996a, 1997, 1998).

Decidual T lymphocytes are mostly localized in Large Lymphoid Clusters (LCC) in the vicinity of endometrial glands, and these are the so-called Intraepithelial Lymphocytes (IEL). This lymphocyte subpopulation is comprised mainly of $\gamma\delta$CD56$^+$, $\alpha\beta$CD4$^+$ and $\alpha\beta$CD8$^+$ cells. Like NK cells, the decidual T cells differ phenotypically and functionally from the T cells of peripheral blood and other tissues. In addition, decidual T cells express a whole range of activator receptors (CD45RA, Kp43, CD69, CD71, CD38, HML-1), and MHC class II antigens (HLA-DR, HLA-DP, HLA-DQ), which makes them similar to the T cells of peripheral blood. In instances of T cell and NK cell activation, when such activation brings about the loss of pregnancy, the decidual immune reaction is located in the LCC region (Mincheva-Nilsson *et al.,* 1994; Geiselhart *et al.,* 1995).

$\gamma\delta$T cells in decidua were shown to specifically colonize the non-pregnant murine endometrium, and to constitute a major cell population in murine, sheep, horse and pig decidua. Human $\gamma\delta$T cells comprise about half of all decidual T cells in early pregnancy. The vast majority of them are Vδ1$^+$, CD4$^-$CD8$^-$, and express CD56. Ultrastructurally, human decidual $\gamma\delta$T cells have the distinctive morphology of large granular lymphocytes. It has been shown that murine decidual $\gamma\delta$T cells specifically recognize a conserved trophoblast antigen and suppress the maternal "anti-fetal" response probably by TGF-β production. Recently, several reports have suggested a regulatory role for the murine decidual $\gamma\delta$T cells in abortions via the release of abortogenic or anti-abortive cytokines. In humans, $\gamma\delta$T cell clones from decidua have been shown to mediate strong non-MHC restricted cytotoxicity (Mincheva-Nilsson *et al.,* 2000; Szekeres-Bartho, 2002; Bubanovic, 2003a). In addition, the percentage of perforin$^+$ decidual T lymphocytes is two times higher than in peripheral blood T lymphocytes (55% vs. 27%), and the prevalent phenotype is CD3$^-$CD4$^-$CD8$^-$CD2$^+$ (95%) CD11c$^+$ (68%). Two different subpopulations of CD8$^+$ decidual lymphocytes exist: CD8$^{bright+}$, which are CD3$^+$ CD56$^-$perforin$^-$ and CD8^{dim+}, which are CD3$^-$ CD56$^+$perforin$^+$ (Rukavina *et al.,* 2000). Classical $\alpha\beta$T cells in the mammalian species have not been considered able to recognize trophoblast because of the atypical expression of MHC class I and class II molecules on human trophoblast cells or MHC expression on murine spongiotrophoblast

(Ljunggren *et al.,* 1990; Sprinks *et al.,* 1993; Kaufmann *et al.,* 1996). γδT
cells, however, can recognize the atypical class Ib MHC molecules such as
HLA-G (Davis *et al.,* 1995). There are data that γδT cells provide a first line
of defence against infectious agents such as bacteria, viruses, and parasites
because of their preferential location in epithelium and underlying stroma at
mucosal surfaces (Chein *et al.,* 1996). Fujihashi *et al.* (1997) have
demonstrated that human gut, like endometrial or decidual, γδT cells are a
potent source of Th1 and Th2 cytokines, and one third of the γδT cells are
producers of both Th1 and Th2 type cytokines, i.e., have a Th0 phenotype. A
significant proportion of γδT cells are thymus-independent and may have a
TCR with germ line determined specificity, lack CD4 and/or CD8, and
recognize HSP as well as specific antigens (Born *et al.,* 1990). The antigens
non-associated with MHC, soluble antigen, and carbohydrates may also be
recognized. On this basis, it has been suggested that γδT cell recognition is
more comparable to antibody recognition than αβT cell recognition of
antigen (Chein *et al.,* 1996).

Table 4.2. Percentage of subpopulations within lymphocyte population of the decidua and
peripheral blood.

Receptors	Decidual lymphocytes (%)	Peripheral blood lymphocytes (%)
CD3$^+$ T	12,2 ± 4,6	71,8 ± 3,5
CD4$^+$ T	5,6 ± 2,1	43,4 ± 6,5
CD8$^+$ T	4,7 ± 1,9	24,9 ± 7,2
CD16$^+$ NK	2,7 ± 2,7	20,3 ± 4,3
CD16$^-$CD56$^{bright+}$ NK	84,8 ± 5,2	0,5 ± 0,3
CD20$^+$ B	0,9 ± 0,3	8,4 ± 2,5

One of the crucial phenomena deciding the fate of pregnancy and
regulating cytotoxic mechanisms of the decidual immune cells is the ratio of
decidual Th1 and Th2 lymphocytes. The number of Th1 cells in
endometrium decreases rapidly as the cycle progresses. Actually, the number
of the cells is the highest in the endometrium in the proliferative phase,
significantly lower in the endometrium in the secretory phase, and the lowest
in the decidua. The Th1/Th2 ratio is over 100 times higher in the
endometrium in the proliferative phase than the early decidua, whereas it is
as little as 4 times higher in the endometrium in the proliferative phase than
in the endometrium in the secretory phase of the cycle. Such decrease of the
Th1/Th2 ratio in the secretory phase could be accounted for by the
immunomodulatory effects of progesterone and estrogens, but a dramatic
decrease in Th1/Th2 ratio during pregnancy is probably the consequence of
the joint immunomodulatory activity of progesterone, Progesterone Induced

Blocking Factor (PIBF), prostaglandine and a large number of trophoblast and decidual cytokines with Th2 inducing potentials (*Figure* 4.1.) (Wegmann *et al.,* 1993; Saito *et al.,* 1999).

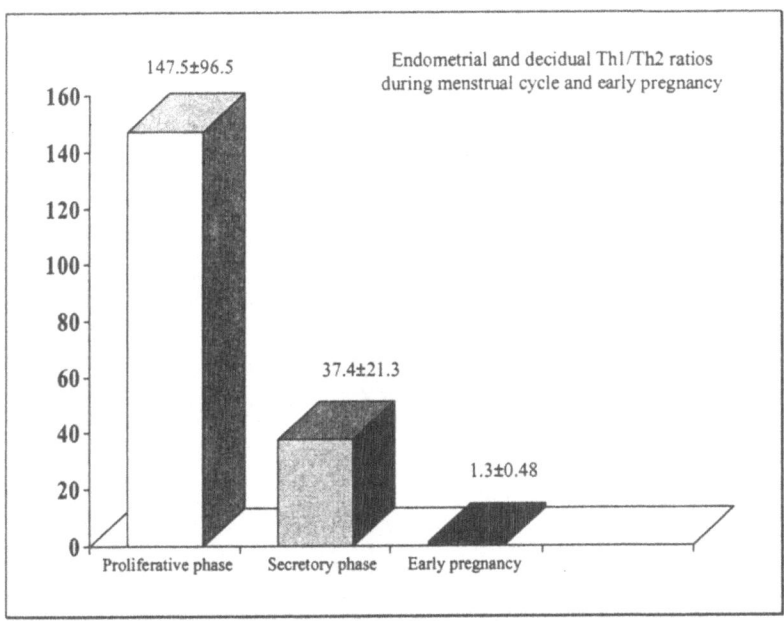

Figure 4.2. Ratios of Th1/Th2 cells in endometrium and early pregnancy decidua.

Th1 type cytokines stimulate abortions, while Th2 type cytokines prevent abortions. In pregnancy, a Th1 Th2 shift is postulated, and data from both mice and humans appear to support this hypothesis (Wegmann *et al.,* 1993; Chaouat, 1994). Indeed, in pregnant mice infected with *Leishmania major* (a parasite that stimulates Th1 response), if the parasite is rejected, abortions occur, whereas if the pregnancies succeed, so also does the infection (Krishnan *et al.,* 1996). Th1 type cytokines stimulate the NK cells and macrophage system that is involved in abortions, whereas Th2 type cytokines suppress the NK cells and macrophage mediated activity. Endotoxin, another potent abortogenic factor, stimulates macrophages; these cells then release TNF in the same way they would if activated by Th1 derived IFN-γ. Surprisingly, immunization to paternal antigens prevents the stress-triggered abortions. Therefore, several different pathways may lead to the activation of NK cells and macrophages involved in the rejection of the embryo. Th2 type immune response appears to be protective, however, it has yet to be resolved whether conventional circulating TCR positive cells that bear recognize and react to the antigens expressed by the trophoblast (Clark, 1993; King *et al.*, 1996a).

Th1 cells produce abortogenic cytokines such as IL-2 and INF-γ which are seen to cause abortion in mice and humans, while Th2 cells however produce IL-3, IL-4 and IL-10 which promote antibody formation and put off the inflammation and NK cell activation. The NK derived INF-γ may activate the macrophages of the feto-maternal interface or other TNF-α secreting cells whereas Th2 cells would suppress this activation (Clark, 1993; Ng *et al.*, 2002; Plevyak *et al.*, 2002).

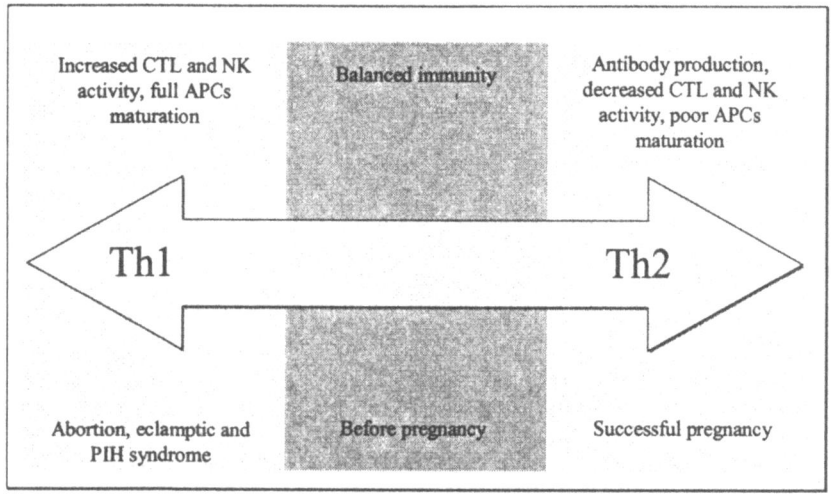

Figure 4.3. The fate of pregnancy depends on pre-existing Th1/Th2 ratios, as well as ratios in pregnancy.

The signals leading to such drastic decrease in the Th1/Th2 ratio most likely come from the blastocyst in apposition and intensify during the invasion of the decidua by trophoblast. A large number of factors stimulating Th2 response like IL-10, TGF-β2 and prostaglandine such as PGE2 have been detected in blastocyst liquid. As well as stimulating Th2 lymphocytes, IL-10, TGF-β2 and PGE2 also act as effective inhibitors of Th1 lymphocytes and the products of their activity (Wood *et al.*, 1988; Wegmann *et al.*, 1993; Clark, 1993; Loke *et al.*, 2000).

4.2.1.4 Decidual Macrophages

Uterine macrophages express the receptors for estrogens and progesterone, so that the hormones are some of the factors controlling the number and activity of endometrial and miometrial macrophages (Miller *et al.*, 1996). The influence of progesterone on decidual macrophages is mostly associated with the stimulatory effects of progesterone on prostaglandine secretion by the macrophages. Together with the trophoblast, these cells are the main site of synthesizing of prostaglandines, whose immunosuppressive role in pregnancy has been well explored and documented (Hunt *et al.,* 1996).

By contrast to CD4$^+$ lymphocytes and NK cells, the number of macrophages in the endometrium in the proliferative and secretory phases is relatively constant, but their number significantly increases in the decidua of the early pregnancy. The concentration of macrophages in periimplantation zone of the endometrium is 45% higher compared to the proliferative or secretory phases of the cycle. Such local increase in the concentration occurs due to these cells being redistributed from other uterine regions. The factors leading to the chemotaxis of macrophages and their staying in periimplantation zone are: (i) trophoblast pro-inflammatory factors present at the site of the invasion of the endometrium; (ii) the local enhancement of steroid hormones and (iii) the increased concentration of steroid hormone induced cytokines, such as CSF-1, GM-CSF, TNF-α, and IL-6. Decidual lymphocytes produce GM-CSF spontaneously, but this production is higher in lymphocytes which are stimulated by paternal antigens, unlike lymphocytes from other tissues (Bulmer *et al.,* 1988; Miller *et al.,* 1996; Hunt *et al.,* 1996).

The macrophages of the decidua capsularis express MHC class II molecules uniformly at all stages of pregnancy. The expression of MHC class I molecules on macrophages of decidua capsularis is rather weak, but can be enhanced by IFN-γ. The degree of MHC expression on macrophages is proportional to the level of their activation, which is most often the result of tissue destruction, IFN-γ and NK or Th1 cell derived cytokines (Yang *et al.,* 1996).

The highest level of macrophage activation is achieved by LPS, so that pregnancy may be compromised by the cytokines released from LPS-activated macrophages of utero-placental interface. First findings concerning the potentially abortogenic role of macrophages and their products were obtained after the treatment of pregnant mice with the supernatant of macrophagal culture stimulated with Freund's adjuvant. Such treatment proved fatal for pregnancies of experimental animals. A subsequent research has shown that macrophages produce abortogenic cytokines only when primed with cell detritus, IFN-γ, the products of NK cells and LPS. The immunomodulatory role of macrophages of the utero-placental interface is affected by secreting factors which inhibit T cell proliferation, such as PGE2 (Wood *et al.* 1988; Hunt, 1989, 1994; Baines *et al.,* 1997). In addition, certain cytokines such as IL-10 and TGF-β reduce macrophage activation as demonstrated by the lower expression of class II and class Ia of MHC molecules and diminished production of TNF-α (Bogdan *et al.*, 1993). In addition, a CSF-1/GM-CSF equilibrium has an important role in regulating the transition between the immunosuppressive and immunostimulatory phenotypes. When this equilibrium is disrupted, implantation is avoided (Tartakovsky *et al.*, 1991).

Ovarian steroids control the effector molecule production from the activated macrophages. Progesterone enhances PGE2 production by human placental macrophages and suppresses the cytotoxicity of human mononuclear phagocytes. In a mouse model, progesterone suppresses the steady state levels of inducible NO mRNA and inhibits NO production from macrophages (Miller *et al.*, 1996). However, it appears that progesterone exerts a direct control over the factor production once their glucocorticoids receptors are engaged and mechanisms not involving specific hormone receptors are activated (Hunt *et al.*, 1996). Perturbed cytokine levels have a negative impact on the fertility and pregnancy maintenance and the loss of regulation of macrophage activity results in a poor reproductive outcome (Tartakovsky *et al.*, 1991).

The receptor phenotype of macrophages (immunosuppressive or immunostimulatory) is variable and depends on the presence of factors such as IL-10, TGF-β2 and PGE2. Under the influence of these factors, decidual macrophages change their receptor's composition to correspond more against immunosuppressive activities i.e. the diminished expression of MHC class II molecules and hindered communication with Th lymphocytes. Some authors are of opinion that decidual macrophages and APCs do not express the costimulatory molecules such as CD80 and CD86, contributing to some extent in developing anergy of decidual CTL or their apoptotic selection. Yet, others claim that decidual APCs only express CD86 costimulatory ligand, thus contributing to the switch towards Th2 type response. The cytokines like IFN-γ and especially IL-12 elicit quite opposite effects on the activity and receptor composition of the decidual macrophages. They inhibit the expression of the CD86 costimulatory molecules and promote the expression of class I and II thus biasing towards CTL and Th1 activation (Tartakovsky *et al.*, 1991; Miller *et al.*, 1996; Hunt *et al.*, 1996).

4.2.1.5   Cytokine Network of the Decidua

The implantation of the human embryo is a multiple paradox, because the feto-placental unit shows the characteristics of successful transplant, parasite and tumor. Many different mechanisms are included in the priming of endometrial environment for successful implantation, but one of most important events is the establishing of optimal cytokine network. Much evidence has suggested that decidual cytokines and chemokines play a very important role in the embryo implantation, endometrial development, and trophoblast growth and differentiation by modulating the immune and endocrine systems. The close correlation between the trophoblast and the decidua are mediated by sex steroid hormones, cytokines and chemokines. As a result of this closely related cross-talk, pregnancy is successfully maintained (Saito, 2001).

Endometrial macrophages, NK cells, dendritic cells, lymphocytes, neutrophils and other decidual cells build a specific cytokine network which regulates the activity of decidual immune cells. The priming of endometrial immune cells and endometrium for the oncoming pregnancy most probably starts with the phenomenon of "seminal priming", followed by the signals coming from the blastocyst, invasion of the decidua by the trophoblast, as well as the enhanced and prolonged progesterone activity of the corpus luteum (Bulmer *et al.*, 1988; Carp *et al.*, 1984; Chaouat *et al.*, 1994).

The cyclic endometrial development regulated by estrogen and progesterone, is characterized primarily by the changes in immune and both the luminal and glandular epithelial cells in the preparation for blastocyst apposition and attachment. If the action of progesterone is antagonized, these changes are inhibited and the uterus is maintained in a pre-receptive state. The secretory phase is in the further modulation of these steroids induced changes by embryonic signals. In the primate, the infusion of chorionic gonadotropins in a manner that mimics blastocyst transit results in the endoreplication and plaque formation in the luminal epithelium. The glandular epithelium responds by the increasing transcriptional and post-translational modifications of secretory proteins and the stromal fibroblasts initiate their differentiation process into a decidual phenotype. The final phase is associated with the trophoblast invasion and remodelling of the endometrial immune and non-immune cellular compartment (Fazleabas *et al.*, 2002).

The domination of Th2 cells and Th2 type of cytokines are preferentially found in healthy, normal decidua. *In vitro* studies have shown that this microenvironmental conditions inhibit the proliferation of cytotoxic and Th1 cells. The secretion of anti-inflammatory cytokines and the inhibition of T and NK cell activation and proliferation, strongly suggest that decidual immune cells create an immunosuppressive environment on the decidual-trophoblast junction. As mentioned above, decidual immune cells produce the cytokine, IL-10. In addition to its functions already mentioned, IL-10 also up-regulates the macrophagal expression of MHC class II molecules. MHC class II molecules bind to Th1 and Th2 receptors to activate both kinds of T cells. However, the presence of IL-10 biases the activation towards Th2 because IL-10 has an inhibitory effect on Th1 cells. The significance of this bias lies in the functions of Th1 and Th2 cells. Anti-inflammatory responses mediated via IL-1R-antagonist, can also down-regulate the inflammatory response and inhibit the activation of Th1 lymphocytes. The growth factors like EGF, TGF, IGF and angiogenetic factors like FGF and VEGF are also included in the preparation of the endometrium for a successful implantation via the regular decidualization and expression of the extracellular matrix molecules such as integrins and cadherin. Moreover, these factors regulate the progressive invasion of the decidua, to the depth of the spiral arteries by the trophoblastic expression of metalloproteases (Hunt *et al.,* 1989; Clark, 1993; Chaouat *et al.,* 1994; Ng *et al.,* 2002).

In addition, TGF-β2 effectively inhibits IL-2 activated NK cells. Treating pregnant mice with anti-TGF-β2 antibodies resulted in the dramatic increase of the fetal resorption rate. The mating of mice strains with the high rates of fetal resorption, significantly lower values of TGF-β2 were found compared to strains with the normal rates of fetal resorption. The pregnancy-protective mechanisms of the decidual Th2/Th3 cytokines are numerous; however, the most important mechanism is considered to be the inhibition of the expression of IL-2Rα, IL-2Rβ and IL-2Rγ receptors on the decidual effector cells (Clark *et al.,* 1992).

A defective or excessive trophoblastic invasion, as well as an aberrant cytokine network of the decidua, probably can result in the complications of pregnancy, as early spontaneous miscarriage, preeclampsia and the growth retardation of vascular origin, placenta accreta or percreta. The decidual immune cells can be activated in a pregnancy-destructive manner. In this case, decidual cytokine network is associated with pro-inflammatory cytokines, which involve Th1 cell activation. Cytokines such as IL-1, IL-2, IL-6, IL-12, IL-15, IL-18, TNF-α and IFN-γ promote the inflammation by increasing the permeability of blood vessels. The increased permeability increases the access of effector cells, thereby increasing lymphocyte and NK cell activation, respectively (Hunt *et al.,* 1989; Clark, 1993; Chaouat *et al.,* 1994; Ng *et al.,* 2002).

The mechanism by which the implanted trophoblast can be killed is largely unknown. As mentioned above, trophoblast is not susceptible to lysis by any immune cells, but probably may be killed by LAK cells and may undergo apoptosis in response to TNF-α and IFN-γ *in vitro.* These cytokines act synergistically to induce apoptosis in human trophoblast cell cultures; however, the cytokine CSF-1 may abrogate the apoptotic effects of TNF-α and IFN-γ (Garcia-Lloret *et al.,* 1996). Administrating TNF-α to pregnant experimental animals resulted in grave losses of fetuses accompanied with significant decreases in the weights of newborns. Anti-TNF-α antibodies and soluble TNF-αR were effective in decreasing fetal resorption rates in strain combinations of the animals characterized by a very high incidence of fetal resorption (Arck, 1997).

Finally, in woman with recurrent miscarriages, the differences in the CD56$^+$ population of cells are seen, and there is evidence for an alteration in the ratio of Th1 and Th2 cytokines produced by the peripheral blood monocytes and clones of decidual CD4$^+$ cells. There is also evidence for the differences in endometrial cytokine production, and in particular decreased production of pro-inflammatory cytokines (Laird *et al.*, 2002).

4.2.2 The Trophoblast

The cells of the compacted 8-cell embryo divide to produce a 16-cell morula. The morula consists of a small group of internal cells surrounded by a larger group of external cells (Barlow *et al.*, 1972). Most of the descendants of the external cells become the trophoblast (trophectoderm) cells. This group of cells produces no embryonic structures. Rather, it forms the tissue of the chorion, the embryonic portion of the placenta. The chorion enables the fetus to get oxygen and nourishment from the mother. It also secretes hormones that cause the mother's uterus to retain the fetus, and produces the regulators of the immune response so that the mother will not reject the embryo as she would an organ graft. The undifferentiated cytotrophoblast of the outer cell layer can develop into the villous syncytiotrophoblast, extravillous anchoring trophoblastic cell columns, or invasive intermediate trophoblast. Within the villi of placenta of most mammals, there always exists a population of cytotrophoblast cells, which remain undifferentiated apparently available for differentiation if necessary (Hertig *et al.*, 1956).

The blastocyst of most mammals hatches from the *zona pellucida* by lysing a hole in it and squeezing through that hole as the blastocyst expands. A trypsin-like protease, s-trypsin, is located on the trophoblast cell membranes and lyses a hole in the fibrillar matrix of the *zona* (Perona *et al.*, 1986; Yamazaki *et al.*, 1989). The released blastocyst soon establishes contact with endometrium. However, for the implantation it is first necessary that the blastocyst and endometrium accomplish a series of receptor and other molecular interactions. The endometrium "catches" the blastocyst on an extracellular matrix containing collagen, laminin, fibronectin, hyaluronic acid, and heparan sulfate receptors. The trophoblast cells also contain integrins that will bind to the uterine collagen, fibronectin, and laminin, and they synthesize heparan sulfate proteoglycan precisely prior to implantation (Carson *et al.* 1993). Once in contact with the endometrium, the trophoblast secretes another set of proteases, including collagenase, stromelysin, and plasminogen activator. These protein-digesting enzymes digest the extracellular matrix of the uterine tissue, enabling the blastocyst to bury itself within the uterine wall (Strickland *et al.* 1976; Brenner *et al.* 1989)

In conclusion, implantation is regulated by means of numerous interludes between the trophoblast and endometrium. While the trophoblast tends to infiltrate itself as deep as possible into the uterine wall, the decidua tends to limit and control the invasion of the trophoblast. Finally, normal implantation and placentation are the results of a well balanced action of the mechanisms which make the trophoblast invasive and decidual mechanisms that limit trophoblast invasion (Tartakovsky *et al.*, 1990; Ng *et al.*, 2002).

4.2.2.1 Trophoblast Expression of MHC Molecules

Restrictiveness and selectiveness are most important attributes of the expression and tissue distribution of MHC molecules on trophoblast tissues. In normal circumstances, the trophoblast does not express MHC class II and Ia molecules, while MHC class Ib expression is limited to only certain trophoblast tissues. For these reasons, $\alpha\beta T$, $\gamma\delta T$ and NK cells recognize the trophoblast cells as "self" tissue (Mincheva-Nilsson *et al.*, 2000; Szekeres-Bartho, 2002; Bubanovic, 2003a).

HLA-G gene transcription and protein expression are strongly activated in trophoblast cells, and have recently been reported in thymus and endothelial cells (Crisa *et al.*, 1997; Blaschitz *et al.*, 1997), while HLA-G transcriptional activity is basal or null in most other cell types (Ulbrecht *et al.*, 1994). The factors responsible for the specific activation of HLA-G genes in trophoblast cells remain to be defined and could include immunomodulatory factors locally secreted by the placenta during pregnancy (Yang *et al.*, 1996; Saito, 2000, 2001; Crisa *et al.*, 1997).

The original type of trophoblast cells constitutes a layer called the cytotrophoblast, whereas the multinucleated type of cell forms the syncytiotrophoblast. The cytotrophoblast initially adheres to the endometrium through a series of adhesion molecules. Moreover, these cells also contain proteolytic enzymes that enable them to enter the uterine wall and remodel the uterine blood vessels so that the maternal blood bathes fetal blood vessels.

The syncytiotrophoblast tissue is thought to further the progression of the embryo into the uterine wall by digesting uterine tissue. The uterus, in turn, sends blood vessels into this area, where they eventually contact the syncytiotrophoblast. Shortly thereafter, the mesodermal tissue extends outward from the gastrulating embryo (Fisher *et al.* 1989).

The syncytiotrophoblast is the superficial layer of villous trophoblast and therefore most exposed to the maternal immune system. The cell of syncytiotrophoblast directly contacts the decidua and maternal blood, however neither the MHC molecules nor mRNA for MHC molecules have been verified on the cells of syncytiotrophoblast, which might be taken to account for the maternal immunotolerance to syncytiotrophoblast (Hammer *et al.*, 1997; Crisa *et al.*, 1997).

Under the syncytiotrophoblast layer, there is a layer of trophoblast cells commonly regarded as the stem-cell cell population. These cells rejuvenate and regenerate the syncytiotrophoblast cells population by means of differentiation process. The cells of villous cytotrophoblast do not express the MHC molecules but verifying mRNA for MHC class I in the cytoplasm could mean that these cells at some circumstances can express the MHC molecules (Ellis *et al.*, 1990; King *et al.*, 2000).

Extravillous trophoblast involves the trophoblast tissue devoid of chorionic villi. The significance of extravillous trophoblast is highest in early phases of placental development and represents the basis for the development of primary chorionic villi. With the development of the placenta completed, the extravillous trophoblast cells diversify into 3 cells groups (i) cell clusters invading the decidua and contacting the maternal immunocompetent cells, (ii) the cells invading spiral arteries and contacting the maternal blood cells and (iii) the cells invading the uterine miometrium, possibly also contacting the maternal immunocompetent cells (Ellis *et al.*, 1990; King *et al.*, 2000; Lodererova *et al.*, 2003). All these subpopulations of villous trophoblast cells react with monoclonal W6/32 antibodies, specific for the constant partitions of HLA, later revealed to be HLA-G. There are reports on verified expression of heavy HLA-C chain on the surface of extravillous trophoblast, but no possible function of the expression of incomplete HLA-C molecules has been revealed so far (King *et al.*, 2000).

Chorionic membrane is in close contact with the decidua capsularis on the large surface of the uterus, except for the surface of placental bed. The layer of chorionic membrane closely contacting the decidua capsularis is covered with cytotrophoblast cells. This cell subpopulation of extravillous cytotrophoblast also expresses HLA-G, and its function on the chorionic membrane is probably identical to the function performed by HLA-G on extravillous trophoblast. HLA-G molecules have also been verified in the superficial cells of fetal membranes and in amniotic fluid like soluble HLA-G molecules (Ellis *et al.*, 1990; Mc Master *et al.*, 1998).

4.2.3   Pregnancy and Cytokines

The trophoblast and placenta display a high cytokine activity. According to this, there are data that implantation is associated with the development of inflammatory reaction, local hyperaemia, increase of the blood flow, lymphocyte and mononuclear infiltration of the environs of implantation zone. The epithelial cells of endometrium secrete LIF, while decidual macrophages secrete IL-1 and TNF-α causing a significant production of prostaglandine. At the same time, T cells produce IFN-γ thus inducing the expression of HLA-DR on decidual immune cells and consequently leading to a stronger communication between the decidual immune cells. Contrary to this, trophoblast cells secrete anti-inflammatory cytokines like TGF-α, IGF II, CSF-1, TGF-β2 and IL-10 (Tartakovsky *et al.*, 1990; Hunt *et al.*, 1996; Arck, 1997).

4.2.3.1 IL-10

Amiot *et al.* (1998) demonstrate the induction of HLA-G molecules by IFN-γ in the monohystiocytic cell line U937, which suggest that inflammatory conditions may also be associated with the increased expression of these molecules. The baseline status of HLA-G expression and its ability to be up-regulated by both, IFN-γ and IL-10, cytokines could thus play a role in regulating immune responses during inflammatory processes. The ability that different cells demonstrate an enhanced expression of HLA-G under the impact of IFN-γ is not so much surprising as well as the fact that IL-10 also supports the expression of HLA-G. This cytokine is known as the strong inhibitor of MHC gene transcription and the expression of MHC molecules (Amiot *et al.*, 1998).

IL-10 is produced by human trophoblasts cells *in vivo* (Roth *et al.*, 1996). The amount of HLA-G mRNA in IL-10 treated trophoblast culture is significantly increased (~7-fold), compared to the untreated cells. In addition, RT-PCR analysis demonstrates an increase in the level of all HLA-G mRNA transcript isotype, including those that encode soluble proteins in IL-10 treated trophoblast culture. IL-10 is therefore a candidate molecule for a strong stimulatory signal for the HLA-G molecules expression on the trophoblast cells, as well as the important factor of pregnancy success. In particular, the reduced IL-10 production has been reported in pathological pregnancies, distinguishing them from normal ones (Hill *et al.*, 1995; Chaouat *et al.*, 1995; Bubanovic, 2003a).

In view of published data, IL-10 production localized in placenta might have immunosuppressive activities, as observed for monocytes. Moreover, in the light of the effect of IL-10 on human monocyte the HLA molecules expression can be postulated that placental IL-10 could have a dichotomous effect on the HLA expression, inducing the HLA-G expression, while down-regulating classical class Ia and class II.

The modulation of MHC class I expression by IL-10 may result from the molecular mechanisms which include the transcriptional regulation of HLA-G expression in trophoblasts and post-transcriptional regulation in monocytes, since the enhancement of the HLA-G transcript expression in monocytes was low after IL-10 treatment (Hill *et al.*, 1995; Marzi *et al.*, 1996; Raghupathy *et al.*, 1997).

Table 4.3. Effects of IL-10 on different immunocompetent cells.

Cells exposed to IL-10	Effects of IL-10
T lymphocytes	Inhibition of Th1 cytokines secretion from Th1 and NK cells. Antagonistic effects in relation to IL-4.
	Inhibition of T cell proliferation and differentiation mediated by Th1 cytokines. Inhibition of IL-2 gene transcription.
	Inhibition of Th cells activity mediated by APCs and DCs. Inhibition of APCs and DCs class II molecules expression.
	Anergy of T lymphocytes
	Inhibition of IFN-γ secretion by CTL.
	Inhibition of apoptosis of Th2 cells.
Macrophages/Monocytes	IL-10 can change morphology and phenotype, as well as type of cytokines secreted by macrophages and monocytes
	The cells are rounded, less adherent and active.
	Inhibition of MHC molecules expression and interruption of communication between cells.
	Inhibition of IFN-γ effects.
	Inhibition of synthesis of IL-1α, IL-1β, IL-6, IL-8 and TNF-α on mRNA level.
	Inhibition of IL-12 effects.
NK cells	Inhibition of IFN-γ secretion.
	Stimulation of NK activity if the cells are already activated by IL-2
	Stimulation of NK activity in relation to NK activity resistant tumors.
B lymphocytes	Stimulation of class II molecules expression.
	Increased resistance to apoptotic signals.
	Stimulation of B lymphocytes differentiation into plasmocytes.
	Stimulation of antibody production.
	Increased susceptibility to IL-2, increasing of number of IL-2R and affinity for IL-2 binding.

The observation of transcriptional down-regulation of classical HLA class I gene transcription in monocytes failed, suggesting a possible effect of IL-10 on post-transcriptional mechanisms. Some authors found that in murine tumor cell lines expressing IL-10 (Salazar-Onfray *et al.*, 1997) or B cells exposed to EBV-encoded IL-10 (Zeidler *et al.*, 1997), this cytokine may down-regulate classical class I expression through TAP proteins. These findings strongly suggest that the constitutive absence of classical HLA class Ia cell-surface expression in villous cytotrophoblasts and syncytiotrophoblast is also likely to be due to the lack of transporters protein.

4.2.3.2   TGF-β

Clark *et al.* (1990) reported that non-T non-B small lymphocytes in decidua of pregnant mice release a potent immunosuppressive factor in vitro which blocks the action of IL-2 and can be neutralized with rabbit anti-TGF-β antibody. In common with murine decidua, it has been found that human decidual suppressor cells released a soluble suppressor activity which was neutralized with an antibody specific for TGF-β. This cytokine which is produced at the human maternal-fetal interface has a role in implantation, possibly by increasing plasminogen activator inhibitors and thus limiting trophoblast invasion (Ditzian-Kadanoff *et al.*, 1993). A deficiency of TGF-β2-containing cells has been reported in first trimester pregnancies destined to abort, although the cell type responsible for TGF-β2 production in human decidua has not been thoroughly characterized (Clark *et al.*, 1990). However, further research demonstrated cytoplasmic immunostaining for TGF-β, predominantly TGF-β2 of decidual CD56⁺CD16⁻CD3⁻ LGL (Clark *et al.*, 1991). Along with these, TGF-β2 is also secreted by endometrial, miometrial, and trophoblast cells, though to a much smaller degree. Apart from CD56⁺CD16⁻CD3⁻ LGL, the significant quantities of decidual TGF-β2 come from the pool of stimulated γδT cells of early and late decidua (Bluestone *et al.*, 1991). In addition, TGF-β2 can be secreted by Th3 cells and may play protective roles during pregnancy, regulating Th1/Th2/Th3 relationship (Raghupathy, 2001).

TGF-β2 exhibits some important effects such as the regulation of cell growth, cell differentiation and migration, production of intercellular matrix, angiogenesis and stimulation of prostaglandine secretion. In addition, TGF-β2 helps implantation and controls the invasion of the trophoblast, as well as delaying the rejection of allotransplant. This effect is evidenced by a strong TGF-β2 mediated suppression of the production of cytokines such as IL-1β, IFN-α, IL-2, TNF-α and TNF-β (Clark *et al.*, 1997). The immunomodulatory effects of TGF-β are partly elicited by directing the differentiation of Th precursors towards Th2 and Th3 cells (Ditzian-Kadanoff *et al.*, 1993; Clark *et al.*, 1997), though some data suggest that TGF-β may also display inhibitory effects on the Th2 cell differentiation and Th2 cytokine secretion (Gorelik *et al.*, 2000).

Progesterone is mentioned as one of the factors regulating the production of TGF-β2 in pregnancy. Also, progesterone has direct inhibitory effects on Th1 cells. Opposite to this, progesterone probably has no direct stimulatory effects on Th2 cells. Rather, stimulation of Th2 cells is mediated by TGF-β2, PIBF and IL-10 (Choi *et al.*, 2000). D'Orazio *et al.* (1998) showed that the presence of APCs to TGF-β2 could significantly enhance the production of IL-10. TGF-β2 also significantly suppresses the production of the pro-inflammatory cytokine IL-12. These data suggest that TGF-β2 contributes to the maintenance of immune privilege by predisposing an APCs to the preferential production of IL-10 cytokine, which in turn might down-regulate or alter certain presenting functions leading to the generation of a suppressive, Th2 type of immune response. In addition, TGF-β as a pleiotropic cytokine with generally anti-inflammatory and immunosuppressive properties inhibits macrophage activation, the generation of CTL, and the expression of MHC class II molecules. Importantly, TGF-β also has clear bimodal effects, with the low concentrations and high concentrations exerting distinct physiologic effects (Gorham *et al.*, 1998). For example, mice deficient in TGF-β1 develop a lethal multiorgan inflammatory immune infiltrate at 3 weeks of age with an increased expression of inflammatory cytokines such as IFN-γ and TNF-α (Shull *et al.*, 1992).

In pregnancy, the early stages of the resorption process induced by LPS were characterized by lymphocyte accumulation in the vicinity of the embryo, preceding any visible embryonic damage. At that time, there is an increased expression of TNF-α in the decidua, which was reduced as the resorption process was completed. In contrast, the TGF-β2 expression was decreased in the decidua, as well as in the glandular epithelium, at all the times assessed. The maternal immunopotentiation with GM-CSF, which controls maternal immune activities supporting normal embryonic development, decreased the resorption rate in LPS-treated mice while normalizing the expression of TNF-α and TGF-β2 in the uterus of these animals throughout the ongoing resorption process. These findings indicate the important role of TNF-α, TGF-β2 and GM-CSF in the mechanisms mediating the early stages of pregnancy loss (Savion *et al.*, 2002).

In mammals, as in other classes of vertebrates, TGF-β2 is an important factor of cell proliferation and tissue differentiation in embryonic development, as well as embryonic survival. Fein *et al.* (2002) found that early embryonic deaths and malformed newborns in diabetic pregnancy are accompanied by an alteration of the TGF-β2 expression. Same authors also found that immunopotentiation with paternal leukocytes can improve the reproductive performance of the diabetic mice, accompanied by a partial normalization of the TGF-β2 expression in embryonic vicinity.

4.2.3.3 IL-4

The activated Th2 cells are the most important pool of IL-4 production. The effects of this cytokine activity are primarily strong bias towards the humoral immune response. The effects of IL-4 on B cells are largely associated with the stimulation of proliferation, differentiation and antibody production. An important function of IL-4 in regulating and promoting the humoral immune responses is its ability to stimulate the differentiation and proliferation of precursor Th cells, biasing towards Th2 type (Ishikawa *et al.,* 1991; Saito *et al.,* 1996).

Accompanied by IL-10, IL-4 strongly suppresses the activity of Th1 lymphocytes, inhibition the already activated macrophages, CTL and NK cells. IL-4 is seen to suppress the expression of receptors for IL-2 on CTL and NK cells, making these cells quite resistant to the activating IL-2 signals (Ishikawa *et al.,* 1991; Saito *et al.,* 1996).

The serum level of IL-4 of the healthy pregnant women decreases progressively, up to the second trimester of pregnancy, which is quite unusual for the cytokines of Th2 group. In the last trimester of pregnancy, the serum concentration of IL-4 is somewhat higher compared to the second trimester, only to drop to its normal value 6 to 11 months after delivery (Shimaoka *et al.,* 2000).

IL-4 has been verified on the trophoblast tissues at all stages of placental development, and unlike IL-3, the trophoblast and syncytiotrophoblast express the receptors for IL-4 (de Moraes, *et al.* 1997). Although the concentration of IL-4 in the sera of healthy pregnant women progressively decreases throughout pregnancy, the PBMC of these women respond to the stimulation by PHA by secreting significantly higher quantities of IL-4, compared to pregnant women suffering from Recurrent Spontaneous Abortion (RSA) syndrome. The patients with RSA who, after undergoing immunopotentiation with paternal leucocytes succeeded in delivering healthy newborns, showed much higher values of serum IL-4 during and after pregnancy than before immunopotentiation (Raghupathy *et al.,* 2000).

However, there are some data that the treatment with IL-4 and consequent acceleration of Th2 immune response may be as fatal for pregnancy as the acceleration of Th1 immune response. Hayakawa *et al.* (2000a) found that treating allopregnant and syngeneic-pregnant Balb/c mice with IL-4 activated splenocytes from virgo Balb/c mice, resulted in developing of fetal resorption and hypertensive-like syndrome accompanied by a massive liver degeneration in all the pregnant animals. Similar effects were obtained after the inoculation of IL-12 stimulated lymphocytes, only in this case some drastic changes occurred on the liver of treated mice.

4.2.3.4   IL-6

IL-6 is one of the most active cytokines in pregnancy. It is produced by a large number of decidual immunocompetent cells, but mostly by Th2, NK and mononuclear cells. Cytotrophoblast and syncytiotrophoblast secrete IL-6 and express receptors for this cytokine. Of all the cytokines stimulating the proliferation of cytotrophoblast, IL-6 exhibits the greatest activity. Being a Th2 cytokine, the level of its secretion significantly increases during pregnancy (Agarwal *et al.,* 2000). The PBMC of pregnant women with RSA show much lower level of secretion of this cytokine, whether of basal secretion or mitogens-induced secretion (Saito, 2000).

Since IL-6 is a strong stimulator of antibody production, the increased secretion of this cytokine may compromise pregnancy. Gutierrez *et al.* (2001) succeeded in clarifying somewhat contradictory data related to IL-6 as a protective factor in pregnancy, and yet a powerful accelerator of differentiation of plasmocytes and antibody production. Namely, the placenta was indeed shown to secrete a specific isotype of IL-6, which also stimulates the differentiation of plasmocytes and production of antibodies. However, the antibodies being secreted under the influence of placental IL-6 were asymmetric, incomplete and almost functionally inadequate compared to the anti-bodies generated after the stimulation of B lymphocytes with IL-6 of non-pregnant women.

IL-6, accompanied by IL-4, stimulates the production of hCG, and therefore the production of progesterone from the corpus luteum. In addition, there is a well documented evidence of IL-6 role in implantation, placentation, pre-delivery cervical dilatation, initiation of maternal contractions and delivery. A large number of studies confirm a significant increase in the concentration of IL-6 in pre-delivery cervical mucus, cervical tissue, fetal membranes, amniotic fluid and the sera of pregnant women, whether in chorioamnionitis-complicated pregnancy, pre-term or normal term delivery. Certainly, the concentration of IL-6 in fetal membrane and amniotic fluid is significantly higher in the presence of chorioamnionitis. There are no reliable data as to which IL-6 isotype take part in the foregoing mechanisms, but it is assumed that they are lymphatic isotype of IL-6 (Saito, 2000; Saji *et al.*, 2000; von Minckwitz *et al.*, 2000).

4.2.3.5 IL-2

IL-2 has been shown to be a powerful proliferative agent for decidual NK cells *in vitro*, as well as an augmenter of cytolytic activity. This, coupled with the observation that decidual NK cells express the IL-2Rβ chain would suggest that IL-2 is responsible for maintaining NK cells in utero. However, subsequent work has failed to detect IL-2 protein in either first-trimester trophoblast or decidual cells or IL-2 mRNA. It therefore appears that IL-2 is not produced anywhere in the normal implantation site (Lala *et al.*, 1990; Ishikawa *et al.*, 1991; Saito *et al.*, 1996). Yet, in pathological pregnancies, decidual immune cells may secrete IL-2, during which the activation of decidual CTL and NK cells occurs followed by the infiltration in placental tissues and initiation of cytolysis and/or apoptosis of trophoblast cells. By penetrating into placental blood-vessels, CTL and NK cells may start the classical reaction of alloimmune rejection of endothelial cells, causing the thrombosis of placental blood vessels and discontinuance of fetoplacental circulation (Lala *et al.*, 1990; Saito *et al.*, 1996; Gorczynski *et al.*, 2002).

During normal pregnancy, the production of IL-2 by PBMC is very low and is maintained on this level up to the delivery day only to increase gradually, reaching physiological values 2-11 months after delivery (Shimaoka *et al.*, 2000). Decidual lymphocytes and PBMC of pregnant women with RSA past synthesize double amounts of IL-2 when stimulated by mitogens com-pared to decidual lymphocytes and PBMC of healthy pregnant women (Raghupathy *et al.*, 1997, 2001). The enhanced decidual production of IL-2 in pregnancies with RSA past certainly is not an isolated phenomenon. The increase of IL-2 production in such patients is only a part of acceleration of the Th1 response and activation of cellular immunity. The reasons for activating Th1 cells in pregnancy are manifold and can be explained by means of various mechanisms ranging from the constitutional peculiarities of Th1/Th2 balance, infections and the so-called adverse combinations of the paternal and maternal MHC phenotype.

4.2.3.6 IL-12

IL-12 is secreted by B lymphocytes and APCs. Its role is the stimulation of the activity of CTL, NK cells and Th1 lymphocytes. IL-12 is seen to enhance the effects of IL-2 activity by many times, while exerting direct stimulatory effects on macrophages and other cells of the innate immunity (Xing *et al.*, 2000). IL-12 is shown to stimulate the production of IFN-γ and TNF-α and via these, enhance the expression of the MHC molecules on immunocompetent and target cells, thus enabling a better communication between immunocompetent cells as well as a more effective recognition of target cells (Saito *et al.*, 1996; Gorham *et al.*, 1998; Xing *et al.*, 2000).

B lymphocytes secrete IL-12 only when acting as antigen-presenting cells, i.e., only when, following the endocytosis of antigens, they present its epitops as MHC class II/peptide complex. This phenomenon could serve to account for the abortion mediated via the immune mechanisms triggered by B cell activation. This mechanism was shown to enhance IL-12 concentration in the sera of women with RSA compared to normal pregnancies. In addition, an enhanced production of IL-12 can be recorded in pregnancies complicated with regard to imminent abortion, PIH, eclamptic, and preeclamptic syndromes. Suggestive of this is the finding that the syngeneic splenocytes of mice which were previously cultured in the presence of IL-12 and injected in allopregnant mice cause a state similar to eclamptic syndrome and are accompanied by massive alterations on the liver (Hayakawa *et al.*, 2000a).

The regulation of IL-12 secretion in pregnancy is mediated mostly by the factors such as PIBF and LIF. Both these factors suppress the secretion of

IL-12 in normal pregnancies whereas decreased concentrations of LIF and PIBF have been verified in the sera of most of endangered pregnancies recording the enhanced concentration of IL-12. PIBF reduces the expression of IL-12 via the inhibition of arachidonic acid metabolism (Par *et al.,* 2000).

4.2.3.7 IL-15

Interleukin IL-15 is secreted by B lymphocytes, NK cells, CTL and Th lymphocytes. Although the effects of this cytokine resemble the effects of IL-2, IL-15 differs in that it regulates the function of a large number of cells, tissues and organs. For example, IL-15 mRNAs have been detected in tissues and cell types, including heart, lung, liver, placenta, skeletal muscle, adherent peripheral blood mononuclear cells, and epithelial and fibroblast cell lines. However, IL-15 mRNA is not detectable in the activated peripheral blood T cells that contain high levels of IL-2 mRNA (Grabstein *et al.,* 1994).

IL-15 has biological activities similar to IL-2 and has been shown to stimulate the growth of NK cells, activated PBMC, TIL, and B cells (Grabstein *et al.,* 1994; Giri *et al.,* 1994, 1995). In addition, IL-15 has also been shown to be a chemoattractant for human blood T lymphocytes and to be able to induce LAK activity in NK cells as well as to be able to induce the proliferation of CTL (Lewko *et al.,* 1995; Armitage *et al.,* 1995).

The placental IL-15 displays a linear increase throughout pregnancy and reaches its maximum in term-placenta. Apart from placental production, an endometrial, i.e. decidual production has also been recorded. The levels of both IL-15 mRNA and protein were significantly reduced in the pre-eclamptic placental tissue compared with the normal controls (Agarwal *et al.,* 2001). In human endometrium, the expression of IL-15 mRNA significantly increased during the secretory phase compared with the proliferative phase, but the most abundant expression of IL-15 mRNA during the menstrual cycle was observed in the midsecretory phase. In the case of pregnancy and decidual transformation of endometrium, the production of IL-15 continues its linear increase until delivery (Okada *et al.,* 2000).

Decidual NK cells (preferably $CD56^+CD16^+$) do not normally display a cytolytic activity towards JEG-3 choriocarcinoma or trophoblast cells. However, the stimulated IL-15 NK cells successfully lyse JEG-3, but not trophoblast cells, unlike IL-2 stimulated NK cells which lyse both JEG-3 and trophoblast cells with equal effectiveness (Verma *et al.,* 2000).

4.2.3.8 IL-18

The regulation of synthesis and secretion of IFN-γ is one of the best controlled mechanisms in the cascade of immune response. IL-18 is the factor controlling synthesis and secretion of IFN-γ as a result of which it has been called IFN-γ inducing factor (IGIF). This cytokine is mostly secreted by macrophages and Kupfer's cells. IL-18 also inhibits the secretion of GMC-SF by B lymphocytes, NK cells, macrophages and T lymphocytes. Moreover, it inhibits the synthesis of IL-10, Th2 lymphocytes activity and decreases the expression of IL-10 receptors (Nakanishi *et al.,* 2001a, 2001b). Since IL-18 helps the secretion of IFN-γ, which necessitates the signal coming from IL-12, it can thus be considered a co-stimulatory factor for the stimulation of IFN-γ secretion (Chang *et al.,* 2000; Eberl *et al.,* 2000). IL-18 has been found to defer the apoptosis of CTL and NK cells without the mediation by other cytokines (Eberl *et al.,* 2000).

The role of IL-18 in the immunopathology of pregnancy has also been verified. The concentration of IL-18 in the sera of pregnant women increases progressively only to reach its peak values with the approaching of the delivery time. The complications like chorioamnionitis and pre-term delivery are also associated with the increase of IL-18 level in the sera of pregnant women (Ida *et al.,* 2000).

4.2.3.9   IFN-γ

IFN-γ is one of the most significant and most active cytokines comprising the so-called major cytokines of Th1 group. As its effects are the strong stimulation of cellular immunity and enhanced MHC expression on target cells, the high decidual IFN-γ activity might result in the expression of paternal MHC molecules on trophoblast tissues, activation of decidual CTL and NK cells and, finally, rejection of the placenta as an allotransplant (Yang *et al.,* 1996; Gorczynski *et al.,* 2002). Although much simplified, the described order of events could be considered as the possible mechanism of strong abortogenic effect of IFN-γ.

Treating allopregnant mice with IFN-γ at early stages of pregnancy will lead to the increased rates of embryo resorption, compared to the same treatment applied in the second half of pregnancy which does not influence the rate of fetal resorption. Nevertheless, in both cases a higher expression of MHC class I molecules has been verified on the trophoblast of treated mice compared to the trophoblast of non-treated animals (Krishnan *et al.,* 1996).

The PBMC of pregnant women with RSA past display a higher activity in producing IFN-γ when stimulated with mitogens, than the PBMC of healthy women (Raghupathy *et al.,* 1997, 2000). During the course of normal pregnancy, the production of IFN-γ is significantly lower. Such decrease in IFN-γ secretion gradually involves all immunocompetent cells, regardless of their location. The degree of suppression of IFN-γ production is highest in the decidual population of immunocompetent cells. The suppression of IFN-γ production concurs with the onset of pregnancy, growing in intensity throughout until the delivery time, taking 2-11 months for the production of IFN-γ to be restored to its normal value (Shimaoka *et al.*, 2000).

The level of IFN-γ in the sera is increased in pregnant women with RSA past. Moreover, lymphocytes in such patients secrete much higher amounts of this cytokine when stimulated with antigens or mitogens. The alloimmunization of the women with RSA past by paternal lymphocytes causes the decrease of the IFN-γ to the level as in the sera of healthy pregnant women while at the same time decreasing the potential of lymphocytes for the secretion of this cytokine (Hayakawa *et al.*, 2000b).

Besides the effects exerted on trophoblast cells by IFN-γ, mediated by the effector cells, *in vitro* studies also indicate the immediate effects of IFN-γ on the trophoblast. IFN-γ may exert direct cytotoxic effects on trophoblast cells. At the same time, IFN-γ significantly increases the susceptibility of cultured trophoblast cells to effector cells and TNF-α. EGF is seen to prevent successfully most of the effects exerted by IFN-γ on the trophoblast, and it is assumed that these mechanisms are associated with the protective role of fibronectin whose synthesis is stimulated by EGF (Pijnenborg *et al.*, 2000).

4.2.3.10 TNF-α and β

TNF-α has been classified as belonging to the group of major Th1 cytokines because of multiple and intensive changes it causes on the immunocompetent cells. Like IL-1, TNF-α raises the level of synthesis and secretion of autocrine growth factors, as well as the sensitivity of immune cells to these factors. Due to this role of TNF-α in the acute phase of inflammation, TNF-α has been grouped as a pro-inflammatory cytokine. It strongly inhibits neutrophils, enhances the production and secretion of antibodies and complement, increases the adhesiveness of thrombocytes and stimulates extravasation of lymphocytes and macrophages. All these effects cause a local inflammatory reaction, whereas, on the system level, TNF-α mostly causes generalized oedema, hypoproteinemia and neutropenia (Beutler, 1999).

TNF-β is an isotype of TNF, but unlike TNF-α, it is secreted by a smaller number of cells, like NK cells, CTL and B lymphocytes. Since the immunocompetent cells are the most significant source of TNF-β, this cytokine is also known as "lymphotoxin". The major stimulating signals for the secretion of TNF-β are the antigenic stimulation of CTL and IL-12. The production of IL-2 by Th1 cells under the influence of TNF-β is identical to the antigenic stimulation of these cells. The effects which TNF-β exerts are mostly seen as the stimulation of terminal differentiation phase of various cell lines, mostly haematopoietic, and stimulation of apoptosis. TNF-β is capable of triggering apoptotic processes on almost all cell types (Ware, 1996; Xing *et al.*, 2000).

Since TNF belongs to the Th1 group of cytokines, with characteristics of a strong apoptotic mediator, any increased activity in pregnancy will lead to a serious compromising of immunoregulation in pregnancy, and complicate further progression of pregnancy. The adverse impact of TNF on the course of pregnancy is generally associated with the fact that TNF is particularly active when inducing apoptotic mechanisms on the proliferative and trophoblast tissue (Krishnan *et al.*, 1996; Arck, 1997; Xing *et al.*, 2000). Probably due to this reason, the synthesis and secretion of TNF in normal pregnancy is very low, and immunocompetent cells show a low potential for the secretion of TNF, even when stimulated by mitogens. A large number of studies indicate the mediatory role of TNF in RSA syndrome or some other forms of compromised pregnancy. Corroborating this is the finding that the PBMC of patients with RSA past show 2-4 times higher ability for the synthesis and secretion of TNF-α compared to healthy pregnant women (Ho *et al.*, 1996; Raghupathy *et al.*, 1997, 2000).

The factors contributing to the enhanced activity of TNF-α and Th1 cytokine in pregnancy are manifold, the most frequent being infections, dysregulation of immunoregulation in pregnancy, poor expression of HLA-G on trophoblast cells, hormonal disorders etc. RSA, eclamptic and PIH syndrome are conditions characterized by an increased decidual production of TNF-α and a significantly higher activity of decidual NK cells, which is in current literature accounted for by the lower expression of membranous HLA-G on placental tissues (Choi *et al.*, 2000; Hayakawa *et al.*, 2000b; Nakanishi *et al.*, 2001; Ng *et al.*, 2002). The conditions such as chorioamnionitis are followed by the sudden increase in the concentration of pro-inflammatory cytokines, especially TNF-α, in the sera of fetus and pregnant woman, fetal membrane and amniotic fluid (Saji *et al.*, 2000).

Apart from the great connection between TNF-α and various other entities of pathological pregnancy, there are some data possibly indicating the important role of this cytokine in the process of trophoblast differentiation and regulating of its invasiveness. In the first trimester of pregnancy, the trophoblast secretes small amounts of TNF-α and expresses receptors for TNF types I and II. The low concentrations of TNF-α do not exert apoptotic effects on the early trophoblast, but mostly serve to facilitate and direct its differentiation. However, higher concentrations of TNF-α, especially accompanied by other pro-inflammatory cytokines possibly modifying the composition of the receptors for TNF-α on trophoblast cells, display a strong apoptotic activity on the placental tissues (Knofler *et al.*, 2000; Ortega *et al.*, 2000).

4.2.4   Activity of Trophoblast TAP Machinery

As in other tissues, trophoblast transporters proteins are essential for the peptide presentation and MHC class I assembly and expression. In first-trimester human placenta, the expression of the TAP1 gene correlated with the expression of HLA class I molecules in trophoblast cells. HLA-G positive extravillous cytotrophoblast cells exhibited a high level of TAP1 mRNA and TAP1 gene products. However, TAP1 mRNA and the protein were absent in HLA class I molecules negative syncytiotrophoblast and villous cytotrophoblast cells (Clover *et al.,* 1995; Roby *et al.,* 1996).

The different trophoblast cell subpopulations have developed efficient regulatory mechanisms to prevent the expression of β2m-associated HLA class Ia molecules at their cell surface. For example, Rodriguez *et al.* (1997) found that the IFN-γ treatment of trophoblast induces a significant enhancement of the transcription of TAP1 and TAP2 rather than an increase of HLA class I or β2m. The same authors found that the constitutive absence of HLA class Ia cell surface expression in term villous cytotrophoblast and syncytiotrophoblast is likely to be due to a lack of TAP1 and TAP2.

The experiments on different animal models showed that in all cells with deficiencies in β2m or TAP expression, MHC class I expression is nearly absent. Likewise, targeted mutations in β2m or TAP genes results in animals with a significant loss of normal MHC class I expression, and their cells are susceptible to NK activity. Because HLA-G and blastocyst MHC molecules resemble MHC class I molecules, such disruption may result in the marked decrease in trophoblast expression of these molecules and fetal susceptibility on NK mediated cytolysis. Indeed, embryonic cells may not normally express TAP genes, and these cells are quite susceptible to killing by IL-2-acti-vated NK cells. However, notwithstanding the possible mechanisms controlling the expression of class Ib molecules via the processing machinery, it is yet surprising that β2m or TAP-deficient animals apparently develop norm-ally (Yokoyama, 1997).

The control of TAP genes transcription in trophoblast cells probably can be provided by prostaglandine, hormones and especially by cytokines. Salazar-Onfray *et al.* (1997) showed that murine tumor cell lines expressing IL-10 or B cells exposed to EBV-encoded IL-10 (Zeidler *et al.,* 1997) may down-regulate classical class I expression through TAP proteins. It has recently been suggested that the constitutive absence of classical HLA class I cell-surface expression in term villous cytotrophoblasts and syncytiotrophoblast is also likely to be due to the IL-10 activity and lack of TAP (Rodriguez *et al.* 1997).

Although TAP molecules apparently participate in the regulation of class Ib expression, their influence on the frequency of class Ib genes transcription is most probably not significant. Therefore, two assumptions seem most logical: (i) TAP molecules take part in the controlling of class Ib expression on the posttranscriptional level and (ii) the expression of class Ib molecules accumulated intracellularly can be regulated by factors other than TAP. TAP1, TAP2, tapasin, and β2m, together with the HLA-G mRNA, are all up-regulated at a very early stage in extravillous cytotrophoblast cell differentiation. In contrast, the accumulation of the HLA-G protein is restricted to the distal portion of extravillous cytotrophoblast cell columns. These results reveal several important aspects of antigen-presenting capability during extravillous cytotrophoblast cell development (Copeman *et al.*, 2000).

1. HLA-G expression is pre-transcriptional and post-transcriptional regulated process.
2. The proteins essential for MHC class I expression on the cell surface are concomitantly expressed during extravillous cytotrophoblast cell differentiation.
3. The HLA-G protein accumulates markedly later than all other such proteins.

HLA-G associates with TAP and binds nonamer peptides derived from a variety of intracellular proteins, suggesting that HLA-G, like the polymorphic HLA class Ia molecules, has the ability to present peptides to the T cell receptor on CD8$^+$ CTL, thereby leading to the elimination of infected cells. Therefore, it has been assumed that class Ib molecules may be involved in protecting of the fetal-placental unit against intrauterine infections (Le Bouteiller, 2000).

4.3   PROSTAGLANDINE

Prostaglandines are tissue hormones present in each anatomical and functional segment of the female genital tract. They play a variety of roles in a variety of biological processes such as endometrial ablation, regulation of ovarian cycle, biosynthesis of sex hormones as well as the influence on periovulatory events and ovulation. They are also included in the process of implantation, placentation and angiogenesis. Finally, prostaglandines play a major role as factors regulating the delivery mechanisms (Wood *et al.*, 1988; Scodras *et al.*, 1990; Brannstorm *et al.*, 1993; Pitzel *et al.*, 1993; Jacobs *et al.*, 1993; Ohno *et al.*, 1994; Norman *et al.*, 1998).

The placenta exerts very a strong prostaglandine activity which tends to grow in intensity during pregnancy (Kankofer *et al.*, 1999). From the start of placental development, prostaglandines are significantly involved as the mediators of both decidualization and placentation. Treating experimental animals with *indomethacine* between day 1 and 6 of pregnancy is seen to inhibit effectively decidualization and implantation (Kankofer *et al.,* 1999; Tessier-Prigent *et al.* 1999). The same authors claim that the activity of inducible COX-2 isoenzyme is more important than COX-1 for the processes of decidualization and implantation. Indicative of this are the findings obtained from COX-2 deficient mice. The line of COX-2 deficient mice is normally characterized by a very weak decidualization, and so depleted offspring. Although PGI2 is critical for the processes of decidualization and implantation, the dominant prostaglandine in the placental tissue are PGF2α as the key mediator for very strong angiogenetic processes, and PGE2 as the main immunosuppressive factor (Norman *et al.,* 1998). Some authors propose PGI2 as a mediator of angiogenesis in the process of placentation. It has been verified to significantly enhance the synthesis and secretion of growth factors such as FGF, VEGF and EGF (Krishnamurthy *et al.,* 1999).

4.3.1   Immunomodulatory Role of Prostaglandine

Decidua is the part of placenta which serves as a site for PGE2 synthesis. Its immunomodulatory role is enabled through the inhibition of IL-2 secretion and inhibition of IL-2R expression on the immunocompetent cells (Lala *et al.,* 1988, 1990). Watson *et al.* (1991) confirmed that the immunosuppressive effects of the supernatant of trophoblast culture on the ConA stimulated lymphocytes might be annulated by *indomethacine* or IL-2.

A series of studies conducted by Lala *et al.* (1988, 1990) based on the continual treatments of allopregnant mice with *indomethacine* and/or IL-2 revealed a high percentage of fetal resorption and spontaneous abortion up to 89%. In subsequent studies, the same group of authors found that decidual NK cells, which are in normal pregnancies suppressed by decidual prostaglandines, are largely responsible for the fetal resorption when stimulated by *indomethacine* and/or IL-2. Unlike normal pregnancies where decidual NK cells display a low level of anti-YAC-1 activity, the decidual NK cells of allopregnant mice pre-treated with *indomethacine* and/or IL-2 display high anti-YAC-1 activity. The same authors claim that it is possible to prevent fetal resorption stimulated with *indomethacine* and/or IL-2 by applying parallel treatments using PGE2.

Vanderbeehen *et al.* (1984) point to similar findings verifying the complete areactivity of maternal lymphocytes which were isolated immediately after delivery in the MLR test against either fetal or paternal lymphocytes. The first signs of the reaction of maternal lymphocytes against paternal or fetal lymphocytes appeared only if maternal lymphocytes were isolated on day 3 after delivery, with their complete reactivity restored on day 6 after delivery. It is possible to abolish all stages of the areactivity of maternal lymphocytes in MLR tests with fetal or paternal lymphocytes by adding anti-CD8 antibodies, *indomethacine* and/or IL-2 i.e. by directly suppressing the activity of T suppressor cells, or by the inhibition of PGE2 synthesis and/or substituting IL-2, the deficit of which is the major outcome of the suppressive activity of prostaglandine (Lala *et al.,* 1988, 1990).

The mechanisms of immunomodulatory action of prostaglandine are not only limited to the inhibition of IL-2 secretion and inhibition of IL-2R expression. Prostaglandines significantly inhibit the secretion of other cytokines of Th1 group as well, also suppressing the activity of effectory cells by enhancing the cAMP level. The effect of stimulation on Th2 cells under the influence of PGE2 has been verified by the enhanced Th2 cytokines secretion. In addition, PGE2 is seen to enhance significantly the concentrations of IL-10 and IL-8 in the maternal blood (Scodras *et al.,* 1990; Denison *et al.,* 1998).

4.4   SEX HORMONES AND IMMUNITY IN PREGNANCY

An example of a physiologic, rather than a pathologic, state in which cytokine biases appear to occur is that of mammalian pregnancy (Wegmann *et al.,* 1993; Clark, 1993; Chaouat *et al.,* 1997; Saito, 2000, 2001). Evidence has long accumulated to indicate that pregnancy is associated with the enhanced humoral and reduced cellular immune activity, consistent with a bias for Th2 cytokines. Steroid hormones, which are secreted in large quantities during pregnancy, are clearly capable of regulating cytokine synthesis in a variety of cell types (Ohno *et al.,* 1994; Choi *et al.,* 2000; Par *et al.,* 2000). Thus, it is possible that they may also play a role in regulating the balance between Th1 and Th2 like activities. For instance, glucocorticoids inhibit IL-2 and IFN-γ (Daynes *et al.,* 1995), induce TGF-β secretion (Bautman *et al.,* 1991), and decrease IL-2R expression in T cells (Paliogianni *et al.,* 1993). Other steroid hormones such as adrenal androgen, dihydroepiandrosterone and 1,25-(OH)2 vitamin D3 selectively evoke, respectively, Th1 or Th2 like cytokine secretion patterns in mice (Daynes *et al.,* 1995, Bubanovic, 2004).

4.4.1   Progesterone

Progesterone is the dominant hormone in pregnancy. The site and control of its synthesis depend mostly on the pregnancy stage and species. In humans, in the first fourth of pregnancy the corpus luteum is the main site for the synthesis of progesterone, while the synthesis itself takes place under the influence of hCG (Kovalevskaya *et al.,* 1999). Starting from the second fourth of pregnancy onwards the main site for this activity is the placenta. The production of progesterone continuously intensifies throughout pregnancy until the delivery time and then begins to drop suddenly, which is presently considered one of the factors responsible for the weakening of the immuno-regulatory mechanisms and initiating of delivery mechanisms (Patel *et al.*, 1999).

The role of progesterone in pregnancy is multiplex: it enables the decidual transformation of endometrium, successful implantation, suppression of maternal immune response as well as the initiation of delivery. In addition, this hormone serves as the main substrate for the synthesis of mineral- and corticosteroids in the fetal and maternal adrenal glands. Progesterone relaxes the smooth musculature of the uterus and blood vessels, contributing thus to a better blood flow on the utero-placental junction (Daynes *et al.* 1995; Kovalevskaya *et al.*, 1999; Patel *et al.*, 1999).

In most mammals, progesterone is essential in maintaining pregnancy, since it can influence various phases of the immune response. The mechanisms of immunomodulatory effects of progesterone can be defined as: (i) direct immunosuppressive effects on immunocompetent cells mediated by mechanisms enhancing cAMP levels, (ii) suppressor effects mediated by factors like PIBF and (iii) joint effects of progesterone and PIBF on the biasing of Th1/Th2 balance towards Th2 response (Szekeres-Bartho *et al.*, 1996, 1999a).

The immunomodulatory effects of progesterone are exerted through the immune cells expressing the receptors for progesterone (PR). The capacity of progesterone binding by the lymphocytes in pregnant women is significantly higher compared to the lymphocytes in non-pregnant women. Specific monoclonal antibodies were used to verify PR on the lymphocytes of pregnant women's peripheral blood, whereas only a small number of these were detected in non-pregnant women (Szekeres-Bartho *et al.*, 1989a, 1990, 1999a). Lymphocyte subpopulation mostly expressing PR belongs to γδT lymphocytes (Polgar *et al.*, 1999). The percentage of PR$^+$ lymphocytes grows throughout pregnancy, unlike RSA and pre-term delivery which are associated with a much lower percentage of PR$^+$ lymphocytes (Szekeres-Bartho *et al.*, 1989a, 1989b).

The regulation of PR expression on lymphocytes primarily depends on the degree of their activation, but is hormone-independent. A very small number of non-activated lymphocytes express PR, unlike ConA, PHA or alloantigen-activated lymphocytes which express a large number of these. An amazingly large number of the lymphocytes in transplant recipients also ex-press PR (Szekeres-Bartho *et al.*, 1989a, 1989b). These data indicate the existence of a possible mechanism for inducing the expression of PR on lymphocytes, based on the continual stimulation by mitogens or alloantigens.

Origin of Anti-tumor Immunity Failure in Mammals

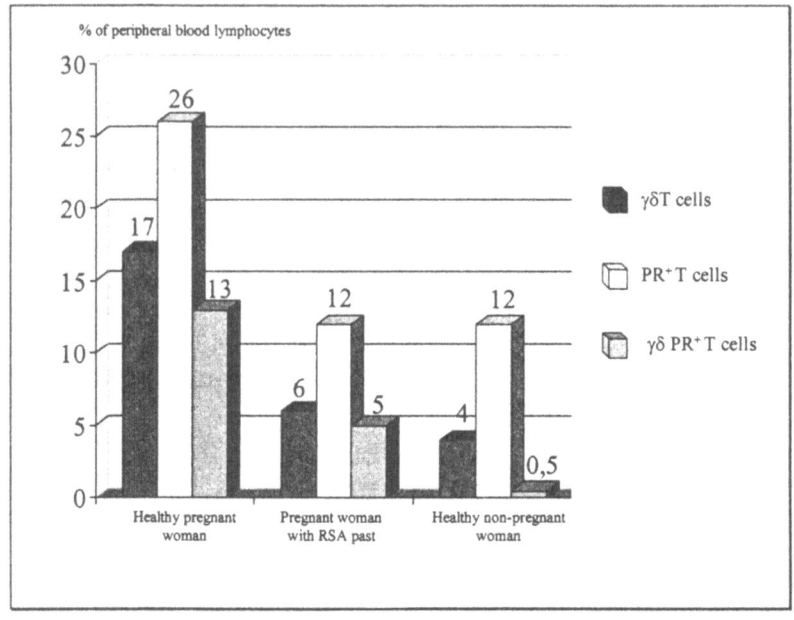

Figure 4.4. The percentage of γδT, PR⁺ T and γδPR⁺ T lymphocytes in the lymphocyte
population of peripheral blood in healthy pregnant women, pregnant women
with RSA past and healthy non-pregnant women (Szekeres-Bartho *et al.,*
1989a, 1999).

Decidual γδT lymphocytes grow in number during pregnancy. Their
number in the uterus is much higher in allogeneic than syngeneic
pregnancies. Most γδT lymphocytes (90%) in pregnant women show a well-
express-ed PR, indicating the activated state of these lymphocytes in
pregnancy. The importance of PR expression on lymphocytes is probably
due to the fact that PR is a powerful mechanism through which it is possible
to influence the lymphocyte activity, especially in pregnancy where the
concentrations of progesterone in the sera are 100-1,000 times higher than in
the non-pregnant state. This simple, logical assumption found a solid
experimental support in the findings of the investigation into the percentage
of γδT PR⁺ lymphocytes in the peripheral blood of women with RSA, in
normal pregnancies and non-pregnant women. Quite unsurprisingly, the
findings showed that in normal pregnant women, the percentage of γδPR⁺
lymphocytes was the highest, in women with RSA significantly lower and
finally in non-pregnant women this number was seen to be the lowest
(Figure 4.4.).
    The subpopulation of Vγ9/Vδ2 lymphocytes produce significantly higher
amounts of TNF-α than Vγ1.4/Vδ1 lymphocytes, possibly because the

activation of lymphocytes via γ1.4δ1 receptors enhances the expression of PR and production of IL-10, without influencing the production of IL-12. On the contrary, the activation of lymphocytes via γ9δ2 receptors enhances the secretion of IL-12 without greatly influencing the degree of PR expression and IL-10 secretion. The ratio of Vγ1.4/Vδ1 and Vγ9/Vδ2 cells in the peripheral blood of healthy pregnant women was 1.38, whereas this ratio is much lower in pregnant women with RSA past - as low as 0.09 (Szekeres-Bartho *et al.*, 1989a, 1999) *(Figure 4.4.)*.

Progesterone has been shown to favour the development of human T lymphocytes producing Th2 cytokines, which inhibit the production of Th1 type cytokines, thus allowing the survival of fetal allograft and therefore a successful pregnancy (Piccinni *et al.*, 1995). More recently, the progesterone induced Th2 bias has been found to stimulate the release of LIF from T lymphocytes, mediated by IL-4. Clinical data showing that women with RSA past have a reduced LIF production suggest that the latter is indeed critical for the implantation and maintenance of fetus in humans (Piccinni *et al.*, 1995, 1998). Also, Piccinni *et al.* (1995) found that the high concentrations of progesterone facilitate the development of Th0 over that of Th1 like cells *in vitro*, in addition to enhancing the secretion of IL-4 and IL-5 and inducing the expression of IL-4 mRNA in the established Th1 clones. Since IL-4 is critical to the development of Th2 cells, progesterone appears to be a good candidate for their promotion.

4.4.1.1 Progesterone Induced Blocking Factor (PIBF)

The immunomodulatory effects of progesterone are mediated partly via protein factor PIBF, which is produced by the lymphocytes of pregnant women, as well as other mammalian species. PIBF is seen to exert strong immunomodulatory effects on pro-inflammatory lymphokine production in the supernatant of the culture of spleen lymphocytes. In addition, PIBF strongly stimulates the secretion of IL-3, IL-4 and IL-10, but has no influence on IFN-γ production. This might indicate the possible influence of PIBF on the switching of Th1/Th2 balance towards Th2 response (Szekeres-Bartho *et al.*, 1985, 1995, 2002; Piccinni *et al.*, 1996).

PIBF is secreted by the PR$^+$ lymphocytes previously exposed to the impact of progesterone. In addition to influencing the Th1/Th2 balance, PIBF effectively suppresses the activity of CTL and NK cells. The concentrations of PIBF in the sera of pregnant women correlate with the outcome of pregnancy. Namely, the high values of serum PIBF indicate normal pregnancy and its successful outcome, whereas the low PIBF concentrations anticipate a spontaneous abortion, pre-term delivery or the presence of fetal anomaly. It has not yet been evidenced whether progesterone is crucial and the only factor deciding the appearance of PIBF (Szekeres-Bartho *et al.*, 1990, 1995).

In conclusion, anti-abortive effects of this factor have been evidenced in humans and experimental animals. These effects of PIBF are largely based on the inhibition of NK activity, the antagonizing of Th1 activity and stimulation of Th2 secretion. The degree of activity of NK cells in the peripheral blood is in reverse correlation with the level of serum PIBF. The mechanism for suppressing NK cells by PIBF is probably based on the inhibition of the production and secretion of Th1 cytokines, especially IL-12 (Szekeres-Bartho *et al.*, 1990, 1995).

4.4.2   Estrogens

Estrogens, estrone (E1), estradiol (E2) and estriol (E3), like other steroids are capable of modulating cytokine synthesis (Correale *et al.*, 1998). Estrogens support Th2 like activities at the high concentrations that are associated with pregnancy. This type of activity may be responsible for the changes in the immune responsiveness that occur during gestation (Wegmann *et al.*, 1993; Huber *et al.*, 1999), and may also provide an explanation for the tendency of some auto-immune disease to remit during pregnancy (Cua *et al.*, 1995; Dalal *et al.*, 1997; Stephanie *et al.*, 2001).

Estrogens and progesterone are most likely to exert a Th2 like influence only at concentrations associated with pregnancy. These doses and higher can enhance IL-4 and IL-10 secretion and inhibit TNF-β production. The lower doses of estrogens alone (<5000 pg/ml) may actually promote Th1 like activity via the enhancement of IFN-γ and TNF-α or β, in the absence of an opposing influence from IL-10, which requires the higher concentrations for the enhancement. E1 and E3 enhance the secretion of IL-10 and IFN-γ in a dose-dependent fashion, almost identical to that of E2. The effect on IL-10 was more potent than the one that occurred with IFN-β. In addition, E1 and E3, like E2, had a biphasic effect on TNF-β secretion, with low concentrations stimulatory and high doses inhibitory. None of the estrogens influenced IL-4 or TGF-β secretion (Correale *et al.*, 1998). In addition, Huber *et al.* (1999) found that testosterone treatment enhanced the Th2 cell lysis while estradiol treatment was protective. *bcl-2* expression, as an anti-apoptotic factor, increases in the cells after the estradiol treatment, but decreases after the testosterone treatment. Hormone induced changes in *bcl-2* expression can possibly explain the selective survival of Th2 cells in pregnancy.

4.5   IMMUNOMODULATORY PROTEINS

TJ6 proteins have significant immunomodulatory roles in both pregnancy-induced tolerance as well as thymic tolerance (Nichols *et al.* 1994; Mandal *et al.,* 1995). Although normally expressed in certain amounts on the cells of uterine tissues, its expression is the strongest in lymph nodes draining the endometrial, i.e. decidual region of the uterus. TJ6 is a protein whose membrane form is regularly expressed on the B lymphocytes of women during a successful pregnancy. The expression of TJ6 on circulating NK cells is associated with an unfavourable outcome of pregnancy. The identification of the cells bearing TJ6 protein in early stages of pregnancy could thus serve as a good prognostic parameter of the outcome of pregnancy (Coulam *et al.,* 1995a, 1995b). The role of TJ6 protein is considered to be crucial during the implantation and early stages of pregnancy. TJ6 protein exerts its immunosuppressive effects by inducing the apoptosis of NK cells in the periimplantation region, as well as by affecting a specific form of their selection.

The regulation of the synthesis and secretion of TJ6 protein is most probably effected via progesterone. The studies concerning the levels of TJ6 protein in patients with insufficient corpus luteum showed the connection between the low levels of TJ6 proteins and low serum values of progesterone. Treating such patients with progesterone would simultaneously cause TJ6 protein to increase. Being biologically highly active molecule, TJ6 protein is thought by some authors to be an unclassified cytokine (Nichols *et al.* 1994; Mandal *et al.,* 1995; Coulam *et al.,* 1995a, 1995b; Beaman *et al.,* 1996).

Pregnancy specific glycoprotein-18 (PSG-18), represents one of the most significant immunomodulatory proteins whose immunomodulatory role has been well explored. This glycoprotein is secreted by trophoblast cells which stimulate IL-10 secretion by the decidual immune cells. For instance, mononuclear cells secrete the high amounts of IL-10 in the presence of PSG-18. Under normal circumstances, the stimulation of mononuclear cells with LPS results in these cells' activation, reflected in the significant secretion of Th1 cytokines. On the contrary, the stimulation of mononuclear cells with LPS in the presence of PSG-18 leads to the multiple enhancement of IL-10 secretion, while the production of Th1 cytokine is adjusted on the basal level. Other immunocompetent cells commonly reacting to antigen stimulation by producing IL-1β, TNF-α and IL-12, but in the presence of PSG-18 the stimulated cells secrete only basal values of these cytokines (Wessells *et al.,* 2000).

4.6   ANTI-TROPHOBLAST IMMUNITY FAILURE BY APOPTOSIS AND/OR ANERGY INDUCTION

Apoptosis is a continuous physiological process associated with embryonic development. This is a highly regulated process induced by specific stimuli. The *Fas-FasL* system is one of the major pathways for the induction of apoptosis in cells and tissues. CD95 *(Fas)* is a type I membrane protein of 45-kDa that belongs to the TNF receptor family of proteins. *FasL*, a type II membrane protein of 37-kDa, belongs to the TNF and CD40 ligand family of proteins. *Fas* is widely expressed in many tissue types; in T and B cells, it is present constitutively in low levels on the surface of resting cells and its expression is enhanced following the lymphocyte activation. In contrast to *Fas*, the expression of *FasL* is reported to be more restricted and often requires the cell activation (Cascino *et al.,* 1996; Nagata, 1997).

Fas are constitutively expressed by a number of tissues such as thymus, liver, heart, and kidney (Nagata, 1997). Also, this molecule is expressed as an inducible protein on the activated T lymphocytes and NK cells, where it participates in the programmed elimination of activated T cells after strong immune challenges. *FasL* is also constitutively expressed by specialized epithelial or endothelial cells associated with the sites of T cell selection like the thymus or peripheral elimination like the eye, inflammatory focuses, testis, brain, decidua and placenta. From these sites of expression, *FasL* is thought to trigger the apoptosis of homing T lymphocytes, thereby enforcing the immune privilege and mediating immune tolerance (Griffith *et al.,* 1995; Streilein, 1996; Bubanovic, 2003a).

There are two main areas of contact between the placenta and maternal tissues: (i) a large surface area formed by the syncytiotrophoblast of the chorionic villi contacting the maternal blood and (ii) the extravillous trophoblast within the decidua (consisting mostly of cytotrophoblast with some syncytial elements) mingling directly with the maternal immune cells. At the trophoblast-blood interface in the first trimester placenta, *FasL* are localized to the cytotrophoblast and the Hoffbauer cells in the mesenchymal core of the villi. At the trophoblast-decidual interface, these molecules are localized to the extravillous trophoblast, while the highest concentration of *FasL* is localized to the invasive trophoblast surrounded by decidual cells and in many cases in immune cells.

Both *Fas* and *FasL* are expressed in the villous placenta. *FasL* expression, initially in the decidua and later in the exchange trophoblast and fetal stroma, has been suggested to limit migration, initially of fetal cells into maternal tissue, then later of maternal cells into fetal tissue. *Fas* is expressed in chorionic trophoblasts and selectively in villous stromal and endothelial cells but has not yet been identified in villous trophoblasts (Runic *et al.,* 1996; Salafia *et al.,* 1996). Payne *et al.* (1999) showed *Fas* expression on the pure populations of term villous cytotrophoblasts, and thus the potential for juxtacrine killing via trophoblast-expressed *Fas* and *FasL*.

Fas-FasL interactions have been implicated as a mechanism for (i) clonal deletion following *in vivo* exposure to super-antigen, peptides and as mechanism of thymic (central) tolerance, (ii) controlling T cell expansion during immune responses and (iii) killing by CTL and NK cells. Studies using either T cell hybridomas or peripheral T cells have demonstrated that TCR induced apoptosis of activated T cells occurs via *Fas-FasL*-mediated autocrine or paracrine suicide pathway. This *Fas-FasL* induced apoptosis plays a central role in both the clonal deletion of autoreactive T cells in the peripheral lymphoid organs and in the elimination of activated T cells following an immune response (Griffith *et al.,* 1995; Streilein, 1996; Nagata, 1997; Bubanovic, 2003a).

There is some evidence that the regulation of apoptosis may be important during the implantation and early embryo development. It has been recently demonstrated that the apoptosis in the endometrial glands can serve as a marker of the receptive endometrium for the implantation. Von Rango *et al.* (1998) detected the apoptosis in the glandular epithelium of the basalis at the beginning of the implantation window which extended to the functionalis in the luteal phase. The proliferation and *bcl-2* expression, which are predominant in the glandular compartment during the proliferative phase, are limited to the stromal compartment during the luteal phase of the menstrual cycle (von Rango *et al.,* 1998; Yamashita *et al.,* 1999). As during the first trimester of pregnancy lymphoid aggregates of large granular lymphocytes CD56[+] NK cells are found in the basal layer of the decidua, authors conclude that apoptosis of these lymphocytes may be the first observation of an immunological preparation of endometrium for the successful implantation. Moreover, the apoptosis may be related to the loss of the protective effect of *bcl-2* which is accompanied by an increased expression of *bax* protein (Akcali *et al.,* 1996; von Rango *et al.,* 1998). However, a decreased expression of *bcl-2* and increased expression of *bax* in the decidua characterizes failing first-trimester pregnancies (Lea *et al.,* 1999).

Data of Yui *et al.* (1994) suggest that a physiological role of TNF-α and of INF-γ expression in the placental villi may be to induce the apoptotic death of cytotrophoblast cells. Both TNF-α receptors localize to the villous trophoblast, however the apoptotic death of primary cytotrophoblast is mediated almost by TNF-α receptor p55 (TNFR-1) and TNF-α receptor p75 (TNFR-2) appears to have a little effect on this process (Yui *et al.,* 1994).

EGF inhibits TNF-α or INF-γ induced apoptosis of primary human trophoblast (Garcia-Lloret *et al.,* 1996). The apoptotic cascade seems to be initiated in the villous cytotrophoblast which in turn promotes syncytial fusion (Huppertz *et al.* 1998) Moreover, data of Ho *et al.* (1999) show that TNF-α and INF-γ stimulate cytotrophoblast apoptosis within those placental villi expressing the low levels of *bcl-2* protein. It has been recently demonstrated that *bcl-2* is present on CD56$^+$ NK cells in the decidual stroma, the decidual glandular epithelium, and in the syncytiotrophoblast of the chorionic villi (Lea *et al.,* 1997).

Chan *et al.* (1999) show the apoptosis in clusters mainly confined to the non-dividing syncytiotrophoblast whereas the proliferative activity is limited to cytotrophoblast and to the stromal cells. Apoptotic DNA-fragmentation has been demonstrated in cytotrophoblast, being most abundant in early placenta. In contrast, *bcl-2* protein expression has been found in syncytiotrophoblast, being less abundant in early placenta. These data indicate that early placenta is characterized by the highly proliferative activity of cytotrophoblast cells associated with the increased occurrence of apoptosis (Mochizuki *et al.,* 1998). Therefore, *bcl-2* may prevent the apoptosis in syncytiotrophoblast (Quenby *et al.,* 1998; Toki *et al.,* 1999).

In contrast to normal pregnancy, apoptotic cells are predominant in the syncytiotrophoblast layer in the cases of spontaneous abortion (Kokawa *et al.,* 1998). Moreover, *bcl-2* expression is consistently lower in syncytiotrophoblast from women undergoing sporadic or recurrent spontaneous abortions (Lea *et al.,* 1997). Placental apoptosis increases as pregnancy progresses that suggest that it is a normal physiological phenomenon throughout gestation (Smith *et al.,* 1997). These findings provide a potential explanation for villous remodelling during placentogenesis. Therefore, the apoptosis seems to participate in the regulation of the extravillous trophoblast invasion.

Several studies indicate that inner cytotrophoblast and outer syncytiotrophoblast layers of anchoring and floating villi of the first trimester express *FasL*. The expression of *FasL* by human trophoblast has been proposed as a mechanism for providing protection against the lytic action of decidual immune cells (Runic *et al.,* 1996). *FasL* is positioned to prevent the exchange of activated immune cells between the mother and the fetus during pregnancy in mice. *FasL* is appropriately positioned first in the uterus and then in the placenta to deter trafficking of activated *Fas*$^+$ immune cells at the maternal-fetal interface (Runic *et al.,* 1996). PHA and IL-2 activated peripheral blood lymphocytes co-cultured with trophoblast undergo apoptosis (Kauma *et al.,* 1999). According to Coumans *et al.* (1999) the frequency of cell death in the peripheral blood CD3$^+$ cell population is higher when the lymphoid cells were co-cultured with the trophoblast expressing *FasL* then when they were cultured alone.

Taken together, these findings suggest that *FasL* expressed by fetal trophoblast cells can induce the apoptosis in activated lymphocytes and provides a mechanism for maternal immune tolerance to the fetus. In this sense, the reduction of placental *Fas* or *FasL* function may be associated with pregnancy loss.

4.6.1   Extrathymic Lymphocyte Maturation (eTLM) and Selection

The decidua is one of the major sites of eTLM. This site of eTLM is temporary and poorly anatomically defined, so that any similarity and/or differences between thymic and decidual eTLM are difficult to demonstrate. However, this pathway of lymphocyte maturation probably have important role in the immune mechanisms of immunomodulation in pregnancy (Bubanovic, 2003a). Pregnancy is the condition of Th2 immunity domination and toleration to proliferative trophoblast. The shift in domination of Th2 cells subpopulation and Th2 immunity happens in the second phase of the menstrual cycle and in early pregnancy. These events are the results of immunosuppressive and immunomodulatory factors activity in pregnancy (Wegmann *et al.*, 1993). However, some authors share the opinion that the shift within endometrial and decidual lymphocyte subpopulations can be the result of eTLM (Abo, 2001).

The majority of conditions with thymic atrophy like pregnancy and malignant tumors are associated with eTLM activation, but the direction of eTLM probably depends on other factors and microenvironmental conditions. Some of the factors that might be polarizing eTLM pathway probably are microbes, costimulatory molecules, *FasL*, prostaglandine, sex hormones, HSP and the level of MHC expression. These factors can probably determine the reactivity of extrathymic derived lymphocytes in the (self)antigen-reactive or antigen-protective way. Some microbes are known to be the promoters of Th1 immunity and might be the factors of eTLM polarization, in the (self)antigen-reactive manner of immune reaction. Concerning this model, (self)antigen-reactive manner of eTLM possibly includes a positive selection of (self)antigen-reactive thymus independent lymphocytes, the activation of thymus dependent lymphocytes like CTL and Th1 cells, the suppression and/or negative selection of Th2 and suppressor cells (Abo, 2001; Bubanovic, 2003a). Oppositely, the domination of Th2 cytokines and other immunosuppressive and/or immunomodulatory factors leads to the development of antigen-protective manner of eTLM. This mechanism of eTLM probably includes the negative selection of antigen-reactive thymus independent lymphocytes, the suppression of thymus dependent lymphocytes like CTL and Th1 cells, the suppression of NK activity, the activation and/or positive selection of Th2 antigen-reactive and suppressor cells (Abo, 2001; Bubanovic, 2003a).

Based on the described models of eTLM, the participation of extrathymic derived lymphocytes in phenomena like auto-immunity, tumor, trophoblast and transplant rejections or escape is probably considerable. In this case, auto-immunity may be the result of thymic involutive factors activity that coexists with the activation of eTLM and microbes, especially viruses. Under these conditions, the focuses of eTLM generate self-reactive thymus independent lymphocytes, probably by positive selection mechanisms. At the same time, the activation of thymus dependent CTL, Th1 and NK cells happens. Since pregnancy and malignancy are associated with the thymic involution, similar processes may be included in the mechanisms of miscarriages and tumor rejection. Unlike these, the mechanisms of tumor escape as well as trophoblast immunosurveillance in successful pregnancy probably are based on the thymic involution, activation of eTLM and interaction with many suppressive/modulatory factors. Under these conditions, the focuses of eTLM generate suppressor lymphocytes, while thymus independent antigen-reactive clones die by apoptosis, probably by the negative selection mechanisms. The inhibition of thymus dependent CTL, Th1 and NK cells also takes place. In addition, the ability of eTLM to participate in the activation and/or positive selection of Th2 and Th3 cells in the mechanisms of trophoblast immunosurveillance and tumor escape occurs as well (Abo, 2001; Bubanovic 2003a).

4.6.2   Failure of Blood-thymus Barrier

The characteristics of both pregnancy and tumor development are the domination of Th2 immunity, lymphocyte clonal anergy and immunotolerance of proliferative tissue (Wegmann *et al.*, 1993). Although tumor and trophoblast immunity failure are mostly the result of local events in tumor and decidual-trophoblast microenvironments, central thymic tolerance in these processes is not excluded (Kim *et al.*, 1995; Bubanovic, 2003b, 2003c). Pregnancy, as well as tumors, is associated with an involution of the thymus accompanied by a massive depletion of the cortical region and alteration in the distribution of thymocytes, with a decrease in $CD4^+CD8^+$, $CD4^+CD8^-$ and $CD4^-CD8^+$ thymocytes. However, $CD4^-CD8^-$ population shows an increase, suggesting impairment in thymocytes differentiation at an early T cell maturation stage (Shanker *et al.*, 2000; Kendall *et al.*, 2000).

The thymus changes dramatically during pregnancy. It shrinks in size, and the cortex is extensively reduced from mid-pregnancy onwards. Other changes associated with pregnancy involve the medullary epithelial cells that undergo an increased level of mitosis. Cell movement through blood vessel walls was clearly observed in mid-pregnancy, but not at other times (Kendall *et al.*, 2000). *In vitro*, trophoblast cells inhibited almost completely the proliferative response of thymocytes whether or not combined with the thymic stroma. Decidual cells were also found to have an inhibitory effect on the thymocytes proliferation while their combination with the thymic stroma decreased the thymocytes proliferation rate almost completely (Savion *et al.*, 1995). Whilst the cortex shrinks, the medulla enlarges and rearranges to create a microenvironment containing the increased numbers of mature thymocytes. Clarke *et al.* (1989) suggests that these recently derived T cells may contribute to the unique populations of cells with suppressive function that appear during pregnancy, and thereby contribute to the immune suppression of the mother to paternal and fetal antigens. In addition, the pregnancy-associated cortical involution of the thymus may reflect the deletion of clones with potential reactivity to paternal and/or fetal antigens.

Pregnancy as involutive factor probably increases the blood-thymus barrier permeability so that it can contribute in the development of central and peripheral tolerance. Many cytokines, oncofetal antigens, soluble receptors, prostaglandine, progesterone, estrogens, trophoblast and other factors which characterize pregnancy, probably can impair the permeability of blood-thymus barrier. As a result of these events, circulating trophoblast-specific or paternal antigens can permeate blood-thymus barrier and participate in the development of acquired thymic tolerance. In view of the tolerance as the phenomenon of selection and anergy of thymocytes, the failure of blood-thymus barrier as a cause of thymic tolerance in mechanisms of trophoblast escape is possible.

4.7 EVOLUTIONARY SOLUTIONS OF VIVIPARITY IN MAMMALS

As is evident from the burgeoning field of reproductive immunology, the intolerance of the feto-placental unit can represent a significant barrier to reproduction even within an apparently genetically cohesive species. In humans, for example, it is estimated that 70% of conceptions fail. Similar estimates of early fetal loss, ranging from 10% to 60%, have been obtained for other mammalian species. In human couples experiencing primary recurrent spontaneous abortion, approximately 45% of cases have been linked to immunological factors (Hill, 1995; Clark, 1999). Paradoxically, matching of MHC genotypes between sexes results in the increased rates of spontaneous abortion in humans (Ober *et al.*, 1998) and other primates (Knapp *et al.*, 1996).

At the interspecies level, immune factors can markedly improve the female's ability to successfully tolerate feto-placental unit in certain extraspecies pregnancies. For example, the pregnancy success rate of donkey embryos gestated in female horses is significantly enhanced by the infusion of serum from mares carrying normal intraspecies horse pregnancies, or the immunization of recipient mares with donkey peripheral blood lymphocytes (Allen *et al.*, 1997). The offspring from interspecies mammalian pregnancies is more an exception than a rule, being mainly possible in equine, cattle and cats. The selective pressure of immune mechanisms in pregnancy could, therefore, be primarily regarded as a barrier to interspecies fertilization. Contrary to this, the evidence about higher incidence of spontaneous abortions in couples with the same or similar MHC genotypes (syngeneic animals) bears out the assumption that the immune mechanisms in pregnancy exert the negative selective pressure on MHC homozygous offspring.

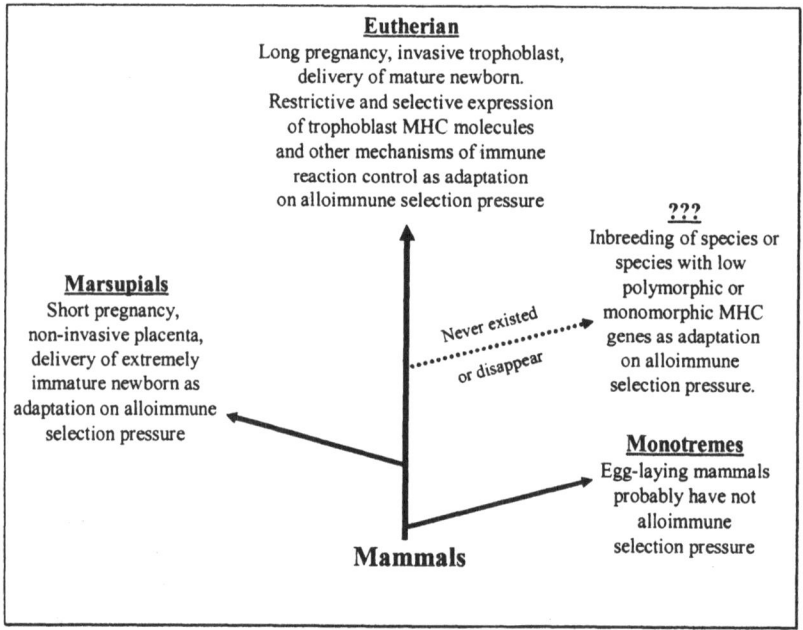

Figure 4.5. Different evolutionary solutions of viviparity in mammals.

Comparative studies carried out on pregnancies of syngeneic and allogeneic mouse and rat strains have shown that allogeneic pregnancies gave more numerous offspring compared with syngeneic pregnancies. In addition, the average weights of fetuses and placentas were higher in allogeneic than syngeneic pregnancies. However, after the stimulation of the maternal immune system with *indomethacine*, the percentage of fetuses in resorption was significantly higher in allogeneic than syngeneic pregnancies. These data could serve to support the hypothesis that the normal immune mechanisms in pregnancy favour MHC heterozygosity. At the same time, the stimulation of the maternal immune system by accelerating Th1 immune response could represent a strong negative pressure on MHC heterozygous offspring (Lala *et al.,* 1988; 1990; Pop Trajkovic *et al.,* 2001).

Considering the assumption that the joint action of alloimmunity and reproductive efficacy can represent, in viviparous vertebrates, a selective pressure, the nature of adaptation mechanisms as a response to selective pressure of alloimmunity could be, thus, one of the crucial issues regarding the origin of many control mechanisms of the immune reaction.

The class of mammals is comprised of three groups, but only two of which are viviparous mammals. Namely, monotremes are the mammals which reproduce themselves by laying eggs, while marsupials and eutherians fall in the group of viviparous mammals. Marsupials are placental mammals characterized by very short pregnancy and primitive placenta, while eutherians are the most numerous and successful group of mammals distinguished by long pregnancy and advanced placenta.

In theory, several adaptation mechanisms could have been involved in the prevention of conflicts between feto-placental unit and maternal immune system in the early evolution of mammals.

1. One of the adaptation mechanisms may be concerned with the duration of pregnancy. The short duration of pregnancy, as in the case of marsupials, could represent the adaptation mechanism against the selective pressure of alloimmunity. Marsupials have short pregnancies despite the fact that they reject the allotransplant slower than eutherian (Stone *et al.* 1997), have less variable MHC genes and develop superficial, almost non-invasive placenta (Mc Kenzie *et al.,* 1997).

2. Other adaptation mechanisms could be based on a decrease or complete absence of MHC variability or inbreeding of species. This could be similar to the process whereby syngeneic strains of experimental animals are obtained. As it requires a high level of precision in inbreeding of interrelated individuals across a large number of generations, the possibility that something similar might have happened spontaneously in natural circumstances, is almost none. In addition, syngeneic experimental animals only have the same alleles of the MHC genes, which do not differ from the MHC genes of other representatives of the species regarding the level of variability. Therefore, "inbreeding of species with monomorphic MHC genes" in natural conditions may differ essentially from the inbreeding of experimental animals. Apart from this

assumption, there are no data about the existence of a group of mammals with monomorphic or low polymorphic MHC molecules. Even if such group of mammals with monomorphic MHC genes did emerge during the evolution, it probably disappeared under a strong pressure of microbes and less effective adoptive immune system.

3. Finally, the adaptation mechanisms found in the most numerous group of mammals, eutherian, base themselves on the advanced mechanisms controlling the immune reaction, such as Th1/Th2 balance, effective immunomodulatory/suppressive decidual cytokine network, the loss of MHC class Ia and II expression on placental tissues, and the expression of monomorphic or low polymorphic of class Ib molecules. These and other adaptation mechanisms for the control of immune reaction probably enabled a "more complete" connection between feto-placental unit and maternal systems, leading consequently to a longer duration of pregnancy and higher invasiveness of the trophoblast, as well as delivering more mature offspring compared to the marsupials.

REFERENCES

Abo T. (2001). Extrathymic pathways of T-cell differentiation and immunomodulation. *Int Immunopharmacol.* 7:1261-1273.

Adkins B., Charyulu V., Sun Q.L. *et al.* (2000). Early block in maturation is associated with thymic involution in mammary tumor-bearing mice. *J Immunol.* 164:5635-5640.

Agarwal R., Loganath A., Roy A.C. *et al.* (2001). Expression profiles of IL-15 in early and late gestational human placenta and in pre-eclamptic *Placenta. Mol Hum Reprod.* 7:97-101.

Agarwal R., Loganath A., Roy A.C. *et al.* (2000). Increased expression of IL-6 in term compared to the first trimester human placental villi. *Horm Metab Res.* 32:164-168.

Akcali K.C., Khan S.A., Moulton B.C. (1996). Effect of decidualization on the expression of bax and bcl-2 in the rat uterine endometrium. *Endocrinol.* 137:3123-131.

Allen W.R., Short R.V. (1997). Interspecific and extraspecific pregnancies in equids: anything goes. *J Hered.* 88:384-392.

Amiot L., Onno M., Drénou B. *et al.* (1998). HLA-G class I expression in normal and malignant hematopoietic cells. Hum. *Immunol.* 59:524-528.

Arck P.C. (1997). Soluble receptors neutralizing TNF-α and IL-1 block stress-triggered murine abortion. *Am J Reprod Immunol.* 37:262-266.

Armitage R.J., Macduff B.M., Eisenman J. *et al.* (1995). IL-15 has stimulatory activity for the induction of B cell proliferation and differentiation. *J Immunol.* 154:483-490.

Baines M.G., Duclos A.J., Antecka E. *et al.* (1997). Decidual infiltration and activation of macrophages leads to early embryo loss. *Am J Reprod Immunol.* 37:471-477.

Barlow P., Owen D.A., Graham C. (1972). DNA synthesis in the preimplantation mouse embryo. *J Embryol Exp Morphol.* 27:431-445.

Batuman O.A., Ferrero A.P., Diaz A. *et al.* (1991). Regulation of TGF-β_1 gene expression by glucocorticoids in normal human T lymphocytes. *J Clin Invest.* 88:1574-1579.

Beaman K., Angkachatchai V., Gilman-Sachs A. (1996). TJ6:The Pregnancy-Associated Cytokine. *Am J Reprod Immunol.* 35:338-341.

Beutler B.A. (1999). The role of TNF in health and disease. *J Rheumatol.* 57:16-21.

Blaschitz A., Lenfant F., Mallet V. *et al.* (1997). Endothelial cells in chorionic fetal vessels of first trimester placenta express HLA-G. *Eur J Immunol.* 27:3380-3385.

Bluestone J.A., Cron R.Q., Barrett T.A. *et al.* (1991). Repertoire development and ligand specificity of murine TCR γδ cells. *Immunol Rev.* 120:5-20.

Bogdan C., Nathan C. *et al.* (1993). Modulation of macrophage function by TGF-β, IL-4 and IL-10. *Ann NY Acad Sci.* 685:713-720.

Born W., Happ M.P., Dallas A. *et al.* (1990). Recognition of heat shock proteins and γδ cell function. *Immunol Today.* 11:40-43.

Brannstorm M., Wang L., Norman R.J. (1993). Effect of cytocines on prostaglandin production and steroidogenesis of incubated preovulatory follicles of the rat. *Biol Reprod.* 48:165-171.

Brenner C.A., Adler R.R., Rappolee D.A. *et al.* (1989). Genes for extracellular-matrix-degrading metalloproteinases and their inhibitor, TIMP, are expressed during early mammalian development. *Genes Dev.* 3:848-859.

Bubanovic I. (2003a). Crossroads of extrathymic lymphocytes maturation pathways. *Med Hypotheses.* 61:235-239.

Bubanovic I. (2003b). Induction of thymic tolerance as possibility in prevention of recurrent spontaneous abortion. *Med Hypotheses.* 60:520-524.

Bubanovic I. (2003c). Failure of blood-thymus barrier as a mechanism of tumor and trophoblast escape. *Med Hypotheses.* 60:315-320.

Bubanovic I. (2004). 1α,25-dihydroxy-vitamin-D3 As New Immunotherapy in Treatment of Recurrent Spontaneous Abortion. *Med Hypotheses.* In press.

Bulmer J.N., Smith J., Morrison L., *et al.* (1988). Maternal and fetal cellular relationship in the human placental basal plate. *Placenta.* 9:237-241.

Carp H.J., Serr D.M., Mashiach S. *et al.* (1984). Influence of insemination on the implantation of transferred rat blastocyst. *Gynecol Obstet Invest.* 18:194-198.

Carson D.D., Tang J.P., Julian J. (1993). Heparan sulfate proteoglycan (perlecan) expression by mouse embryos during acquisition of attachment competence. *Dev Biol.* 155:97-106.

Cascino I., Papoff G., De Maria R. *et al.* (1996). Fas/Apo-1 (CD95) receptor lacking the intracytoplasmic signaling domain protects tumor cells from fas-mediated apoptosis. *J Immunol.* 156:13-17.

Chan C.C., Lao T.T., Cheung A.N. (1999). Apoptotic and proliferative activities in first trimester placentae. *Placenta.* 20:223-227.

Chang J.T. *et al.* (2000). The costimulatory effect of IL-18 on the induction of antigen-specific IFN-γ production by resting T cells is IL-12 dependent and is mediated by up-regulation of the IL-12 receptor β2 subunit. *Eur J Immunol.* 30:1113-1119.

Chaouat G. (1994). Synergy of lipopolysaccharide and inflammatory cytokines in murine pregnancy: alloimmunization prevents abortion but does not affect the induction of preterm delivery. *Cell Immunol.* 157:328-340.

Chaouat G., Assal Meliani A., Martal J. (1995). IL-10 prevents naturally occurring fetal loss in the CBAxDBA/2 mating combination, and local defect in IL-10 production in this abortion-prone combination is corrected by in vivo injection of IFNTau. *J Immunol.* 154:4261-4266.

Chaouat G., Tranchot Diallo J., Volumenie L. *et al.* (1997). Immune suppression and Th1/Th2 balance in pregnancy revisited: a (very) personal tribute to Tom Wegmann. *Am J Reprod Immunol.* 37:427-434.

Chien Y.H., Jores R., Crowley M.P. (1996). Recognition by T cells. *Annu Rev Immunol.* 14:511-517.

Choi B.C., Polgar K., Xiao L. *et al.* (2000). Progesterone inhibits *in vitro* embryotoxic Th1 cytokine production to trophoblast in women with recurrent pregnancy loss. *Hum Reprod.* 1:46-59.

Clark D.A., Drake B., Head J.R. et al (1990). Decidua-associated suppressor activity and viability of the individual implantation sites of allopregnant C3H mice. *J Reprod Immunol.* 17:253-264.

Clark D.A., Lea R.G., Dehburg J. *et al.* (1991). TGF-β related suppressor factor in mammalian pregnancy decidua; homologies between the mouse and human in successful pregnancy and in recurrent unexplained abortion Clloque. *INSERM.* 212:171-179.

Clark D.A. *et al.* (1992). Role of unique species TGF-β in preventing rejection of the conceptus during pregnancy. In: Gergely J. *et al.* Progress in immunology VII. Budapest: Springer-Verlag. pp.841-852.

Clark D.A. (1993). Cytokines, decidua and early pregnancy. *Oxf Rev Reprod Biol.* 15:83-111.

Clark D.A., Vince G., Flanders K.C. *et al.* (1994). CD56+ lymphoid cells in human first trimester pregnancy decidua as a source of novel TGFβ 2-related immunosuppressive factors. *Hum Reprod.* 9:22070-2277.

Clark D.A., Merali F.S., Hoskin D.W. et al (1997). Decidua associated suppressor cells in abortion-Pron DBA/mated CBA/2mated CBA/J mice that release bioactive TGFβ2-related immunosuppressive molecules express a bone marrow-derived natural suppressor cell marker and γδ T-cell receptor. *Biol Reprod.* 56:1351-1360.

Clark D.A. (1999) Signaling at the fetomaternal interface. *Am J Reprod Immunol.* 41:169-173.

Clarke A.G., Kendall M.D. (1989). Histological changes in the thymus during mouse pregnancy. *Thymus.* 14:65-78.

Clover L.M., Sargent I.L., Townsend A. *et al.* (1995). Expression of TAP1 by human trophoblast. *Eur J Immunol.* 25:543-553.

Copeman J., Robin N.N., Caniggia I.H. *et al.* (2000). Posttranscriptional Regulation of Human Leukocyte Antigen G During Human Extravillous Cytotrophoblast Differentiation. *Biol Reprod.* 62:1543-1550.

Correale J., Arias M., Gilmore W. (1998). Steroid Hormone Regulation of Cytokine Secretion by Proteolipid Protein-Specific CD4+ T Cell Clones Isolated from Multiple Sclerosis Patients and Normal Control Subjects1. *J Immunol.* 161:3365-3374.

Coulam C.B., Beaman K.D. (1995a). Reciprocal alteration in circulating TJ6+ CD19+ and TJ6+ CD56+ leukocytes in early pregnancy predicts success or miscarriage. *Am J Reprod Immunol.* 34:219-224.

Coulam C.B., Stern J. (1995b). Effect of seminal plasma on implantation rates. *Early Pregnancy.* 1:33-36.

Coumans B., Thellin O., Zorzi W. *et al.* (1999). Lymphoid cell apoptosis induced by trophoblastic cells: a model of active foeto-placental tolerance. *J Immunol Methods.* 224:185-196.

Crisa L., McMaster M.T., Ishii J.K. *et al.* (1997). Identification of a thymic epithelial cell subset sharing expression of the class Ib HLA-G molecule with fetal trophoblasts. *J Exp Med.* 186:289-293.

Croy B.A. *et al.* (1995). Characterization of murine decidual natural (NK) cells and their relevance to the success of pregnancy. *Cell Immunol.* 93:315-326.

Croy B.A., Guimond M.J., Luross J. *et al.* (1997). Uterine natural killer cells do not require IL-2 for their differentiation or maturation. *Am J Reprod Immunol.* 37:463-470.

Cua D.J., Hinton D.R, Stohlman S.A. (1995). Self-antigen-induced Th2 responses in experimental allergic encephalomyelitis (EAE)-resistant mice: Th2-mediated suppression of auto-immune disease. *J Immunol.* 155:4052-4057.

Dalal M., Kim S., Voskuhl R.R. (1997). Testosterone therapy ameliorates experimental auto-immune encephalomyelitis and induces a T helper 2 bias in the autoantigen-specific T lymphocyte response. *J Immunol.* 159:3-7.

Davis M.M., Chien Y. (1995). Issues concerning the nature of antigen recognition by α beta and γδ T-cell receptors. *Immunol Today.* 16:316-318.

Daynes R.A., Araneo B.A., Hennebold J. *et al.,* (1995). Steroids as regulators of the mammalian immune response. *J Invest Dermatol.* 105:14S.

de Moraes Pinto M.I., Vince G.S. Flanagan B.F. *et al.* (1997). Localization of IL-4 and IL-4 receptors in the human term placenta, decidua and amniochorionic membranes. *Immunol.* 90:87-91.

Denison F.C., Kelly R.W., Calder A.A. *et al.* (1998). Cytokine secretion by human fetal membranes, decidua and placenta at term. *Hum Reprod.* 13:3560-3565.

Ditzian-Kadanoff R., Garon G., Verp M.S. *et al.* (1993). γδT cells in human decidua. *Am J Obstet Gynecol.* 168:831-836.

D'Orazio T.J., Niederkom J.Y. (1998). A novel role for TGF-beta and IL-10 in the induction of immune privilege. J Immunol. 160:2089-2098.

Eberl M. *et al.* (2000). IL-18 potentiates the adjuvant properties of IL-12 in the induction of a strong Th1 type immune response against a recombinant antigen. *Vaccine.* 18:2002-2008.

Ellis S.A., Palmer M.S, McMichael A.J. *et al.* (1990). Human trophoblast and the choriocarcinoma cell line BeWo express a truncated HLA class I molecule. *J Immunol.* 144:731-736.

Fazleabas A.T., Strakova Z. (2002). Endometrial function: cell specific changes in the uterine environment. *Mol Cell Endocrinol.* 186:143-147.

Fein A., Magid N., Savion S. *et al.* (2002). Diabetes teratogenicity in mice is accompanied with distorted expression of TGF-β2 in the uterus. *Teratog Carcinog Mutagen.* 22:59-71.

Ferry B.L., Starkey P.M., Sargent I.L. *et al.* (1990). Cell populations in the human early pregnancy decidua: natural killer activity and response to IL-2 of CD56-positive large granular lymphocytes. *Immunol.* 70:446-452.

Fisher S.J., Cui T.Y., Zhang L. *et al.* (1989). Adhesive and degradative properties of human placental cytotrophoblast cells in vitro. *J Cell Biol.* 109:891-902.

Fujihashi K., Kweon M.N., Kiyono H., *et al.* (1997). A T cell/B cell/epithelial cell internet for mucosal inflammation and immunity. *Springer Semin Immunopathol.* 18:477-494.

Garcia-Lloret M.I., Yui J., Winkler-Lowen B. *et al.* (1996). Epidermal growth factor inhibits cytokine-induced apoptosis of primary human trophoblasts. *J Cell Physiol.* 167:324-332.

Geiselhart A., Dietl J., Marzusch K. *et al.* (1995). Comparative analysis of the immunophenotypes of decidual and peripheral blood large granular lymphocytes and T cells during early human pregnancy. *Am J Reprod Immunol.* 33:315-322.

Giri J.G., Ahdieh M., Eisenman J. *et al.* (1994). Utilization of the beta and γ chains of the IL-2 receptor by the novel cytokine IL-15. *EMBO J.* 13:2822-2830.

Giri J.G., Kumaki S., Ahdieh M. *et al.* (1995). Identification and cloning of a novel IL-15 binding protein that is structurally related to the α chain of the IL-2 receptor. *EMBO J.* 14:3654-3663.

Gorczynski R.M., Hadidi S., Yu G. *et al.* (2002). The same immunoregulatory molecules contribute to successful pregnancy and transplantation. *Am J Reprod Immunol.* 48:18-26.

Gorelik L., Fields E.P., Flavell A.R. *et al.* (2000). TGF-β Inhibits Th Type 2 Development Through Inhibition of GATA-3 Expression. *J Immunol.* 165:4773-4777.

Gorham D.J., Güler L.M., Fenoglio D. *et al.* (1998). Low Dose TGF-β Attenuates IL-12 Responsiveness in Murine Th Cells1. *J Immunol.* 161:1664-1670.

Grabstein K.H., Eisenman J., Shanebeck K., *et al.* (1994). Cloning of a T cell growth factor that interacts with the beta chain of the IL-2 receptor. *Science.* 264:965-268.

Griffith T.S., Brunner T., Fletcher S.M. *et al.* (1995). Fas ligand-induced apoptosis as a mechanism of immune privilege. *Science.* 270:1189-1192.

Hammer A. *et al.* (1997). Amnion epithelial cells, in contrast to trophoblast cells, express all classical HLA class I molecules together with HLA-G. *Am J Reprod Immunol.* 37:161-171.

Hayakawa S. *et al.* (2000a). Murine fetal resorption and experimental pre-eclampsia are induced by both excessive Th1 and Th2 activation. *J Reprod Immunol.* 47(2):121-138.

Hayakawa S., Karasaki-Suzuki M., Itoh T. *et al.* (2000b). Effects of paternal lymphocyte immunization on peripheral Th1/Th2 balance and TCR Vβ and Vγ repertoire usage of patients with recurrent spontaneous abortions. *Am J Reprod Immunol.* 43:107-115.

Haynes M.K., Flanagan M.T., Perussia B. *et al.* (1995). Isolation of decidual lymphocytes from chorionic villus samples: phenotypic analysis and growth in vitro. *Am J Reprod Immunol.* 33:190-199.

Hertig A.T., Rock J. (1956). A description of 34 human ova within the first 17 days of development. *Am J Anat.* 98:435-494.

Hill J.A. (1995). Immunological factors in recurrent spontaneous abortion. In: Kurpisz M, Fernandez N, ed. Immunology of human reproduction. Oxford: Bios Scientific Publishers; pp.:401-424.

Hill J.A., Polgar K., Anderson D.J. (1995). T-helper 1Type immunity to trophoblast in women with recurrent spontaneous abortion. *J Am Med Ass.* 273:1933-1938.

Ho H.N., Chao K.H., Chen C.K. *et al.* (1996). Activation status of T and NK cells in the endometrium throughout menstrual cycle and normal and abnormal early pregnancy. *Hum Immunol.* 49:130-136.

Ho S., Winkler-Lowen B., Morrish D.W. *et al.* (1999). The role of Bcl-2 expression in EGF inhibition of TNF-α/IFN-γ-induced villous trophoblast apoptosis. *Placenta.* 20:423-430.

Huber S.A., Kupperman J., Newell M.K. (1999). Estradiol prevents and testosterone promotes Fas-dependent apoptosis in CD4+ Th2 cells by altering Bcl 2 expression. *Lupus.* 8:384-387.

Hunt J.S. (1989). Cytokine networks in the uteroplacental unit: macrophages as pivotal regulatory cells. *J Reprod Immunol.* 16:1-11.

Hunt J.S. (1994) Immunologically relevant cells in the uterus. *Biol Reprod.* 50:461-470.

Hunt J.S., Robertson S.A. (1996). Uterine macrophages and environmental programming for pregnancy success. *J Reprod Immunol.* 32:1-25.

Huppertz B., Frank H.G., Kingdom J.C. *et al.* (1998). Villous cytotrophoblast regulation of the syncytial apoptotic cascade in the human *Placenta. Histochem Cell Biol.* 110:495-508.

Ida A., Tsuji Y., Muranaka J. *et al.* (2000). IL-18 in pregnancy; the elevation of IL-18 in maternal peripheral blood during labour and complicated pregnancies. *J Reprod Immunol.* 47:65-74.

Ingman V., Robertson V. (2002). Defining the actions of TGFβ in reproduction. *Bioessays.* 24: 904-914.

Ishikawa T., Uchiyama T. *et al.* (1991). IL-4 down-regulates IL-2 receptor p75 by accelerating its endocytosis. *Int Immunol.* 3:517-525.

Jacobs A.L., Carson D.D. (1993). Uterine cell secretion of IL-1α induces prostaglandin E_2 (PGE_2) and $PGF_2α$ secretion by uterine stromal cells *in vitro. Endocrinol.* 132:300-308.

Kankofer M., Wiercinski J. (1999). Prostaglandin E2 9-keto reductase from bovine term placenta. *Prost Leukot Essent Fatty Acids.* 61:29-32.

Karre K. (1995). Express yourself or die: peptides, MHC molecules, and NK cells. *Science.* 267:978-979.

Kaufmann S. (1996). γδ and other unconventional T lymphocytes: what do they see and what do they do? *Proc Natl Acad Sci USA.* 93:2272-2279.

Kauma S.W., Huff T.F., Hayes N. *et al.* (1999). Placental Fas ligand expression is a mechanism for maternal immune tolerance to the foetus. *J Clin Endocrinol Metab.* 84:2188-2194.

Kendall M.D., Clarke A.G. (2000). The thymus in the mouse changes its activity during pregnancy: a study of the microenvironment. *J Anat.* 197:393-411.

Kelly R.W. *et al.* (1997). Immunomodulation by human seminal plasma: benefit for spermatozoon and pathogen. *Hum Reprod.* 12:2200-2207.

Kim J., Modlin R.L., Moy R.L. *et al.* (1995). IL-10 production in cutaneous basal squamous cell carcinoma - A mechanism for evading the local T cell immune response. *J Immunol.* 155:2240-2247.

King A., Balendran N., Wooding P. *et al.* (1991). CD3- leukocytes present in the human uterus during early placentation: phenotypic and morphologic characterization of the CD56++ population. *Dev Immunol.* 1:169-190.

King A., Jokhi P.P., Burrows T.D. *et al.* (1996a). Functions of human decidual NK cells. *Am J Reprod Immunol.* 35:258-260.

King A., Boocock C., Sharkey A.M. *et al.* (1996b). Evidence for the expression of HLA-C class I mRNA and protein by human first trimester trophoblast. *J Immunol.* 156:2068-2076.

King A., Loke Y.W., Chaouat G. (1997). NK cells and reproduction. *Immunol Today.* 18:64-66.

King A., Burrows T., Verma S. *et al.* (1998). Human uterine lymphocytes. *Hum Reprod.* 4:480-485.

King A., Hiby S.E., Gardner L. *et al.* (2000). Recognition of trophoblast HLA class I molecules by decidual NK cell receptors-a review. *Placenta.* 21 Suppl A:S81-85.

Klonoff-Cohen H.S., Savitz D.A., Cefalo R.C. *et al.* (1989). An epidemiologic study of contraception and preeclampsia. *JAMA.* 262:3143-3147.

Knapp L.A., Ha J.C., Sackett G.P. (1996). Parental MHC antigen sharing and pregnancy wastage in captive pigtail macaques. *J Reprod Immunol.* 32:73-88.

Knofler M., Mosl B., Bauer S. *et al.* (2000). TNF-α/TNFRI in primary and immortalized first trimester cytotrophoblasts. *Placenta.* 21:525-535.

Kokawa K., Shikone T., Nakano R. (1998). Apoptosis in human chorionic villi and decidua during normal embryonic development and spontaneous abortion in the first trimester. *Placenta.* 19:21-26.

Kovalevskaya G., Birken S., Kakuma T., *et al.* (1999). Evaluation of nicked human chorionic gonadotropin content in clinical specimens by a specific immunometric assay. *Clin Chem.* 45:68-74.

Krishnamurthy P., Bird I.M., Sheppard C. *et al.* (1999). Effect of angiogenetic groeth factor on endothelium-derived prostacycline production by ovine uterine and placental arteries. *Prostaglandins Other Lipid Mediat.* 57:1-12.

Krishnan L., Guilbert L.J., Wegmann T.G. *et al.* (1996). T helper 1 response against Leishmania major in pregnant C57BL/6 mice increases implantation failure and fetal resorptions. Correlation with increased IFN- γ and TNF and reduced IL-10 production by placental cells. *J Immunol.* 156:653–662.

Laird S.M., Tuckerman E.M., Cork B.A. *et al.* (2003). A review of immune cells and molecules in women with recurrent miscarriage. *Hum Reprod Update.* 9:163-174.

Lala P.K., Kennedy T.G., Parhar S. (1988). Suppression of lymphocyte alloreactivity by early gestational human decidua. II. Characterization of suppressor mechanisms. *Cell Immunol.* 116:411-422.

Lala P.K., Scodras J.M., Graham C.H. *et al.* (1990). Activation of maternal killer cells in the pregnant uterus with chronic indomethacin therapy, IL-2 therapy, or a combination therapy is associated with embryonic demise. *Cell Immunol.* 127:368-381.

Lauritzsen G.F., Hofgaard P.O., Schenck K. *et al.* (1998). Clonal deletion of thymocytes as a tumor escape mechanism. *Int J Cancer.* 78:216-222.

Le Bouteiller P. *et al.* (2000). HLA-G in the human placenta: expression and potential functions. *Biochem Soc Trans.* 28:208-212.

Lea R.G., Al- Sharekh N., Tulppala M. *et al.* (1997). The immunolocalization of bcl-2 at the maternal-foetal interface in healthy and failing pregnancies. *Hum Reprod.* 12:153-158.

Lea R.G., Riley S.C., Antipatis C. *et al.* (1999). Cytokines and the regulation of apoptosis in reproductive tissues: a review. *Am J Reprod Immunol.* 42:100-109.

Lee N., Llano M., Carretero M. *et al.* (1998). HLA-E is a major ligand for the natural killer inhibitory receptor CD94/NKG2A. *Proc Natl Acad Sci USA.* 95:5199-5204.

Lewko W.M., Smith T.L., Bowman D.J. *et al.* (1995). IL-15 and the growth of tumor derived activated T-cells. *Cancer Biother.* 10:13-20.

Lodererova A., Honsova E., Viklicky O. (2003). Detection of HLA-G on human extravillous cytotrophoblast and skeletal muscle with a new monoclonal antibody MEM-G/1. *Folia Microbiol.* 48:239-242.

Loke Y.W., King A. (2000). Immunology of implantation. *Baillieres Best Pract Res Clin Obstet Gynaecol.* 14:827-837.

Ljunggren H.G., Kärre K. (1990). In search of the "missing self": MHC molecules and NK cell recognition. *Immunol Today.* 11:237-244.

Maeurer M.J., Lotze M.T. (1998). IL-7 knockout mice. Implications for lymphopoiesis and organ-specific immunity. *Int Rev Immunol.* 16:309-322.

Maki K., Sunaga S., Komagata Y. (1996). IL 7 receptor-deficient mice lack γδ T cells. *Proc Natl Acad Sci USA.* 93:7172-7177.

Mandal M., Nichols T.C. Beaman K.D. *et al.* (1995). Purification and characterization of a pregnancy associated protein: TJ6. *Am J Reprod Immunol.* 33:60-65.

Martin-Villa J.M., Luque I., Martinez-Quiles N. *et al.* (1996). Diploid expression of human leukocyte antigen class I and class II molecules on spermatozoa and their cyclic inverse correlation with inhibin concentration. *Biol Reprod.* 55:620-629.

Marzi M., Vigano A., Trabattoni D. *et al.* (1996). Characterization of type 1 and type 2 cytokine production profile in physiologic and pathologic human pregnancy. *Clin Exp Immunol.* 106:127-131.

Mc Kenzie M., Cooper W. (1994). MHC class II variability in a marsupial. *Reprod Fertil Dev.* 6:721-726.

Mc Master M., Zhou Y., Shorter S. *et al.* (1998). HLA-G Isotype Produced by Placental Cytotrophoblasts and Found in Amniotic Fluid Are Due to Unusual Glycosylation1. *J Immunol.* 160:5922-5928.

Miller L., Hunt J.S. (1996). Sex steroid hormones and macrophage function. *Life Sci.* 59:1-14.

Mincheva-Nilsson L., Baranov V., Yeung M.M. *et al.* (1994). Immunomorphologic studies of human decidua-associated lymphoid cells in normal early pregnancy. *J Immunol.* 152:2020-2032.

Mincheva-Nilsson L., Nagaeva O., Sundqvist K.G *et al.* (2000). γδT cells of human early pregnancy decidua: evidence for cytotoxic potency. *Internat Immunol.* 12:585-596.

Mochizuki M., Maruo T., Matsuo H. *et al.* (1998). Biology of human trophoblast. *Int J Gynaecol Obstet.* 60:21-28.

Murray F.A., Grifo A.P. Jr, Parker CF. (1983). Increased litter size in gilts by intrauterine infusion of seminal and sperm antigens before breeding. *J Anim Sci.* 56:895-900.

Nagata S. (1997). Apoptosis by death factor. *Cell.* 88:355-365.

Nakanishi K. *et al.* (2001a). IL-18 is a unique cytokine that stimulates both Th1 and Th2 responses depending on its cytokine milieu. *Cytokine Growth Factor Rev.* 12:53-72.

Nakanishi K. *et al.* (2001b). IL-18 regulates both th1 and th2 responses. *Annu Rev Immunol.* 19:423-474.

Ng S.C., Gilman-Sachs A., Thaker P. *et al.* (2002). Expression of intracellular Th1 and Th2 cytokines in women with recurrent spontaneous abortion, implantation failures after IVF/ET or normal pregnancy. *Am J Reprod Immunol.* 48:77-86.

Nichols T.C., Kang J.A., Angkachatchai V. *et al.* (1994). Expression of membrane form of the pregnancy associated protein TJ6 on lymphocytes. *Cell Immunol.* 155:219-223.

Norman S.J., Poyser N.L. (1998). Prostaglandin production by guinea-pig placenta and other uterine tissues during mid-pregnancy. *Placenta.* 19:631-641.

O W.S., Chen H.Q., Chow P.H. (1988). Effects of male accessory sex gland secretions on early embryonic development in the golden hamster. *J Reprod Fertil.* 84:341-344.

Ober C, Aldrich CL. (1997). HLA-G polymorphisms: neutral evolution or novel function? *J Reprod Immunol.* 36:1-21.

Ober C., Hyslop T., Elias S. *et al.* (1998). Human leucocyte antigen matching and fetal loss: a result of a 10-year prospective study. *Hum Reprod.* 13:33-38.

Ohno Y., Kasugai M., Kurauchi O. *et al.* (1994). Effect of IL 2 on the production of progesterone and prostaglandin E₂ in human fetal membranes and its consequences for preterm uterine contraction. *Eur J Endocrinol.* 130:478-484.

Okada S., Okada H., Sanezumi M. *et al.* (2000). Expression of IL-15 in human endometrium and decidua. *Mol Hum Reprod.* 6:75-80.

Ortega F.V., Vergara M.A. (2000). Apoptosis in trophoblast of patients with recurrent spontaneous abortion of unidentified cause. *Ginecol Obstet Mex.* 68:122-131.

Paliogianni F., Ahuja S.S., Balow J.P. *et al.* (1993). Novel mechanism for inhibition of human T cells by glucocorticoids: glucocorticoids inhibit signal transduction through IL-2 receptor. *J Immunol.* 151:4081-4085.

Pang S.F., Chow P.H., Wong T.M. (1979). The role of the seminal vesicles, coagulating glands and prostate glands on the fertility and fecundity of mice. *J Reprod Fertil.* 56:129-132.

Par G., Bartok B., Szekeres-Bartho J. (2000). Cyclooxygenase is involved in the effects of progesterone-induced blocking factor on the production of IL-12. *Am J Obstet Gynecol.* 183:126-130.

Patel F.A., Clifton V.L., Chwalisz K. *et al.* (1999). Steroid regulation of prostaglandin dehydrogenase activity and expression in human term placenta and chorio-decidua in relation to labor. *J Clin Endocrinol Metab.* 84:291-299.

Paul P., Rouas-Freiss N., Khalil-Daher I. *et al.* (1998). HLA-G expression in melanoma: A way for tumor cells to escape from immunosurveillance. *Proc Natl Acad Sci USA.* 95:4510-4515.

Payne G.S., Smith C.S., Davidge T.S. *et al.* (1999). Death Receptor Fas/Apo-1/CD95 Expressed by Human Placental Cytotrophoblasts Does Not Mediate Apoptosis1. *Biol Reprod.* 60:1144-1150.

Peitz B., Olds-Clarke P. (1986). Effects of seminal vesicle removal on fertility and uterine sperm motility in the house mouse. *Biol Reprod.* 35:608-617.

Perona R.M., Wassarman P.M. (1986). Mouse blastocyst hatch in vitro by using a trypsin-like proteinase associated with cells of mural trophectoderm. *Dev Biol.* 114:42-52.

Piccinni M.P., Giudizi M.G, Biagiotti R. *et al.* (1995). Progesterone favors the development of human T helper cells producing Th2Type cytokines and promotes both IL-4 production and membrane CD30 expression in established Th1 cell clones. *J Immunol.* 155:128-136.

Piccinni M.P., Beloni L., Livi C., *et al.* (1998). Defective production of both leukemia inhibitory factor and type 2 T-helper cytokines by decidual T cells in unexplained recurrent abortions. *Nat Med.* 4:1020-1025.

Piccinni M.P., Romagnani S. (1996). Regulation of fetal allograft survival by hormone-controlled Th1- and Th2Type cytokines. *Immunol Res.* 15:141-148.

Pijnenborg R. *et al.* (2000). Cytotoxic effects of TNF-α and IFN-γ on cultured human trophoblast are modulated by fibronectin. *Am J Reprod Immunol.* 43:107-115.

Pitzel L., Jarry H., Wuttke W. (1993). Effect and interactions of prostaglandin F$_2$α, oxytocin and cytokines on steroidogenesis of porcine luteal cells. *Endocrinol.* 132:751-756.

Plevyak M., Hanna N., Mayer S. *et al.* (2002). Deficiency of decidual IL-10 in first trimester missed abortion: a lack of correlation with the decidual immune cell profile. *Am J Reprod Immunol.* 47:242-250.

Polgar B., Barakonyi A., Xynos I. *et al.*, (1999). The role of γ/δ T cell receptor positive cells in pregnancy. *Am J Reprod Immunol.* 41:239-244.

Pop Trajkovic Z., Najman S., Kamenov B. *et al.* (2002). Effects of indometacin on allogeneic and singeneic pregnancy. *Facta Universitatis.* 9:166-170.

Porter B.O., Malek T.R. (2000). Thymic and intestinal intraepithelial T lymphocyte development are each regulated by the γc-dependent cytokines IL-2, IL-7, and IL-15. *Semin Immunol.* 12:465-474.

Quenby S., Brazeau C., Drakeley A. *et al.* (1998). Oncogene and tumour suppressor gene products during trophoblast differentiation in the first trimester. *Mol Hum Reprod.* 4:477-481.

Raghupathy R. (1997). Th1Type immunity is incompatible with successful pregnancy. *Immunol Today.* 18:478-482.

Raghupathy R., Makhseed M., Azizieh F. *et al.* (2000). Cytokine production by maternal lymphocytes during normal human pregnancy and in unexplained recurrent spontaneous abortion. *Hum Reprod.* 15:713-718.

Raghupathy R. (2001). Pregnancy: success and failure within the Th1/Th2/Th3 paradigm. *Semin Immunol.* 13:219-227.

Reid B.L. (1966). The fate of the nucleic acid of sperm phaged by regenerating cells. *Aust NZJ Obstet Gynaecol.* 6:30-34.

Robertson S.A. *et al.* (1997). Cytokine-leukocyte network and the establishment of pregnancy. Am J *Reprod Immunol.* 37:438-442.

Robertson S.A., Ingman W.V., O'Leary S. *et al.* (2002). TGFα mediator of immune deviation in seminal plasma. *J Reprod Immunol.* 57:109-128.

Robertson S.A., Bromfield J.J., Tremellen K.P. (2003). Seminal 'priming' for protection from pre-eclampsia-a unifying hypothesis. *J Reprod Immunol.* 59:253-265.

Roby K.F., Gershon D., Hunt J.S. (1996). Expression of the transporter for antigen processing-1 (Tap-1) Gene in subpopulations of human trophoblast cells. *Placenta.* 17:27-32.

Rodriguez A.M., Mallet V., Lenfant F. *et al.* (1997). IFN-γ rescues HLA class Ia cell surface expression in term villous trophoblast cells by inducing synthesis of TAP proteins. *Eur J Immunol.* 27:45-54.

Roth I., Corry D.B., Locksley R.M. *et al.* (1996). Human placental cytotrophoblasts produce the immunosuppressive cytokine IL 10. *J Exp Med.* 184:539-542.

Rukavina D., Rubesa G., Gudelj L. *et al.* (2000). Characteristics of perforin expressing lymphocytes within the first trimester decidua of human pregnancy. *Am J Reprod Immunol.* 33:394-404.

Runic R., Lockwood C.J., Ma Y. *et al.* (1996). Expression of Fas ligand by human cytotrophoblasts: implications in placentation and fetal survival. *J Clin Endocrinol Metab.* 81:3119-3122.

Saito S., Nishikawa K., Morii T. *et al.* (1994). A study of CD45RO, CD45RA and CD29 antigen expression on human decidual T cells in an early stage of pregnancy. *Immunol Lett.* 40:193-197.

Saito S. *et al.* (1996). IL-4 blocks the IL-12 induced increase in natural killer activity and DNA synthesis of decidual CD16-CD56bright+ NK cells by inhibiting expression of the IL-2Rα, β and γ. *Cell Immunol.* 170:71-77.

Saito S., Tsukaguchi N., Hasegawa T. *et al.* (1999). Distribution of Th1, Th2, and Th0 and the Th1/Th2 cell ratios in human peripheral and endometrial T cells. *Am J Reprod Immunol.* 42:240-245.

Saito S. (2000). Cytokine network at the feto-maternal interface. *J Reprod Immunol.* 47:87-103.

Saito S. (2001). Cytokine cross-talk between mother and the embryo/placenta. *J Reprod Immunol.* 52:15-33.

Saji F., Samejima Y., Kamiura S. *et al.* (2000). Cytokine production in chorioamnionitis. *J Reprod Immunol.* 47:185-196.

Salafia C.M., Mill J.F., Ossandon M. (1996). Markers of regulation of apoptosis and cell proliferation in preterm and term placental villi. *J Soc Gynecol Invest.* 3:80A.

Salazar-Onfray F., Charo J., Petersson M. *et al.* (1997). Down-regulation of the expression and function of the transporter associated with antigen processing in murine tumor cell lines expressing IL-10. *J Immunol.* 159:3195-3200.

Savion S., Toder V. (1995). Pregnancy-associated effect on mouse thymocytes in vitro. *Cell Immunol.* 162:282-287.

Savion S., Zeldich E., Orenstein H. *et al.* (2002). Cytokine expression in the uterus of mice with pregnancy loss: effect of maternal immunopotentiation with GM-CSF. *Reproduction.* 123:399-409.

Scodras J.M., Parhar R.S., Kennedy T.G. *et al.* (1990). Prostaglandin-mediated inactivation of natural killer cells in the murine decidua. *Cell Immunol.* 127:352-367.

Shanker A., Singh M., Sodhi A. (2000). Ascitic growth of a spontaneous transplantable T cell lymphoma induces thymic involution. 2. Induction of apoptosis in thymocytes. *Tumour Biol.* 21:315-327.

Shimaoka Y., Hidaka Y., Tada H. *et al.* (2000). Changes in cytokine production during and after normal pregnancy. *Am J Reprod Immunol.* 44:143-147.

Shimizu T., Kawamura T., Miyaji C. *et al.* (2000). Resistance of extrathymic T cells to stress and the role of endogenous glucocorticoids in stress associated immunosuppression. *Scand J Immunol.* 51:285-292.

Shull M.M., Ormsby I., Kier A.B, *et al.* (1992). Targeted disruption of the mouse TGF-β1 gene results in multifocal inflammatory disease. *Nature.* 359:693-695.

Smith S.C., Baker P.N., Symonds E.M. (1997). Placental apoptosis in normal human pregnancy. *Am J Obstet Gynecol.* 177:57-65.

Sprinks M.T., Sellens M.H., Dealtry G.B. *et al.* (1993). Fernandez H. Preimplantation mouse embryos express Mhc class I genes before the first cleavage division. *Immunogenetics.* 38:35-40.

Stephanie M., Liva R., Rhonda R.V. (2001). Testosterone Acts Directly on CD4+ T Lymphocytes to Increase IL-10 Production. *J Immunol.* 167:2060-2067.

Stone W.H., Manis G.S., Hoffman E.S. *et al.* (1997). Fate of allogeneic skin transplantations in a marsupial *(Monodelphis domestica). Lab Anim Sci.* 47:283-287.

Streilein J.W. (1996). Peripheral tolerance induction: lessons from immune privileged sites and tissues. *Transplant Proc.* 28:2066-2070.

Strickland J.E., Saviolakis G.A., Fowler A.K. *et al.* (1976). Impaired estrogen-mediated production of type C viral DNA polymerase in aged NH swiss mouse uteri. *Proc Soc Exp Biol Med.* 153:63-69.

Swat W., Ignatowicz L., von Boehmer H. *et al.* (1991). Clonal deletion of immature CD4+CD8+ thymocytes in suspension culture by extrathymic antigen-presenting cells. *Nature.* 351:150-153.

Szekeres-Bartho J. *et al.* (1985). Progesterone treated lymphocytes of healthy pregnant women release a factor inhibiting cytotoxicity and prostaglandin synthesis. *Am J Reprod Immunol Microbiol.* 9:15-19.

Szekeres-Bartho J. *et al.* (1989a). Lymphocytic progesterone receptors in human pregnancy. *J Reprod Immunol.* 16:239-247.

Szekeres-Bartho J. *et al.* (1989b). Progesterone receptors in lymphocyte of liver-transplanted and transfused patients. *Immunol Letters.* 22:259-261.

Szekeres-Bartho J. *et al.* (1990). The effect of a progesterone induced immunologic blocking factor on NK-mediated resorption. *Am J Reprod Immunol.* 24:273-283.

Szekeres-Bartho J. *et al.* (1995). Progesterone-induced blocking factor (PIBF) in normal and pathological pregnancy. *Am J Reprod Immunol.* 34:342-348.

Szekeres-Bartho J., Wegmann T.G. (1996). A progesterone-dependent immunomodulatory protein alters the Th1/Th2 balance. *J Reprod Immunol.* 31:81-95.

Szekeres-Bartho J., Barakonyi A., Polgar B. *et al.* (1999a). The role of γ/δ T cells in progesterone-mediated immunomodulation during pregnancy: a review. *Am J Reprod Immunol.* 42:44-48.

Szekeres-Bartho J. *et al.* (1999b). Nonspecific Immunological mechanisms and hormones. In: Gupta S.K. Reproductive Immunology. Kluwer&Norsa Publishing House, Delhi pp.:218-224.

Szekeres-Bartho J. (2002). Immunological relationship between the mother and the fetus. *Int Rev Immunol.* 21:471-495.

Taga K., Mostowski H., Tosato G. (1993). Human interleukin-10 can directly inhibit T-cell growth. *Blood.* 81:2964-2999.

Tartakovsky B., Ben-Yair E. (1991). Cytokines modulate preimplantation development and pregnancy. *Dev Biol.* 146:345-352.

Tessier-Prigent A., Willems R., Lagarde M. *et al.* (1999). Arachidonic acid induces differentiation of uterine stromal to decidual cells. *Mol Endocrinol.* 13:1005-1017.

Thaler CJ. (1989) Immunological role for seminal plasma in insemination and pregnancy. *Am J Reprod Immunol.* 21:147-150.

Toki T., Horiuchi A., Ichikawa N. *et al.* (1999). Inverse relationship between apoptosis and Bcl-2 expression in syncytiotrophoblast and fibrinType fibrinoid in early gestation. *Mol Hum Reprod.* 5:246-251.

Uchikawa R., Matsuda S., Yamada M. *et al.* (1997). Nematode infection induces Th2 cell-associated immune responses in LEC mutant rats with helper T cell immunodeficiency. *Parasite Immunol.* 19:461-468.

Ulbrecht M., Rehberger B., Strobel I. *et al.* (1994). HLA-G: expression in human keratinocytes in vitro and in human skin in vivo. *Eur J Immunol.* 24:176-180.

Vanderbeehen Y., Vleighe V., Duchateau J. *et al.* (1984). Suppressor T-lymphocytes in pregnancy. *Am J Reprod Immunol.* 5:20-24.

Verma S., Hiby S.E., Loke Y.W. *et al.* (2000). Human decidual natural killer cells express the receptor for and respond to the cytokine IL 15. *Biol Reprod.* 62:959-968.

von Minckwitz G., Grischke E.M., Schwab S. *et al.* (2000). Predictive value of serum IL-6 and -8 levels in preterm labor or rupture of the membranes. *Acta Obstet Gynecol Scand.* 79:667-672.

von Rango U., Classen-Linke I., Krusche C.A. *et al.* (1998). The receptive endometrium is characterised by apoptosis in the glands. *Hum Reprod.* 13:3177-3189.

Ware C.F. (1996). Apoptosis mediated by the TNF-related cytokine and receptor families. *J Cell Biochem.* 60:47-55.

Watson E.D., Zanecosky H.G. (1991). Regulation of mitogen-induced and TCFG-induced lymphocyte blastogenesis by prostaglandins and supernatant from equine embriyos and endometrium. *Res Vet Sci.* 51:61-65.

Wegmann T.G., Lin H., Guilbert L., Mosmann R.T. (1993). Bidirectional cytokine interactions in the maternal-fetal relationship: is successful pregnancy a Th2 phenomenon? *Immunol Today.* 14:353-356.

Wessells J., Wessner D., Parsells R. *et al.* (2000). Pregnancy specific glycoprotein 18 induces IL-10 expression in murine macrophages. *Eur J Immunol.* 30:1830-1840.

Wood G.W. Kamel S. Smith K. (1988). Immunoregulation and prostaglandin production by mechanically-derived and enzyme-derived murine decidual cells. *Cell Immunol.* 116:411-422.

Xing Z., Zganiacz A., Santosuosso M. *et al.* (2000). Role of IL-12 in macrophage activation during intracellular infection: IL-12 and mycobacteria synergistically release TNF- and nitric oxide from macrophages via IFN- induction. *J Leuk Biol.* 68:897-902.

Xu C., Mao D., Holers V.M. *et al.* (2000). A critical role for murine complement regulator Crry in fetomaternal tolerance. *Science.* 287:498-501.

Yamashita H., Otsuki Y., Matsumoto K. *et al.* (1999). Fas ligand, Fas antigen and Bcl-2 expression in human endometrium during the menstrual cycle. *Mol Hum Reprod.* 5:358-364.

Yamazaki K., Kato Y. (1989). Sites of zona pellucida shedding by mouse embryo other than muran trophectoderm. *J Exp Zool.* 249:347-349.

Yang Y., Chu W., Geraghty D.E. *et al.* (1996). Expression of HLA-G in human mononuclear phagocytes and selective induction by IFN-γ. *J Immunol.* 156:4224-4229.

Yokoyama M.W. (1997). The mother-child union: The case of missing-self and protection of the fetus. *Proc Natl Acad Sci USA.* 94:5998-6000.

Yui J., Garcia-Lloret M., Wegmann T.G. *et al.* (1994). Cytotoxicity of TNF-α and γ-IFN against primary human placental trophoblasts. *Placenta.* 15:819-835.

Zeidler, R., Eissner, G., Meissner, P. *et al.,* (1997). Downregulation of TAP1 in B lymphocytes by cellular and Epstein-Barr virus-encoded IL-10. *Blood.* 90:2390-2397.

Chapter 5

Tumors in Mammals and Non-mammalian Classes of Vertebrates

5.1   COMPARATIVE ONCOLOGY

In vertebrates, neoplasia is a disease in which genetically altered cells escape from the normal growth regulation and monitoring of the immune system. Several morphological features distinguish neoplasms from normal tissues and from other lesions such as inflammation. In all vertebrates, neoplastic growth is not controlled by the same mechanisms controlling adult tissues, but the mechanisms of neoplastic growth can be similar or nearly same with the mechanisms controlling embryonic tissues. This results in a persistent, expanding or infiltrating growth without the architecture of the normal tissue (Wellings, 1969; Ruddon, 1995; Schwemmler, 1998). The term "tumor" is the most common synonym for "neoplasia" and stands for a swelling as a result of the accumulation of a new, fast-growing abnormal tissue. However, grossly visible mass may not necessarily characterize the tumors. In the cases of certain types of lymphomas and leukemia, "neoplasia" exists without macroscopically visible tumors (Kieser *et al.*, 1991). In all vertebrates, neoplasms have varying degrees of abnormality in cellular appearance and growth rates, and functional differences are usually apparent between neoplastic tissue and related normal tissue (Schumberger *et al.*, 1948; Schwemmler, 1998).

Comparative oncology is a branch of comparative pathology that is relatively new biomedical discipline. The need for a comparative research into tumors across different groups of living beings has arisen from a relatively old notion that all multicellular organisms may develop tumors. This apparently very simple fact indicate the possible nature of tumor growth, which can be associated with the basic features of the cells of multicellular organisms such as the division, development, growth and differentiation of cells and tissues. Regarding this, tumor growth is commonly defined as a fundamental disorder in the regulation of cell division, growth, differentiation and cell socialization (Stewart, 1972; Harshbarger, 1996).

One of the tasks of comparative oncology is the identification of oncogenic factors and mechanisms of their influence on the cells of different, phylogenetically distant multicellular organisms, as well as of differences in sensitivity of these cells to oncogenic factors. Even though there are no reliable data regarding the incidence of tumors in different groups of multicellular organisms, some studies indicate the possibility that tumor incidence might be growing along with the growing complexity of observed species and their place on the evolutionary scale (Stewart, 1972; Harshbarger, 1974, 1996, Bubanovic *et al.*, 2004). The rationale behind such phenomenon could be the stability of genome and the reliability of mechanisms controlling the cell cycle. For example, simple multicellular organisms, such as yeasts, may have simpler and more reliable machinery and more effective repair mechanisms for damaged DNA sequences. Unlike these, higher multicellular organisms, like vertebrates, have a more complex but less stable genome, as well as more complex surveillance mechanisms, which could be taken into account for a higher tumor incidence in vertebrates in comparison with the lower multicellular organisms. Yet, there are data that significant differences regarding tumor incidence might be present across different classes of vertebrates. Thus, Effron *et al.* (1977) presented the rate of neoplasia at necropsy of captive wild animals of the Zoological Society of San Diego collection. Neoplasia was present at necropsy in 2.75% of 3,127 mammals, 1.89% of 5,957 birds, and 2.19% of 1,233 reptiles. Interestingly, neoplasms were not detected during 198 necropsies of amphibians. The same authors argue that the most common types of tumors differ greatly across vertebrate classes as well. Notably, lymphosarcoma was the most common tumor registered in birds and reptiles while various types of tumors, such as adenomas, hepatoma and different lines of carcinomas, were registered in mammals. Unfortunately, the research did not include tumor incidence in cartilaginous fish and bony fish, though there are data indicating a very low tumor incidence in the former and particularly in the latter (Hendricks *et al.*, 1980). Namely, Hendricks *et al.* (1980) failed to prove the presence of tumors on 144 necropsies in brown bullheads *(Ictalurus nebulosus)*, but tumor incidence in the same type of fish taken from polluted waters ranged 30% on the sample of 532.

Neoplasms do appear in all non-mammalian vertebrates and this is important in and of itself since these animals can serve as models to understand the behaviour and trajectory of such tumors in mammals and humans. An inter-disciplinary effort must begin among collectors, zookeepers, veterinarians, comparative pathologists in order to obtain comprehensive documentation which would help us to understand neoplasms more thoroughly not only in non-mammals, but also in mammals and humans.

Varying in frequency, sporadic neoplasia involving any body system can be seen in all non-mammal vertebrate species. For example, neoplasia is common in snakes, less common in lizards and amphibians, and rare to uncommon in chelonians such as turtles and crocodilians. In snakes, neoplasia such as lymphoid malignancies, oral and cutaneous fibrosarcoma, as well as malignancies of the reproductive tract are most common. In addition, some species may be predisposed for certain neoplasia. For example, colonic adenocarcinoma is common in *Corn snakes, San Francisco Garter Snake* usually get cutaneous melanomas, while oral fibrosarcomas appear commonly in *Pythons,* etc. Lizards are most likely to suffer from lymphoid malignancies, while the most common tumor in chelonians is herpesvirus-associated papillomas. At the same time, cutaneous squamous cell carcinoma and lymphoid malignancies are rarely seen in turtles (Zwart *et al.,* 1972; Jacobson *et al.,* 1980, 1981b; Schultze *et al.,* 1999).

Amphibians are mostly predisposed for lymphoid and myeloproliferative disorders. Melanomas, papillomas, ovarian, hepatic and renal neoplasia are occasionaly seen in this vertebrate group (Schumberger *et al.,* 1948; Ruben *et al.,* 1977, 1997).

In birds, neoplasia is better documented in captive and domesticated species than wild birds. *Psittacines* may be predisposed for cutaneous papilloma as caused by papillomavirus. Also, tumors like the syndrome of cloacal carcinoma, biliary and pancreatic adenocarcinoma and alimentary squamous cell carcinoma are possibly viral. Cutaneous and coelomic lipomas are very common in budgies, like *Amazons cockatoos.* Xanthoma, as a benign tumor is probably induced by trauma and possibly associated with hypercholesterolemia. Fibrosarcomas are more common on wings, but less so on trunk. Squamous cell carcinoma of uropygial gland, oral cavity and skin are documented in canaries. Renal adenocarcinomas, nephroblastoma, Sertoli cell tumors, seminomas, pituitary adenomas are tumors discovered in different species of birds. Also, ovarian neoplasms, granulosa cell tumors and adenocarcinomas are common in all *Psittacines.* Other neoplasms like hemangiosarcoma, lymphoid malignancies, viral papillomas, retrovirus associated lymphoma and nephroblastoma, fibrosarcoma, hemangiosarcoma, cutaneous squamous cell carcinoma and adenocarcinomas are also documented in birds. In captured raptor birds, considering that these birds are fairly long lived in captivity and closely observed, neoplasia is sporadic with few if no trends (Effron *et al.*, 1977; Reece, 1992; Wilson *et al.*, 2000; Forbes *et al.*, 2000; Garner, 2001).

Although some neoplasms are directly hereditary, genetic predisposition is only one of factors affecting the occurrence of all neoplasms. The tendency of certain species to develop particular types of tumors is a well-known aspect of comparative oncology (Shlumberger, 1957). Apart from a great similarity in cell organization across vertebrates, there is an opinion that DNA and mechanisms regulating the cell cycle in lower vertebrates are more stable and more resistant to the influence of various oncogenes (Ruben *et al.*, 1997). This explanation is quite acceptable from the aspect of control mechanisms of the cell cycle and DNA stability as being "the first line of defence" from malignant cell alteration. The immune system could represent "the second line" of anti-tumor defence. Due to the possible significance of the immune system in anti-tumor protection, differences in anti-tumor efficacy among different groups of multicellular organisms could also be of great influence for the incidence of manifesting tumors (Bubanovic *et al.*, 2004).

It is very difficult to make a viable conclusion as to how much the failure of the "first", particularly "the second" line of defence really contribute to even greater differences in tumor incidence across various groups of multicellular organisms. This dilemma may be solved by future research in the field of comparative oncology, particularly by developing disciplines such as comparative tumor genetics and comparative tumor immunology.

Anti-tumor immunity in invertebrates is based solely on the activity of the innate immune system, relatively poorly developed cytokine network, a small number of effector cells and proportionally poorly developed mechanisms regulating immune response. Therefore, one of the primary tasks of comparative tumor immunology may be the identification of mechanisms of anti-tumor immunity in invertebrates and comparing them with their counterparts in vertebrates. On the other hand, all vertebrates are characterized by both innate and adaptive immunity, displaying great differences related to the advancement stage of the adaptive immunity, its correspondence with innate immunity. In addition, differences in the mechanisms controlling the immune reaction, number of cytokines and their role in the immune response are also important.

The identification and comparative analysis of the foregoing mechanisms in invertebrates and vertebrate classes may contribute significantly in clarifying the mechanisms of anti-tumor immunity failure in mammals, as well as tracing the possible link between these and immunoreproductive mechanisms. Furthermore, this could be a way of introducing new, more effective adjuvant and/or immunotherapeutic procedures.

5.1.1 Genetic Properties of Tumors in Vertebrates

In one-cell systems, each division of the cell generates a new individual. Here, the speed and number of cell division depend largely on the conditions of the external environment. Unlike one-cell organisms, the cell division in multicellular organisms enables not only the reproduction but also growth and regeneration of tissues and organs. Undoubtedly, therefore, cell division is a critical event for the survival of multicellular organisms. Yet, an uncontrolled cell division may significantly impair the homeostasis of multicellular organisms and eventually lead to a death outcome. Such phenomenon is commonly called tumor growth (Hartwell *et al.,* 1989; Ruddon, 1995; Schwemmler, 1998).

From this and other reasons, it is clear that nature must control the amount of cell proliferation that occurs. The story of the factors involved in cell cycle regulation lead to the concept of "checkpoints" in the cell cycle. Checkpoints are feedback controls that monitor and regulate various steps in the cell division, growth, maturation, differentiation and death. The feedback controls of every checkpoint have three components: (i) a sensor that monitors the completion of critical events, (ii) a signal that is generated by the sensor and (iii) the response machinery that halts or delays the checkpoint event (Hartwell *et al.,* 1989). In eukaryotic cells, there are checkpoints (i) at the decision to enter S phase, (ii) at the decision to enter mitosis and (iii) at the decision to exit from mitosis. For example, the lack of completion of DNA replication prevents cells from entering mitosis.

When provided with plentiful nutrients, favourable temperature, and other factors, one-cell organisms proliferate uncontrollably. Since nutrients are plentiful in the tissues of the body, the cells must refrain from proliferating in circumstances where bacterium would proliferate readily, so that most cells in the adult are not growing or dividing but instead are in a resting state, performing their specialized function while retired from the division cycle. There are many data that for the cells of a multicellular organism, nutrients are not enough: in order to grow and divide, a cell must receive specific positive signals from other cells (Laiho *et al.,* 2003; Malumbres *et al.,* 2003). Many of these signals are actually proteins called *growth factors,* which take effect by interacting with complementary receptors. Growth factors are positive signals (i.e. for cell division) as opposed to negative intracellular signals (the inhibitors of cell division) which function as mechanisms regulating the cell cycle. The phenomenon of positive extracellular and negative intracellular signals is universal and common to all multicellular organisms indicating a high level of evolutionary conservation of the genes regulating these mechanisms, and also the importance of regulating the cell cycle (Ruddon, 1995; Schwemmler, 1998; Laiho *et al.*, 2003; Malumbres *et al.*, 2003).

In all multicellular organisms, cells have evolved an intricate defence network to maintain genomic integrity by preventing the fixation of permanent damage from endogenous and exogenous mutagens. Cell cycle checkpoints, a major genomic surveillance mechanism, exist at the G1/S and G2/M transitions that are regulated in response to DNA damage (Hartwell *et al.*, 1989; Laiho *et al.*, 2003; Malumbres *et al.*, 2003). These are called the G1, G2 and M checkpoints. Upon reaching a checkpoint, a cell determines whether to continue "cycling" or to remain in a particular phase of the cycle. The G1 checkpoint regulates the transition from G1 to S phase. Once a cell enters the S phase, it commits itself to divide. A cell will enter S only if it reaches a certain threshold size, an ample quantity of nutrients is available, and growth factors are present. The G2 checkpoint regulates the transition from G2 to M. A cell must have attained a certain threshold size and have undergone a proper DNA division before passing through the G2 checkpoint. Finally, the M checkpoint occurs at the metaphase of mitosis, so the cell does not continue mitosis if any of the spindle fibers is unattached to its kinetochore (Schwemmler, 1998; Laiho *et al.*, 2003; Malumbres *et al.*, 2003).

Nowadays, most biologists believe that several pre-conditions are necessary for an unobstructed cell cycle:

1. *Cytoplasmic volume* - The amount of cytoplasm in a cell appears to be an important determinant of cyclin synthesis. A cell will divide once it reaches a critical size, and not before. Indeed, experimental manipulations where a cell is cut (but survives) delays entry into S phase until the cell grows. How a cell "knows" it has reached a critical size is not known, but some biologists posit the ratio of the amounts of DNA to cytoplasm may play a role (Hartwell *et al.*, 1989; Laiho *et al.*, 2003; Malumbres *et al.*, 2003).

2. *Growth factors* - One important function of growth factors is to regulate protein synthesis and thus the rate at which cells grow. Most factors that stimulate cell proliferation also stimulate cell growth, but the correspondence is not always exact. Some factors will make cells of a given type grow but do not get them past the G1 checkpoint in their cycle, while other factors will get them past the G1 checkpoint but do not make them grow. It seems that in mammals there is not so rigid a rule coupling cell size and cell division as there is in yeasts. Growth factors are proteins that induce mitosis. These molecules were first detected in solutions containing platelets, the cell fragments involved in blood clotting. Platelet-derived growth factor from one species of eukaryote has

been used to induce mitosis in many other eukaryotes, suggesting this mechanism of inducing mitosis is ancient and highly conserved. Growth factors interact with other molecules to cause the expression of genes coding for cyclin proteins. Many different kinds of growth factors have been identified, and it appear that most cells need various combinations to divide properly (Schwemmler, 1998; Laiho *et al.*, 2003).

3. *Density-dependent inhibition* - When cells dividing on the opposite sides of a wound meet, cell cycling is inhibited. This process is called density-dependent inhibition. It is not exactly known how this mechanism inhibits cyclin synthesis (Laiho *et al.*, 2003; Malumbres *et al.*, 2003).

Finally, defects in the mechanisms controlling the rate of cell cycling lead to an uncontrolled cell growth. In most cases, the results of this frenzy of cell divisions are tumors.

5.1.1.1   Oncogenes

There are data that retroviral oncogenes or their precursors were naturally present in the genomes of virtually all normal vertebrate cells. Molecular evidence suggesting the normal cellular origin of retroviral oncogenes was first obtained by showing that radiolabelled DNA from the avian retroviral oncogenes *src* hybridized specifically to the normal uninfected avian cellular DNA as well as to the normal mammalian DNA and even normal human DNA. All retroviral oncogenes are now known to have hybridizing analogs, that is, close relatives, in the genomes of virtually all normal vertebrate cells. The normal host cellular analogs of *v-onc* genes are called *c-onc* genes or proto-oncogenes (Gonda *et al.*, 1982).

The retroviral oncogene analogs found in normal DNA have the usual structure of genes, possessing both *exons* and *introns*, whereas the oncogenes in retroviruses do not have *introns*. It has been concluded that the oncogene analogs found in normal DNA represent the native DNA and are not of viral origin. It appears, then, that retroviral oncogenes are copies (allowing for subtle differences in gene sequences) of normal cellular genes which were picked up and transduced into the retroviral genome by pre-existing retroviruses. Nevertheless, retroviral oncogenes usually show structural mutations or changes in expression relative to the corresponding proto-oncogenes. Since proto-oncogenes are highly conserved in vertebrates and are demonstrable in the cellular genomes of virtually all metazoans, their roles in cellular functions are apparently of fundamental importance. Many gene products of proto-oncogenes, by analogy to the corresponding oncogene, have been identified as proteins (kinases, growth factors, growth-factor receptors) with known functions in normal cells (Gonda *et al.*, 1982; Dornburg, 2003; Mikkers *et al.*, 2003).

The direct evidence for the activation of proto-oncogenes to transforming genes *in vivo*, independently of any retroviral gene participation, has been obtained by isolating proto-oncogenes, attaching them to the promotor or enhancer sequences, and introducing this DNA into one-cell mouse embryos (fertilized eggs) and thus ultimately the germ cells of the so-called transgenic mice. The promotor causes a high rate of transcription ("activation") of the introduced proto-oncogene resulting in a high incidence of malignant tumors in some of the progeny mice. In other experiments, "weak" retroviruses which lack a separate viral oncogene have been shown to produce tumors only when, by chance, the proviral DNA is inserted into the cellular genome next to a cellular proto-oncogene, which is then activated through the effect of the inserted viral promoter (Dornburg, 2003; Mikkers *et al.*, 2003).

In summary, the current oncogene theory postulates that the oncogenes of retroviruses are derived from normal cellular genes (proto-oncogenes); and that an increased expression of proto-oncogenes, or an inappropriate expression of a mutated, functionally altered form of proto-oncogenes brought about by the action of carcinogenic agents or occurring spontaneously throughout life, contribute to neoplastic transformation and the development of malignant tumors. Finally, it is now known that at least four different classes of cellular genes are involved in the several steps of neoplastic transformation: oncogenes; tumor supresor genes; mutator genes; and genes that mediate apoptosis (Ruddon, 1995; Schwemmler, 1998; Dornburg, 2003; Mikkers *et al.*, 2003).

5.1.2   Phenomenon of Cell Socialization in Embryogenesis vs. Oncogenesis

Various hypotheses have been made with a view to explaining the mechanisms of carcinogenesis; and they do it more or less successfully. One of well founded hypotheses starts by assuming a close connection between the carcinogenesis and physiological processes of embryogenesis, as well as tissues regeneration. The development of a tumor cell is, according to this model, the result of its ontogenic regression, i.e. the activation of one group or a whole set of genes regulating the embryogenesis. This causes the otherwise normal cell to undergo de-differentiation, leading thus to the failure of the mechanisms regulating the cell cycle. From this aspect, the malignant alteration of the cell can be understood as its ontogenic regression, in most cases only partial but sometimes even complete, into the embryonic or embryonic-like stage (Wohlgemuth, 1957; Schwemmler, 1991, 1998).

Division, growth, differentiation, functioning and apoptosis are considered the five most significant phases in the life cycle of any cell, from the embryonic stage to its death. The regulating mechanisms of each of these are based on the antagonism between the activating and suppressive genes. A number of factors take part in the regulation of their activity, including hormones, cytokines, growth factors, prostaglandine, membranous and soluble receptors etc. In the matter of its ability to proliferate, growth, course of differentiation, functioning and apoptosis, the behaviour of the cell is largely determined by the environmental influence and genome activity (Schwemmler, 1991, 1998).

The onset of ontogenesis is characterized by rapid cell divisions in the embryonic stage. Soon, the mechanisms regulating cell growth and proliferation engage in the process. These are known to suppress cell division, but promote cell differentiation and socialization. Finally, a normal cell ends its life by the programmed cell death. According to the embryonic model of carcinogenesis, the activation of the genes responsible for the regulation of embryogenesis under normal circumstances, which co-occurs with a partial deactivation or complete inactivation of the mechanisms regulating cell division, differentiation, socialization and apoptosis, is regarded as malignant alteration of the cell. Once altered, the cell continues its new life as a tumor cell (Schwemmler, 1991, 1998).

The mitotic index of embryonic cells is extremely high. At the same time as the mechanisms of differentiation and socialization activate, their proliferative potential begins to drop significantly. The frequency with which most of the tumor cells divide is not as nearly high as that of embryonic cells. Nevertheless, it is kept relatively constant due to their resistance to various signals coming from the environmental cells and the failure of control mechanisms of the cell cycle (Ruddon, 1995; Schwemmler, 1991, 1998).

Following a sequence of divisions, under normal conditions, embryonic cells enter the differentiation phase. This means not only a gradual adjustment to the surrounding microenvironment, both in anatomical and functional sense, but also the point whereof they begin to take part in regulating the differentiation of undifferentiated cells. The process through which undifferentiated embryonic cells transform and begin the new life phase as a part of the tissue, organ and organism is known as the phenomenon of cell socialization. Considered from this point of view, tumor cells are rather unsociable and quite autochthonous. In the course of time, as the embryonic cells reach maturity, they steadily grow ever more sensitive to apoptotic signals. These mechanisms are either weakened or completely absent in tumor cells during and after alteration (Schwemmler, 1998; Laiho *et al.,* 2003; Malumbres *et al.,* 2003).

The transplantation of syngeneic embryonic cells into mature organisms, performed in the course of some experimental studies, has shown that these cells begin to differentiate after a while into the cells of the surrounding tissue, which points to their preserved ability to differentiate and socialize. The transplantation into any of the tissues of the adult animal will not result in differentiation or socialization of tumor cells, but in their rapid growth and further dissemination. The above findings would lead a geneticist to conclude that oncogenesis might be, in fact, an incomplete/aberrant ontogenic regression, i.e. the activation of only those genes which make the altered cells similar but not identical to embryonic cells, especially regarding socialization, the ability to differentiate and apoptosis (Ruddon, 1995; Schwemmler, 1991, 1998).

5.1.3   Selection of Tumor Cells as Micro-evolutionary Process of Tumor Development

We can regard the emergence and development of tumors as a micro-evolutionary process accompanied by all, or almost all, elements that characterize evolution in the Darwinian sense. The evolution of tumors in all vertebrates progressed on a time scale of months and years within a limited population of altered cells, involving all the phenomena observed in the long-term evolution of species. These phenomena include random changes in the genome of altered cells, various forms of selection pressure and selection of tumor cells. In the micro-evolutionary process of tumor cell selection, the role of the factors of selection pressure is played by (i) genetic mechanisms which regulate cell socialization (ii) normal cells which surrounding tumor cells, (iii) the mechanisms of differentiation, (iv) nutrients and (v) the cells of the immune system. Regarding tumors as micro-evolutionary process under different forms of selection pressure, the role of a "predator" is most likely ascribed to the cells of the immune system.

In all vertebrates, tumor cells are defined by two heritable properties: they and their progeny (i) reproduce in defiance of the normal restraints and (ii) invade and colonize territories normally reserved for other cells. The combination of these two properties makes tumor cells dangerous for the host, i.e. malignant (Tomlinson *et al.*, 1996, 1999; Rubin, 2001).

There is now much evidence corroborating the fact that tumors develop mainly from only one cell, the so-called "stem" tumor cell. Many scientific studies conducted on this purpose have shown that all malignant cells in a chronic myelocytic leukemia display the same chromosome anomaly called the Philadelphia chromosome. At the same time, the aberrant cells of certain tumor produce an identical immunoglobulin type or express only one particular isoenzyme type (Ruddon, 1995). Like in mammals, tumors of other vertebrates originate from a single primary tumor cell, even when a cancer is metastasized. The evidence for this phenomenon comes mainly from the studies analyzing the DNA of tumor cells. In most cases, almost all tumor cells display the same type of DNA damage, whether in the form of mutations, translocations or some other DNA impairment. This was first detected on the tumor cells of mammals and verified on most tumor cell types of other vertebrates later on. The initial genetic aberration which leads to cell alteration and represents the basis of carcinogenesis is, however, not sufficient for the model of carcinogenesis as a micro-evolutionary process to be applicable. For a successful tumor growth, it is first necessary that altered cells not only pass on genetic aberration to their daughter cells, but also that daughter cells are able to make new genetic changes (Ruddon, 1995; Tomlinson *et al.,* 1996, 1999; Dornburg, 2003; Mikkers *et al.,* 2003).

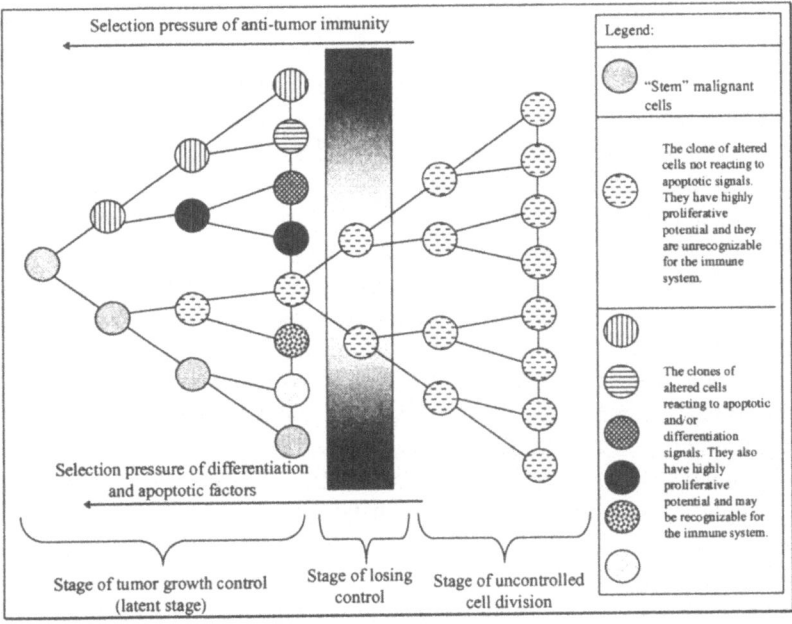

Figure 5.1. Oncogenesis as micro-evolutionary process.

Such genetic changes would have to occur in every, or almost every, successive tumor cell generation, giving each generation a new "quality" which would enable the survival of at least some cells under the strong selection pressure. Finally, such model of carcinogenesis would include two mechanisms related to genetic events and the stability of the genome: (i) initial genetic or chromosome aberration as a basis for the alteration of the "stem" tumor cell and (ii) the permanent mutagenesis (the permanent production of a change in the DNA sequences) possibly responsible for the production of new generations of altered cell clones which, in a true

Darwinian manner, find solutions for the survival and development of a manifesting tumor (Tomlinson *et al.,* 1996, 1999).

Clearly, drastic as one mutation or event may be, they probably can not lead to cell alteration. Furthermore, there is an opinion that the mutations of some important genes end in the rapid death of a cell due to the remarkable changes of its homeostasis and a relatively easy way in which these mutations are identified by the cells of the immune system. With regard to this, carcinogenesis is most commonly referred to as a long-term genetic aberration of low intensity, which continually destabilizes the genome (Young *et al.,* 1993; Rubin, 2001).

Since a single mutation is not enough to convert a typical healthy cell into a tumor cell that proliferates without restraint, the genesis of a tumor as a rule requires that several independent rare accidents occur together in one cell. To that effect, there is opinion that tumors develop in slow stages from mildly aberrant cells by a process in which an initial population descendants of a single mutant ancestor evolves, from slightly aberrant to malignant cells, through the successive cycles of mutation and natural selection. Albeit tumor cells defy the normal controls on cell division, there are other requirements which maintain tumor development, such as stimulation of the development of blood vessels to bring the nutrients and oxygen they require for growth. However, it seems that the most important factor for the transformation from mild to tumor alteration depends on the frequency of mutations. The mutation rate may be high because of mutagens in the environment or because of the intracellular defects in the machinery governing the replication, recombination, and repair of DNA (Young *et al.,* 1993; Tomlinson *et al.,* 1999; Rubin, 2001).

In conclusion, a given tumor cannot be blamed entirely on a single event or a single cause: tumors as a rule result from the chance occurrence in one cell of several independent accidents, with cumulative effects. Speaking in strict terms, the "stem" tumor cell theory cannot explain the differences among the cells of one tumor population regarding antigen phenotype, the degree of dysplasia, sensitivity to apoptosis, etc. These could only be explained by the fact that tumor cells are genetically less stable and literally pass through circles of evolutional selection during the sub-clinical phase of disease. Tumor instability and growth are counterbalanced by such selective mechanisms as anti-tumor immunity and factors which regulate proliferation, differentiation and apoptosis, so that its future will depend largely on the activity of these. It is interesting that during the monthly and yearly development of a tumor within tumor cell population a lot of malignant sub-varieties start to appear. It means the ability to proliferate, ability to differentiate and sensitivity to apoptosis, antigen phenotype and receptors' range, among others. The varieties displaying a decreased proliferation along with a relative sensitivity to apoptotic signals are eliminated over time by the activity of immune system cells. Eventually, it is only those tumor cell varieties with the highest proliferative potential, the lowest sensitivity to apoptotic signals and the expression of MHC and tumor associated antigens that remain in the tumor cell population. Judging by this, we can conclude that tumor development is not only represented by the malignant alteration of cells but also by the further selection of those most capable of escaping all regulating and immune mechanisms (Young *et al.,* 1993; Schwemmler, 1998; Tomlinson *et al.,* 1999; Rubin, 2001).

Genetic, immune and other microenvironmental regulating mechanisms are responsible for the altered cell selection. Once their quality declines due to ageing or an influence from the external environment, the protective barrier is disrupted and the malignant process gets out of control. Another possibility is that certain malignant cell patterns might be produced as a result of variability of the altered cells' genotype and phenotype that the immune system cannot recognize as such and eliminate, which consequently leads to the onset of malignant disease. Finally, the altered cells may steadily weaken the protective "barrier" in the pre-malignant stage when the regulation is still possible, until the moment when they get out of control (Schwemmler, 1998; Tomlinson *et al.,* 1996; Rubin, 2001) (*Figure* 5.1.).

5.1.4 Factors Influencing Oncogenesis in Vertebrates

The researches in the field of fundamental biology of tumors and oncogenesis are most commonly carried out on mammals. However, the similar research on non-mammalian vertebrate classes show a clear link between oncogenesis in all vertebrates with such factors as ageing, chemical carcinogens, radiation and, especially, viral infections. As with mammals, sex, environmental conditions, infections, chronic and long-term stress (life in captivity) can significantly contribute to oncogenesis in non-mammalian vertebrate classes (Effron *et al.,* 1977; Harshbarger, 1974, 1977, 1984; Baumann *et al.,* 1990).

5.1.4.1 Age
Neoplasms typically become more common in older vertebrates (Harshbarger, 1974, 1977). This relationship between age and tumor frequency also occurs in wild, domestic and laboratory animals exposed to chemical and other carcinogens (Anisimov, 2003). Ageing is, in itself, a factor of oncogenesis, but becomes particularly significant when accompanied by one or more factors such as chronic viral infections and chemical agents. Most authors are of opinion that the ageing of organisms and cells is a factor of the genome instability, as well as the factor of

accumulation of mutations resulting in oncogenesis. From the immunological point of view, ageing represents the weakening of immune monitoring and anti-tumor immune response. Yet, tumor incidence in some vertebrate groups does not seem to confirm such an opinion. Namely, tumor frequency in younger snakes and lizards is significantly higher compared to that in older samples of long-living reptiles such as crocodiles and turtles. There is no clear explanation of this phenomenon, but it may be connected with genetic, immune and evolutionary factors of tumor development, as well as factors controlling oncogenesis (Anisimov, 2003; Dall'Ara, 2003).

5.1.4.2 Temperature

Environmental temperature is an important factor in any aspects of vertebrate pathology, because the temperature of most cold-blooded vertebrates is essentially the same as that of the surrounding medium (air or water). In most cold-blooded vertebrates exposed to carcinogens, low temperature, in most cases, reduces the incidence of neoplasms and metastasis. Yet, there are examples where low temperature favours oncogenesis. For instance, the melanoma of *Xiphophorus* develops entirely in the fishes which live in water temperatures between 26 and 27.5° C, but this type of tumors never develops in the fish living in warmer waters (31 to 32° C) (Perlmutter *et al.*, 1988).

Environmental temperature can influence the incidence of virally induced tumors in cold-blood vertebrates by two possible mechanisms: (i) the immune system may be more effective in the recognition and elimination of tumor cells at certain temperature and (ii) the virulence and effects of the viruses on genome stability and/or cell division may depend on temperature. If the second hypothesis is correct, the close relationship between the viral infections and the excessive growth indicates that those lesions are probably not neoplasms, but rather hyperplastic lesions. The activity of animals during the periods with higher temperatures may be also significant because of their higher mobility and easier dissemination of viruses. In addition, an enhanced metabolic activity at certain temperature might be associated with a higher incidence of mitosis and genome instability (Curtis *et al.*, 1995).

Environmental temperature probably has no significant influence on tumor frequency in warm-blooded vertebrates (birds and mammals), though constant body temperature in these could be related to a more intense, metabolic and mitotic cell activity throughout the year, unlike cold-blooded vertebrates in which these processes depend largely on external temperature.

5.1.4.3 Gender

Despite many studies which prove that males and females of some vertebrate groups are more prone to certain types of tumors, there is yet no solid evidence that gender significantly determines the overall tumor frequency (Ruddle, 1973; Bremermann, 1987). For instance, the female of *Japanese hatcheries* are more susceptible to liver tumors (Takashima, 1976), while the males of swordtails *(Xiphophorus helleri)* are prone to melanoma (Siciliano *et al.,* 1971).

In humans, tumors are more likely to appear in males than females. This can be explained by a more dynamic lifestyle, higher percentage of smokers among men, etc. From the immuno-endocrinological point of view, a higher tumor incidence in men could be explained by the immunosuppressive effects of testosterone, unlike the immunomodulatory effects of sex steroids in women. Therefore, the women's immune system is more active and more vigorous which, on the other hand, accounts for a significantly higher incidence of auto-immune disease (Ruddle, 1973; Bremermann, 1987; Ruddon, 1995).

5.1.4.4 Chemical Carcinogens

Many chemicals that are carcinogens in mammals are also known to cause neoplasms in other classes of vertebrates. A variety of chemicals can alter the course of oncogenesis in vertebrates by acting as carcinogens, promoters or anti-carcinogens (Balls *et al.,* 1964; Bailey *et al.,* 1996; Mitsumori, 2002). Chemicals and pollutants seem to be involved in increasin g the prevalence of neoplasms in vertebrates, but in many cases, it is not known whether these chemicals act as carcinogens, promoters, co-carcinogens, or as the activator of oncogenic viruses. Some chemicals are probably both carcinogens and promoters. The maternal transfer of liposoluble chemicals to offspring, especially in egg laying species, can also affect oncogenesis. Some of chemical carcinogens act directly on the target cells, but many others take effect only after they have been changed to a more reactive form by metabolic processes - notably by a set of intracellular enzymes known as the cytochrome P-450 oxidases. These enzymes normally help to convert ingested toxins and foreign lipid-soluble materials into harmless and easily excreted compounds, but they fail in this task with certain substances, converting them instead into direct carcinogens. Although the known chemical carcinogens are highly diverse, most of them have at least one property in common - they cause mutations (Bailey *et al.,* 1996; Mitsumori, 2002).

As with ageing and exposure to radiation, chemical carcinogens may be also responsible for the weakening of anti-tumor immune mechanisms. Such effects are probably mediated by the changes in immune cells' metabolic activity, the quality of their communication and mutagen effects on the immune cells (Wohlgemuth, 1957; Schwemmler, 1991; Ruddon, 1995; Dall'Ara, 2003).

5.1.4.5   Radiation

A large number of studies have pointed to a strong correlation between various types of radiation, genetic damage and oncogenesis. A common mutagen is ultraviolet radiation (UVR) from sunlight. Overexposure causes damage to skin tissues and numerous mutations in exposed cells. At the same time, repair mechanisms protect most of the damaged cells and those which escape repair are usually controlled by the immune system. However, the extensive exposure overwhelms both the repair and defensive mechanisms, producing skin tumor in most vertebrates. Some evidence suggests that the exposure to UVR could be associated with the weakening of the immune response. Specifically, the experiments with UVR induced skin tumors in laboratory mice revealed that the mice developing tumor after the exposure to UVR do not reject the tumor, whereas those not exposed to UVR reject transplanted tumors (Setlow *et al.,* 1989; Jhappan, 2003; Smith, 2003).

X-rays also act as strong mutagens and carcinogens in all vertebrates. Which tumor type will develop after the exposure to x-rays depends largely on exposition, the dose of absorbed energy and body area exposed to radiation. High doses of radiation affecting the whole organism soon lead to malignant diseases. The mechanism of high carcinogenic x-ray potential is reflected in a variety of mutations, resulting from overexposure, their rapid accumulation up to a number necessary for a complete destabilization of the genome and high susceptibility of the immune system to x-rays.

An overexposure to x-rays causes the extreme shrinking of the total mass of lymphoid tissues, severe impairment of the mechanisms of immune reaction and probably complete incapacitation of anti-tumor monitoring. Joined together, the foregoing factors soon lead to malignant alterations in a majority of vertebrates (Setlow *et al.,* 1989; Jhappan, 2003; Smith, 2003).

5.1.4.6   Viruses

Oncogenic viruses can produce tumors in mammals, birds, and other vertebrates, or transform cultured cells to a neoplastic state and comprise DNA viruses and RNA viruses. Retroviruses probably cause most infectious neoplasms in vertebrates. These neoplasms are diverse and include lymphoproliferative disease, leukemia, dermal carcinomas and sarcomas, papilloma, leiomyosarcoma, fibroma and neural tumors. Viruses causing these diseases are difficult to isolate in cell culture, but the transmission of the disease by the cell-free inoculum and the presence of reverse transcriptase activity provide evidence that retroviruses are the etiological agents causing certain neoplasms of vertebrates (Jacobson *et al.,* 1980; Borwn *et al.,* 1986; Hedrick *et al.,* 1987; Cui *et al.,* 1991; Gould *et al.,* 1993; Hosel *et al.,* 2003).

Of the many families of RNA viruses, the only members of the retrovirus family are capable of inducing animal tumors and transforming cultured cells. The oncogenic RNA viruses are a subfamily of the retroviruses and are a cause of naturally occurring tumors and leukemias in a wide range of vertebrate animals, including mammalian, avian, and reptilian species. The RNA tumor viruses are classified according to their natural host, such as avian, murine, feline, and primate leukemia/sarcoma virus species (Jacobson *et al.,* 1980; Borwn *et al.,* 1986; Hedrick *et al.,* 1987; Cui *et al.,* 1991; Gould *et al.,* 1993; Hosel *et al.,* 2003).

Most oncogenic retroviruses have a gene known as a transforming gene or oncogene and termed *v-onc.* Under the influence of the viral promoter sequence, the *v-onc* gene is transcribed along with other viral genes and is responsible for the neoplastic transformation of the cell. All rapidly transforming retroviruses possess one, or rarely two, unique oncogenes of which more than 20 have been isolated and characterized (Gould *et al.,* 1993; Hosel *et al.,* 2003).

Tumors in vertebrates can be also caused by the so-called oncogenic DNA viruses. In general, the infection of cells with an oncogenic DNA virus may result either in a productive lytic infection with cell death and release of newly formed virus or in cell transformation to the neoplastic state with little or no virus production but with the integration of viral genetic information into the cell DNA. Finally, the mechanisms of DNA virus oncogenesis indicate that oncogenesis is an attribute of the viral DNA, that the viral DNA is integrated into the host cell DNA, and that the protein products of viral genes maintain transformation to the neoplastic state. Viral DNA is integrated into the tumor cells in many studied cases of vertebrate's tumors, but the additional agents or factors may be involved at different stages of the progression to invasive tumor (Jacobson *et al.,* 1980; Borwn *et al.,* 1986; Hedrick *et al.,* 1987; Cui *et al.,* 1991; Gould *et al.,* 1993; Hosel *et al.,* 2003).

5.2 TUMORS IN NON-MAMMALIAN CLASSES OF VERTEBRATES

Non-mammalian oncology is generally important for comparative oncology because non-mammalian species can be good models for our understanding of the nature of neoplasia in mammals. Furthermore, non-mammalian species are useful in the evaluation of chemicals, radiation and viruses as carcinogens and as the indicators for the presence of environmental carcinogens. Also, it might be a model for the determination of genetic factors that regulate oncogenesis and immunological mechanisms as the most important anti-tumor defence (Dawe, 1990; Bailey *et al.,* 1996, Bubanovic, 2003a; Bubanovic *et al.,* 2003d, 2004).

5.2.1   Tumors in Cartilaginous fish

Although frequency of neoplasm varies in different types of fishes, there are no taxa known to be insusceptible (Harshbarger, 1974; Harshbarger *et al.*, 1981). Sharks and their relatives, the skates and rays, have enjoyed tremendous success during their nearly 400 million years of existence on earth. Sharks and rays do get tumors (Harshbarger, 1974; Smith *et al.*, 1969; Wellings, 1969), but their incidence is probably much lower than among the other vertebrates (Rosen *et al.*, 1980).

The comparative statistical data on tumor incidence in various vertebrate classes are limited and, for some reasons, hardly available. However, Smithsonian Institution in Washington owns a large tissue collection from various groups of vertebrates and invertebrates from all possible sources throughout the world, which could possibly represent tumor tissues, including the samples inadequately identified as tumors. Not surprisingly though, only 25 of the total 10,000 sample tissues originate from cartilaginous fish, such as sharks and rays. Most of these are classified as fibrous responses to wounds, parasites, or enlarged thyroid glands sometimes developed by sharks in captivity, leaving only 8 to 10 legitimate tumors among all the shark and rays tissues examined (Smith *et al.*, 1969; Harshbarger, 1974, 1977, 1984; Harshbarger *et al.*, 1981).

There are only few published reports of neoplasms in cartilaginous fish, due to the relatively small number of animals kept in captivity and infrequent experimental procedures with sharks and rays. However, certain experiments performed on nurse shark and clearnose skate as laboratory animals, indicate a significantly higher resistance of these fishes to the effects of various carcinogens than such mammalian species of experimental animals such as mice, rats, guinea pigs and rabbits. In some studies, the sharks and skates were exposed to powerful carcinogenic chemicals by placing the chemicals in their food or surrounding tank water, or by a direct injection into muscle. Although the metabolic response in cartilaginous fish which works to eliminate carcinogens resembles that of mammals, no representative of either sharks or rays developed a malignant disease (Bodine *et al.*, 1985, 1989; Rast *et al.*, 1997).

The confirmation of the assumption that cartilaginous fish indeed have developed the most effective anti-tumor mechanisms, will have to wait for controlled comparative studies to be carried out on a large number of species of different vertebrate classes. Nevertheless, many scientists familiar with this issue agree that possible mechanisms ensuring low tumor incidence in sharks and rays might be contained within the genome stability and/or remarkably efficient immune response. Compared to the mammalian immune system, which is quite specialized, the shark immune system appears to be less advanced but remarkably effective. Sharks apparently possess immune cells with the same functions as those of mammals, but the shark immune cells appear to be generated and stimulated differently. Furthermore, the variety of immune cells is associated with the mammalian immune system, while sharks have only several kinds of effectory cells, as well as cytokines. Another difference lies in the fact that sharks, skates, and rays lack a bony skeleton, and so do not have a bone marrow. In mammals, immune cells are produced and mature in the bone marrow, the thymus and other sites, and, after a brief lag time, these cells are mobilized to the bloodstream to fight invading microbes and tumor cells. In sharks, the immune cells are produced in the spleen, thymus and epigonadal organ (unique tissues associated with the gonads). Some maturation of these immune cells occurs at the sites of cell production, as with mammals. But some studies have determined that a significant number of immune cells in sharks actually mature as they circulate in the bloodstream (Rosen *et al.,* 1980; Bodine *et al.,* 1985, 1989; Rast *et al.,* 1997).

As with bony fish, malignant tumors are in cartilaginous fish far less invasive in comparison with the tumors of the similar or same origin in mammals. The results of such tumor-host relationships are: (i) less malignant potential of the neoplasm; (ii) relatively infrequent and slow metastasing and (iii) a considerably slower growth of tumor tissue (Hayes *et al.,* 1989). A number of factors may be included in the control of tumor growth in cartilaginous fish, such as: body temperature, microenvironmental factors, the absence of lymph drainage and lymph nodes, genetic factors, immune surveillance, anti-tumor immunity etc. It is difficult to say exactly to what degree is each involved in creating the phenomenon of low neoplasm incidence in cartilaginous fish.

5.2.2 Tumors in Bony fish

Many reports indicate the possibility of tumor development in bony fish. These are related both to benign and malignant tumors of different localization and histological origin. For example, it has been verified that bony fish are prone to such neoplasms as lymphoma, nephroblastoma, melanoma and hepatoma (Harshbarger, 1974; Masahito *et al.,* 1985; Kaiser, 1989). Lucke' (1942) and Harshbarger *et al.* (1990) later on described tumors such as neurofibromas and neurilemomas in snipers.

Although the number of publications relating to tumors in bony fish is incomparably larger than tumor related studies of cartilaginous fish, yet it appears that the characteristics of the tumors in these classes of vertebrates are almost identical. Namely, many common neoplasms of fish are relatively well differentiated, and this could be related to their weakly malignant behaviour. Other reasons for a relatively low malignant potential of bony fish neoplasms could include effective anti-tumor immunity, the stability of the genome, and a good control of the mechanisms of cell cycle, as well as low body temperature and the absence of lymph drainage and lymph nodes (Kaiser, 1989).

Because of a relatively low degree of malignancy, tumors in bony fish are commonly followed by secondary complications, such as the low mobility of the host, increased susceptibility to predation, infections and difficult feeding. For these reasons, tumors in bony fish are a direct cause of death outcome only in cases of plasmocytoid leukemia, i.e. in *Onchorchyncus tshawytscha* (shinook salmon) (Kent *et al.*, 1990).

As with other vertebrates, chemical carcinogens may also induce the developing of various neoplasms in bony fish. There is evidence indicating that the incidence of neoplasms in freshwater bony fish is proportional to the concentration of carcinogens and length of the exposure of population (Hendricks *et al.*, 1980; Baumann *et al.*, 1990; Grizzle, 1988). Etoh *et al.* (1983) postulate that the susceptibility of bony fish to developing neoplasia is proportional to their age. This is particularly evident in older bony fish samples which have been exposed to chemical carcinogens. In addition, the stage of development at which fish are exposed to carcinogens can also affect carcingenicity. For instance, the percentage of rainbow trout *(Oncorhynchus mykiss)* with neoplasms developed 10-12 months after a prehatching exposure to aflatoxin B1 is higher if embryos are exposed after, rather than before, they have reached the stage when the liver is present as a discrete organ. Incidence is even greater if yolk-sac larvae are exposed (Wales *et al.*, 1978; Hendricks *et al.*, 1980). Hendricks *et al.* (1980) found that river bullheads of combined ages 4 and 5 had a significantly greater prevalence of biliary carcinomas (35.5%) than those of ages 2 and 3 combined (18.4%). Biliary carcinoma was significantly more prevalent than hepatocellular carcinoma in age 4 fish (sexes combined) and in males of ages 3 and 4. In addition, Maccubin *et al.* (1991) present data regarding age-related tumor prevalence in captured fish. Within a year, 36% of the fish examined for external tumors and 23% of those examined for liver tumors were age two or younger. Fish two years or younger in that study had an external tumor prevalence of 2% or less and a liver tumor prevalence of 5% or less, while by age four they had a prevalence of about 12% for liver tumors and 22% for external tumors. By age of five both external and liver tumor prevalence exceeded 40%.

Faisal *et al.* (1991) found that lymphocytes from *Leiostomus xanthurus* display a weaker response to the stimulation by PHA and LPS if the fish are from chemically polluted waters. The strongest response to these agents was recorded in the lymphocytes of the fishes from non-polluted waters. These data indicate the possibility that apart from direct carcinogen effects, which may also affect the quality of anti-tumor immunity and thus increase the incidence of visible neoplasms in fish.

Apart from the evidence that the age and period of the activity of carcinogens greatly affecting the incidence of neoplasms in bony fish, the data about the connection between sex and certain neoplasms are sometimes uncorrelated; still, the research indicates an overall prevalence of these in females. For example, the neoplasia such as hepatocellular carcinoma occurs more commonly in the females of *Japanes hatcheries* than males (Takashima, 1976). Likewise, Baumann *et al.* (1990) propose that hepatocellular carcinomas in *Brown bullheads* are frequenter in females, but the incidence of cholangicarcinomas is equal in both sexes. According to some authors a higher incidence of neoplasms in bony fish females could be brought in connection with the metabolic activity of estradiol (Hendricks *et al.*, 1980; Nunez *et al.*, 1989).

Season can also influence apparent tumor prevalence in bony fish. During the summer when fish are more metabolically active, tumors develop more quickly and become noticeable through histopathology. During the winter many older fish (particularly those with tumors) die, due to the tumor prevalence is usually lower in spring and higher in autumn.

As body temperature of almost all fishes depends on the temperature of surrounding water, the assumption that the water temperature influence tumor rate and malignant potential of the neoplasms of bony fish seems quite logical. Certain evidence suggests that the incidence of tumors, particularly those chemically induced, is lower in bony fish which live in cold waters (Egami *et al.*, 1981; Hendircks *et al.*, 1984; Curtis *et al.*, 1995). Yet, other evidence points to the incidence of melanoma in *Xiphophorus* being greater if the fish are kept in cold water (Perlmutter et al, 1988).

The susceptibility of certain bony fish species to developing apparent tumors may be inherited. Classic examples of such genetic proneness are melanoma affecting *Xiphophorus* (Malitschek *et al.*, 1995), tumors of pigment cells in the so-called M-clone of *Poecilia formosa* (Shartl *et al.*, 1997) and gonadal tumors in hybrids of goldfish *(Carassius auratus)* and common carp *(Cyprius carpio)* (Down *et al.*, 1990).

As with all other vertebrates, radiation and virus-induced infections significantly increase the incidence of neoplasms in bony fish. UV radiation and X-rays can increase the incidence of melanoma (Setlow *et al.*, 1989), tumors of thyroid glands (Hart *et al.*, 1977) and neuroblastoma (Getchell *et al.*, 1998) in hybrids of *Xiphophorus* and *Poecilia formosa* even 2-10 times. There is a possibility that virus infections, especially those induced by retroviruses, are the cause of most common neoplasms in bony fish. The percentage of such neoplasms possibly outnumbers all other tumors caused by aetiological factors such as chemical carcinogens, radiation, genetic factors and ageing. The most frequent neoplasms related to retroviral infections in bony fish are mainly sarcoma (dermal sarcoma, fibrosarcoma, leiomyosarcoma and lymphosarcoma), leukemia and tumors of nervous tissue. Unlike mammals, carcinomas rarely appear in bony fish (Harada *et al.*, 1990). The same authors found that it is difficult to isolate viruses from the cell cultures of fish neoplasms, but the transmission of the disease by cell-free inoculums and the presence of reverse transcriptase activity provide evidence that retroviruses are the aetiological agents causing certain neoplasms of fish (Harada *et al.*, 1990). As well as retroviruses, the group of viruses known as *Herpesviridae*, plays a role in developing neoplasms in bony fish, especially cutaneous carcinoma (Hedrick *et al.*, 1987).

All the foregoing carcinogens may also cause non-neoplastic but neoplasia-like lesions in bony fish, so that an accurate differentiation between neoplastic and neoplasia-like lesions continues to be a problem (Harshbarger, 1977, 1981). This problem is probably associated with the fact that mammals are the most explored vertebrate group both in experimental and clinical sense, so that most of the standards differentiating between benign and malignant lesions have been established on the basis of knowledge gained from mammalian models. Hence, many phenomena, axioms, experimental models even prejudices may easily impede the interpreting of events in the oncology of fishes and other non-mammalian vertebrates.

The rapid regression of malignant tumors in bony fish is relatively common, drawing attention and detailed investigation. This phenomenon is rarely seen among higher vertebrates and could be associated with the abrupt activation of anti-tumor immune response. Other, less viable explanations of such sudden regression of neoplasms in bony fish are mainly associated with the assumptions about the loss of proliferative potential of tumors as an adopted resistance of tumor cells to necessary growth factors.

5.2.3   Tumors in Amphibians

Spontaneous tumors in *Urodelea* amphibians have been considered uncommon, and this resistance has sometimes been associated with the natural regenerative capacity of tissues in such species or with the resistance of DNA to the effects of carcinogens. Ruben *et al.* (1997) instigate that neoplastic cells in amphibian lose apoptotic capacity, thus any unique aspects of programmed cell death may also play a role in cancer resistance in this vertebrate class.

Although Effron *et al.* (1977) isolated no animals affected with neoplasms during the necropsy on 198 samples of amphibians, there are reports indicating the presence of neoplasms in this vertebrate group. Pfeiffer *et al.* (1979) describes spontaneous, non-pigmented, benign epitheliomas which were found in 44 of 1586 (2.8%) ageing, captured adult newts *(Cynops pyrrhogaster)*. Spontaneous neoplasms have been reported in all major organ systems in both anuran and urodele amphibians, but with less frequency in the urodeles. The tumors of the integument are well represented, but the neoplasms of the haematopoietic cells are rare (Sleeman *et al.,* 1999). In frogs, there is a seasonal change in tumor prevalence with tumors being most common in early spring as frogs emerge from hibernation. Additionally, there are evidence that some virus-induced tumors in frogs can be connected to the activity of these viruses which depends on the metabolic activity of the host, i.e. external temperature. For example, a common tumor in anurans is the herpesvirus induced renal adenocarcinoma named Lucke's tumor. For example, Lucke's tumor affecting the northern leopard frog *(Rana pipiens)* was caused by a herpesvirus. The transmission of the disease is probably during the time that adults occupy breeding ponds. The maturation of the virus occurs after tumor-bearing frogs enter hibernation and the release of the virus depends on the warming temperatures of spring (Letcher, 1992). In addition, malignant lymphomas have been reported in South African clawed frogs *(Xenopus laevis)*. These neoplasms were transmitted to unaffected clawed toads by injection of cell-free extract from the tumors, suggesting a viral etiology (Robert *et al.,* 1995; Gregory *et al.,* 2000).

There are data that amphibian are resistant to carcinogen-induced neoplasia. Powerful direct carcinogen, e.g. the nitrosamines, with demonstrable lymphotoxicity, affect immune reactivity, but rarely, if ever, initiate cancer. Special amphibian physiological factors may affect cancer development, e.g. their regenerative capacity, dramatic metamorphosis, poor tolerance to altered-self antigenicity and the capacity for apoptosis. However, cancer cells, or chemical carcinogens introduced into these tissues, induce normal accessory structures to form, but not cancer.

Antigen-specific tolerance is absent until metamorphic climax, probably because T cell education is limited in the thymus. With metamorphosis, a rising glucocorticoids titter produces T cell anergy in the periphery, due to the early development then, amphibian may be particularly susceptible to oncogenesis. Antigen-specific tolerance in late metamorphosis and in adults relies on thymic education and peripheral suppressor function. As *Xenopus* are not easily tolerized to single epitopic changes in self-antigenicity, they will reject, rather than tolerate, altered-self cells (Ruben et al, 1997).

While spontaneous tumors may develop in inbred and isogeneic strains of *Xenopus laevis,* the South African clawed frog, they are extremely rare in wild-type populations of all amphibians (Ruben *et al.,* 1997). Interestingly,

Xenopus laevis lymphoid tumor cells of the *ff* genotype grow after transplantation in inbred *ff* tadpoles or young post-metamorphic animals, but do not grow in fully grown *ff* adults. The ability to grow is lost progressively after metamorphosis and is apparently due to an immune response of the adult host. The resistance of the host against transplanted tumor cells rises during the post-metamorphic development in parallel with the second histogenesis observed in the thymus, the expression of MHC class II by peripheral T cells and the recovery of T cell effector functions such as MLR, and can be abrogated by a sub-lethal irradiation. These results suggest that the lack of tumor rejection by larvae results from an incomplete effector function rather than an absence of recognition. Full responsiveness cannot be elicited before adulthood (Robert *et al.*, 1995).

As well as other vertebrates, amphibian, too, show a higher incidence of neoplasms in the circumstances of exposure to the increased radiation and UV rays. Thus, Alford *et al.* (1999) found that the increased exposure to UV radiation may reduce the survival rates of adult amphibians by damaging their DNA, increasing the probability of developing tumors, and causing the suppression of amphibian immune systems. Neoplasms are more common in older samples, especially if these were exposed to carcinogens effects of chemicals, radiation and particularly viruses. The nature of neoplasms as appearing amphibian is very similar to that in bony fish. Namely, the former show a higher degree of differentiation compared to the corresponding neoplasms in mammals. They are less invasive and metastases less frequently. As in bony fish, external temperature considerably affects the speed of growth and invasiveness of neoplasms in amphibians (Balls *et al.*, 1964; Ruben *et al.*, 1977; Effron *et al.*, 1977; Letcher, 1992; Alford *et al.*, 1999).

5.2.4   Tumors in Reptiles

The descriptions of neoplasia in reptiles are uncommon in comparison to the prevalence reported for mammals or birds (Jacobson, 1981a, 1981b). Within the class *Reptilia*, lymphoid neoplasias are more frequently found in snakes (Effron *et al.*, 1977; Jacobson *et al.*, 1980, 1981b) although there are also descriptions in lizards (Zwart *et al.*, 1972; Effron *et al.*, 1977; Romagnano *et al.*, 1996; Schultze *et al.*, 1999) crocodilians (Scott *et al.*, 1927) and terrestrial chelonians (Harshbarger, 1974). However, some authors are of opinion that the statements that reptiles appear to have a lower incidence of cancer than other vertebrates remain unsubstantiated (Fitzgerald, 1995). Although Shlumberger *et al.* (1948) point to as little as 25 described tumors in reptiles; Jacobsona (1981) description includes 159 tumors of this vertebrate class.

All of the major groups of neoplasms found in mammals have been reported in reptiles except primary neoplasms involving the central nervous system. Most neoplasms appear to occur more commonly in old, long-term captive reptiles, but the interesting observation is that viral origin has been identified in the majority of known tumors in this vertebrate group. For instance, papillomas involving the periorbital tissues and integument in the axis of the hindlimbs and forelimbs are common in wild and mariculture-reared sea turtles and a herpesvirus has been identified in these tumors. Virus particles also have been found associated with other reptile neoplasms. A C-type RNA virus was identified by electron microscopy to be budding from spleen tissue cultures from a *Russell's viper* diagnosed as having a precardial myxofibroma (Zeigel *et al.*, 1969). Electron microscopic examination of an embryonal rhabdomyosarcoma in a *corn snake* revealed the presence of virus-like particles in the cells of neoplasm (Lunger *et al.*, 1974). Finally, similar particles have also been identified in the majority of known tumors in reptiles, notably in neoplastic lymphoid cells of a *California king snake* affected by lymphosarcoma (Jacobson *et al.*, 1980).

As in other vertebrates, ageing considerably influences tumor incidence in snakes, so that fibrosarcomas, renal cell carcinomas, granulocytic sarcomas and lymphomas are the most frequent tumors in old snakes (Jacobson, 1981; Stanley, 1999). Except for fibropapillomatosis, neoplastic disease is very infrequently seen in sea turtles. There are only a few reports in the literature describing neoplastic diseases of turtles. A fibroadenoma of the lung was reported in a *Horsfield's turtle,* as well as a lymphoblastic lymphosarcoma which involved all major organs or lymphoblastic lymphoma a loggerhead sea turtle (*Caretta caretta*) (Oros *et al.*, 2001). In addition, Frye *et al.* (1975) reported about parathyroid adenoma was identified in a *red-footed turtle,* as well as report of Effron *et al.* (1977) about adenomatous proliferation of the intrahepatic bile-ducts was reported in a male African pancake turtle *(Malacochersus tornierii).* Squamous cell carcinomas are rarely described in all reptiles and very infrequently in chelonians. There are descriptions of this type of tumor in a Ceylon terrapin *(Geoemyda trijuga)* (Cowan, 1968) and a European pond turtle *(Emys orbicularis)* (Harshbarger, 1974, 1976). These cases are the only two descriptions of carcinomas in turtles.

Despite the fact that crocodilians belong to long-living reptiles, a relatively small number of neoplasms have been detected in this particular reptilian group. Scott (1927) described the sarcoma of small round cells in a salt water crocodile. The author observed tumor of ventral cerebellum surface as well as portal hepatic loci as being the primary site of neoplasms. In their revision of previous display of little round cell tumor, Schumberger *et al.* (1948) suggest that the neoplasm could, in fact, be lymphosarcoma.

There are only a few other reports on neoplasms in crocodilians, one of which being a big-sized seminoma in alligators (Wadsworth *et al.*, 1956).

Being very popular as a home pet, and thus available for observations, iguanas are frequently reported with a relatively small number of neoplasms. Most of the neoplasms registered in these reptiles can be found in other vertebrates as well. Stolk (1964) reported on two cases of hepatoma in common iguana *(Iguana iguana)*. The reports of other authors are related to neoplasms such as cholangioma (Well *et al.*, 1992), biliary adenocarcinoma (Frye, 1991), teratoadenocarcinoma (Harshbarger, 1974), ovarian teratoma as a rapidly growing tumor with metastasis to the pancreas, interrenal carcinoma, equivalent to the mammalian adrenal cortical carcinoma (Fitzgerald *et al.*, 1995), and several reports about lymphocytic leukemias (Harshbarger 1974; Frye 1991; Romagnano *et al.*, 1996).

There are a variety of other pathologic inflammatory and non-inflammatory changes that may grossly resemble neoplastic disease. These pseudoneoplasms must always be considered in differential diagnosis. Bacteria, fungi, and metazoan parasites typically result in a granulomatous inflammatory response in reptiles, and the clinician must be able to differentiate these lesions from true neoplasms (Harshbarger 1974, 1984).

Fitzgerald (1995) is of opinion that neoplasm incidence in reptiles is not as low as assumed, as necropsy, biopsy or pathological evaluation are rarely carried out. These could, indeed, be the valuable resources for making an overall impression about a low incidence of neoplasms. The research carried out by Effron *et al.* (1977) and Harshbarger (1974, 1984) indicates, however, that reptiles, along with cartilaginous fish and amphibian, could be a vertebrate class displaying a remarkably high degree of resistance against various types of tumors.

As in bony fish, the most common tumors in reptiles are liver tumors and lymphoproliferative neoplasms, the activity of which depends largely on external temperature. These neoplasms are quite well differentiated, with a low proliferative potential, low metastizing capacity and low malignant potential. In addition, the majority of malignant tumors in reptiles have virus origin or has been induced by chemical carcinogens.

We know little about the anti-tumor immunity and immunity in general, of extinct reptilian groups, such as *Dinosaurian*. The research of dinosaur fossil remnants revealed that this widely-spread vertebrate group is susceptible to tumors. To that effect, during the summer of 2003, first evidence was obtained indicating the presence of brain tumor in fossilized predator called *Gorgosaurus,* which lived 75 million years ago.

5.2.5 Tumors in Birds

Tumors are occasionally found in most avian species affecting any area or tissue of the body. As time passes, more and more types of tumors are described and identified in captured and wild birds, most bearing a close similarity in structure to those reported in other vertebrates. Many tumour types have been identified in birds including the following: fibrosarcoma, lymphoma, adenocarcinoma (affecting the preen gland), papilloma, squamous cell carcinoma, myxofibroma, fibroma, histiocytic sarcoma, leiomyoma, epidermoid carcinoma, haemangioma and mast cell tumours. Papilloma often affects the cloaca, choana or feet and on occasions caused by herpes virus. Those affecting the cloaca or choana often progress to cause a fatal bile duct carcinoma (Reece, 1992; Taylor *et al.*, 2001). In addition, feather plucking may occur over the site of a skin cancer or the trauma of repeated plucking may cause such a lesion (most associated with chronic ulcerative dermatitis). A wide range of cutaneous neoplasms also has been documented in various species together with suggested treatment regimes (Forbes *et al.*, 2001; Wilson *et al.*, 2000).

Apart from the notion that almost all forms of neoplasms affecting mammals can be also found in birds, the neoplasms of haematopoietic system take a prominent place in the oncology of birds. The research so far has pointed to retroviruses as being the most common cause of neoplasms (Payne *et al.*, 1992; Gould *et al.*, 1993). However, retroviruses participate equally in inducing other forms of neoplasms as well. In Budgerigars, haematopoietic neoplasms are rare, while the most common neoplasms are carcinomas of the kidney, ovary, and testis. The carcinomas of the genitourinary tract and fibrosarcomas in chickens are a part of a spectrum of neoplasms that may be caused by infectious type C retroviruses (avian leukosis/sarcoma viruses). In addition, two similar forms of hematopoietic neoplasia (myelocytomatosis and myeloblastosis) have been observed in chickens. These neoplastic diseases are associated with avian myeloblastosis virus and often result in hepatomegaly and splenomegaly from neoplastic cell infiltration (Reece, 1992; Payne *et al.*, 1992; Gould *et al.*, 1993).

The avian leukosis/sarcoma viruses are closely related and, depending on their genetic makeup, cause a variety of neoplasms with short to' long latencies. Some viral strains such as avian myeloblastosis virus, avian erythroblastosis virus, and the sarcoma viruses contain specific viral oncogenes that cause a rapid neoplastic transformation of target cells with a subsequent tumor development within a few days or weeks (Payne *et al.*, 1992; Gould *et al.*, 1993).

Marek's disease virus (MDV), a highly cell-associated avian herpesvirus, is the etiological agent of Marek's disease (MD), a malignant T cell lymphoma (Churchill *et al.*, 1967; Nazerian *et al.*, 1968). The MDV genome codes for several unique proteins, some of which have been associated with the oncogenicity of the virus. *Meq* is the most extensively studied gene of MDV and it codes for a protein that shares significant homology to the *jun/fos* family of transcriptional factors. *Meq* is consistently expressed in all MDV-transformed cells, suggesting that it may play an important role in transformation (Jones *et al.*, 1992). In addition, pp38 is a phosphoprotein expressed in both lytically infected and tumor cells (Cui *et al.*, 1991). The function of this protein is still not clear but has been suggested to be involved in the maintenance of malignant transformation (Xie *et al.*, 1996).

Rous sarcoma is a connective tissue tumor caused by the Rous sarcoma virus (RSV), an oncogenic RNA virus. Tumors develop after the injection of the virus into susceptible chickens. The tumors may regress or progress depending on the level of the anti-tumor immune response produced by the MHC (Collins *et al.*, 1977; Schierman *et al.*, 1977). The variation in RSV tumor outcome among the identical MHC genotypes from the crosses of inbred lines differing at the MHC (Collinis *et al.*, 1977; Brown *et al.*, 1984), among different inbred lines identical at the B complex (Medarova *et al.*, 2002), or among the crosses of non-inbred lines (Collins *et al.*, 1985) implicated a role for non-MHC genes. For example, non-MHC T lymphocyte alloantigens, Ly-4 and Th-1, and the B lymphocyte alloantigen Bu-1 interacted to alter the response against RSV tumors in the crosses of B2xB2 inbred lines (Gilmour *et al.*, 1986).

All birds, particularly commercial egg laying chicken commonly suffer from malignant and benign ovarian and oviduct neoplasms. Virus aetiology of these neoplasms has not been proved, though they are assumed to have a certain connection with frequent ovulations (25-28 h) typical for the commercial lines of chickens (Fathalla, 1971). The incidence of ovarian and oviduct neoplasms displays a linear growth in parallel with ageing, and correlates directly with the number of ovulations. Fredrickson (1987) reported that developed ovarian and oviduct cancer was verified in 32.4% hens aged between 2 and 7. In another study, ovarian neoplasm was found in 15.6% hens, whereas this percentage was 16.7% in the necropsies of already dead birds. A histological and immuno-histological research of ovarian neoplasms in hens revealed a certain similarity with ovarian neoplasms in mammals. Ilchmann *et al.* (1975) showed in their study that oviduct neoplasms in various bird strains are highly-differentiated in 95%, whereas the presence of poorly-differentiated and highly-invasive carcinoma has been verified in as little as 5% of affected birds.

In conclusion, as reported by Effron *et al.* (1977), birds and mammals display the highest incidence of neoplasms compared with all other vertebrates. Like other vertebrates, neoplasms in birds are mostly virus-induced, though chemical carcinogens also play an important role in the carcinogenesis of this vertebrate group. A number of studies showed that the incidence of manifesting neoplasms in birds can be closely related to the efficacy of anti-tumor response. Collins *et al.* (1977) reports that spontaneous regression of Rous sarcoma and percentage of survival of affected birds is likely to be related with the expression of certain alleles of MHC genes. To that effect, in chickens percentage of survival of the animals with haplotype B12, infected with Marek's virus is over 95%, while the other haplotypes show significantly lower incidence of survival in comparison with the haplotype B12 (Plachy *et al.*, 1992).

5.2.6   Some Important Conclusions

Progress has been made regarding the causes of neoplasia in non-mammalian vertebrates, but many questions are still open. Based on the recent research it is possible to make some important conclusions that could represent the ground and guidelines for the future research in the field of comparative oncology, tumor immunology, genetics and immunogenetics.

1. In vertebrates, tumors are classified according to the tissue and cell type from which they arise. Tumors arising from epithelial cells are termed carcinomas; those arising from connective tissue or muscle cells are termed sarcomas. Neoplasms that do not fit in either of these two broad categories include the various leukemia and lymphoprpliferative diseases, derived from haemopoiesis cells, and tumors derived from cells of the nervous system. Mammals are most susceptible to carcinoma, unlike other vertebrates which are most prone to lymphoproliferative neoplasms, leukemia and sarcomas.
2. Like in mammals, both oncogenic viruses and chemical carcinogens appear to be the common causes of neoplasia in non-mammal. However, the interactions between viruses of non-mammalians and homoeothermism and poikilothermism or environmental factors, like temperature, chemical carcinogens, radiation and food, as well as ageing and the mechanisms of anti-tumor immunity have not been adequately considered.
3. The differentiation between neoplastic and non-neoplastic, but neoplasia-like lesions continues to be a problem. Neoplasia in non-mammals, as well as in mammals, has been defined almost exclusively by histological examination. However, one must ask whether histological diagnosis, which is based on the knowledge originating from mammals, adequately interprets the neoplasms in non-mammals. Gross and cytologic similarities between certain non-neoplastic conditions and neoplasms in lower animals including fish, amphibians,

and reptiles have invited misinterpretations and contested interpretations. The major categories of neoplasia-like lesions, illustrated by specific examples from material accessioned into the Registry of Tumors in Lower Animals, include infections by foreign organisms like alga and amoebas. Among these are: algal protothecosis, amebic pseudotumors, giant islets of endocrine pancreas (Brockmann bodies) in liver, atypical sites of hematopoietic tissue, non-parasitic hyperplasia of thyroid gland, erythroblastic proliferation suggestive of pernicious anemia, adenofibrosis, parasite-induced hyperplasia like trematode-induced fibrosis, ciliate-induced monocytic leukocytosis, trematode-induced melanosis, glochidiosis, dysmorphogenesis (teratoid anomalies); virus-induced hypertrophy of lymphoid tissues and reactive lesions like metaplasia, regeneration and inflammation (Harshbarger, 1974, 1984).

4. Temperature has an important effect on development and regression of neoplasms. Although temperature is a major factor in all aspects of poikilothermic physiology and pathology, these mechanisms involved in temperature changes in the behaviour of fish, amphibian and reptile neoplasms have not been adequately explored yet.

5. The regression of non-mammalian neoplasms, including some considered malignant, needs additional genetic and immunologic studies. Frequent and rapid regression suggests that the immune system in many circumstances can be very efficacious in preventing of tumor escape.

6. Important genetic, immunogenetic and immunological differences that have been verified in non-mammalian classes of vertebrates, especially those concerning control mechanisms of the immune reaction, may provide a viable explanation for a more effective anti-tumor immunity of these animals in contrast to mammals.

7. The importance and usefulness of the transplantation of neoplastic tissue need to be clarified. A successful transplantation of tumors within the species of non-mammals has occasionally been used as the evidence of the neoplastic nature of the lesion. However, such experiments are limited by an insufficient number of inbred strains within non-mammalian vertebrate classes. Basic information regarding transplant rejection and factors that affect growth of normal tissue when transplanted to syngeneic animals need to be determined.

8. Non-mammalian neoplasms metastasize less often and less aggressively than do similar tumors in mammals. The rationale behind this phenomenon is the absence of lymph drainage and lymph nodes However, the majority of non-mammalian neoplasms are possibly differentiated better than the corresponding neoplasms in mammals, and also the cells of these express a greater number of adhesive molecules.

Most neoplasms of wild non-mammals do not appear to affect the size of species population; however, shifts in age distribution can occur. Additional consideration should be given to the potential long-term effects if high frequencies of neoplasms occur for several generations.

9. Non-mammals offer several advantages over mammals in screening for carcinogenicity, but additional refinement is needed for test procedures. Factors that need consideration include the relative advantages of various species within classes, route of exposure to carcinogens, genetic mechanisms of carcinogenesis and anti-tumor immunity.

5.3 ANTI-TUMOR IMMUNITY AS AUTO-IMMUNITY

Tumor development is frequently accompanied by the immune response against "self" and altered antigens expressed by tumor cells, because these antigens on vertebrate tumors are the most prevalent molecules recognized by the immune system (Houghton, 1994; Jäger *et al.*, 2001). This reflects the fact that tumor arise from the hosts' own tissues, and are not truly "foreign", except in the cases when tumor cells express the so-called fusion proteins and/or viral peptides. Thus, in some respects, the immune recognition of tumor appears to be different from the immune recognition of bacteria, and typically more akin to auto-immunity. In addition, the immune reaction to virally infected cells showing no malignant alterations, displays some characteristics of auto-immune reaction. This inevitably activates the regulatory mechanisms which prevent a complete destruction of tissues and organs. From these reasons, the recognition of "self" antigens on tumor cells in most circumstances presents problems for the host immune system. First, the immunity to tumor may not develop because all vertebrates pass across the embryonic phase of establishing of specific immune tolerance on "self" molecules. Second, even when the immune system can recognize and respond to tumor antigens, immunity may not be sufficient to reject cancers, due to the activation of the mechanisms which control auto-immunity. Finally, if immunity to "self"-tumor antigens develops, there are potential auto-immune sequelae, which may also result in the activation of the control suppressor/modulatory mechanisms of the immune reaction.

Auto-antibodies specific to different "self" molecules have been found in the sera of tumor bearers, which could be taken as an evidence for frequent joint activity of anti-tumor immunity and auto-immunity. This emphasizes the idea that tumor patients can mount tumor immunity which could be, in part, auto-immunity. In contrast to patients with auto-immune diseases, in the majority, if not all, tumor patients the immune system is unable to combat tumor growth.

Tumors seem to find ways to generate tolerance in the immune system by activating the control mechanisms of auto-immune reaction responsible for the tolerance against "self" molecules. These mechanisms include a down-regulation of MHC class I molecules and cellular constituents involved in the antigen processing and presentation pathways (Salih *et al.*, 2001). Tumors can also induce several different biochemical defects in physiology of T lymphocytes. In addition, the immune response against tumors is hindered by the functional hierarchy in the immunogenicity of T and B cell determinants, abnormalities occurring in the communication between the cells of innate and adaptive immunity, as well as the inadequate cytokine network (Stavely-O'Carrol *et al.*, 1998).

In line with Burnett's theory of clonal selection, T-cell clones specific to dominant determinants of tumor antigens are probably deleted during embryonic development in the process of negative selection. This could possibly continue into an adult stage as a central (thymic) deletion of tumor-specific clones (Bubanovic, 2003b), or even as a peripheral deletion in the course of extrathymic lymphocyte maturation (Bubanovic, 2003c). Thus, most of the tumor determinants are expected to be immunologically silent; hence effective tumor immunity cannot be induced via "self"-vaccination. Additionally, as tumor accumulates antigens during transformation they also gradually induce tolerance in T cells against these antigens.

Notwithstanding these and other escape mechanisms, in few cancer patients a spontaneous regression of malignant tumors was observed (Palo *et al.*, 1977; Paul *et al.*, 1994). Data about potential coupling of auto-antibodies and prolonged/sustained survival or even spontaneous tumor regression corroborate the previous observation. Breast cancer patients with a natural humoral response to MUC-1 and/or hsp90 exhibited a better outcome (von Mensdorff-Pouilly *et al.*, 1996; Conroy *et al.*, 1996). Similar to immunological events in some auto-immune diseases, tumor in regression exhibited mainly a Th1 type response, as well as non-pathogenic auto-antibodies, but thus the form of auto-immunity did not always develop into the auto-immune disease. There is data that about the potential coupling of tumor immunity with auto-immunity has been suggested by the clinical observation that the patients with metastatic melanoma who develop vitiligo have a better prognosis (Bystryn *et al.*, 1987). In addition, there are observations that support a possible protective role for the auto-immune diseases in cancer patients. In this respect, the mortality rate of cancer patients with multiple sclerosis was found to be significantly lower than that of cancer patients in general (Palo *et al.*, 1977). This could be associated with the activation of control anti-auto-immune mechanisms which may also inhibit auto-immunity and anti-tumor activity of the immune system.

Differentiation antigens are one prototype of "self" antigens expressing by tumor cells of all vertebrates (Houghton *et al.*, 1982; Vijayasaradhi *et al.*, 1990; Brichard *et al.*, 1993). A differentiation antigen distinguishes a cell lineage from other cell types, and is typically expressed at specific stages of differentiation (Boyse *et al.*, 1969). The immune recognition of mammalian malignant tumors has been intensively investigated in melanoma model, a cancer arising from melanocytes (Houghton *et al.*, 1994). The studies of melanoma have surprisingly shown that the immune system commonly recognizes the products of genes that are specifically expressed by melanocytes, particularly genes that are involved in the synthesis of pigment (Houghton *et al.*, 1994; Sakai *et al.*, 1997). Examples include tyrosinase, the critical enzyme required for the synthesis of the pigment melanin, and tyrosinase-related proteins that determine the type of melanin synthesized (Vijayasaradhi *et al.*, 1990; Wang *et al.*, 1996). In general, differentiation antigens can be recognized by antibodies and by $CD8^+$ and $CD4^+$ T cells, and thus can be broadly seen by the immune system (Vijayasaradhi *et al.*, 1990; Clynes *et al.*, 1998; Overwijk *et al.*, 1999).

The immunization of mice against the tyrosinase family antigen induces tumor immunity and auto-immunity that is mediated by autoantibodies (Weber *et al.*, 1998; Clynes *et al.*, 1998). Immunity against tyrosinase antigen led to tumor protection and to depigmentation that was indistinguishable from auto-immunity induced by immunization (Weber *et al.*, 1998). Similar results were observed by Naftzger *et al.* (1996) and Overwijk *et al.* (1999) after the immunization with syngeneic tyrosinase antigen expressed by baculovirus in insect cells or in vaccinia virus, respectively. In differentiation antigen systems, tumor immunity and auto-immunity were mediated by autoantibodies, and tumor immunity depended on NK cells and functional Fc receptors; neither tumor immunity nor auto-immunity required $CD8^+$ T cells (Naftzger *et al.*, 1996; Weber *et al.*, 1998; Clynes *et al.*, 1998; Overwijk *et al.*, 1999).

In conclusion, the potential coupling of tumor immunity with auto-immunity has been suggested by the clinical observation that patients with metastatic tumor who develop auto-immune phenomena have a better prognosis and are more likely to respond to therapy (Bystryn *et al.*, 1987; Rosenberg *et al.*, 1996). The differences in mechanisms underlying tumor immunity and auto-immunity could be a consequence of fundamental differences in effector mechanisms used to kill tumor cells versus normal cells. The activation of the auto-immune process in parallel with an effective anti-tumor response could mean the failure of protective control mechanisms of the immune reaction that may be responsible for the prevention of auto-immune diseases. On the other hand, the activation of suppressor/modulatory mechanisms possibly accompanied by the activation of anti-tumor auto-immune-like immune response could be a factor of anti-tumor immunity failure in all vertebrates (*Figure* 1.5.).

5.4   PREGNANCY AS A "SUCCESSFUL" TUMOR

The development of the placenta in eutherian mammals involves rapid invasion of the uterine wall by the trophoblast cells, a process with certain similarities to tumor cell invasion (Enders, 1991). Unlike tumor cell invasion of the host's healthy tissue, the unique interaction between genetically diverse tissues such as trophoblast and endometrial cells is a sequence of well-coordinated events, both in terms of time and place. During early pregnancy, fetal chorionic villi that contact the uterine wall, give rise to columns of mononuclear cytotrophoblasts that penetrate the superficial portion of the uterus (Bell *et al.*, 1986; Enders, 1991). From these columns emanate cytotrophoblasts that invade into the uterus and its arterioles (Zhou *et al.*, 1997). This process is quite similar to the tumor cell invasion of the host's healthy tissue, but unlike the penetration of neoplastic cells, trophoblast invasion consists of more control levels, including immune control mechanisms. The precise mechanisms controlling the invasion of trophoblast cells into maternal tissue are unknown; however, there are several lines of evidence to suggest that the interaction between steroid hormones, leukocytes and locally produced cytokines, and growth factors are involved (Clark, 1991; Bubanovic, 2003a, Bubanovic *et al.*, 2003d, 2004).

Upon blastocyst adherence, the stromal cells of the uterine endometrium undergo a decidual transformation. This important step in creating the conditions for transplantation is called a decidualization of the endometrium (Bell *et al.*, 1986; Enders, 1991). Decidualization is a very complex process which, apart from the changes occurring on the endometrial cells, as well as their metabolism and receptor repertoire, also includes a complete recombination of the immune cells of the endometrium. The re-subpopulation of immune cells daring decidualization entails the changes in the number and mutual relationships between certain lymphocyte subsets, APCs and NK cells. These events are assumed to enable the establishing of a "friendly setting" for the invading trophoblast cells expressing paternal antigens (Clark, 1991; King et al, 1991, 1997; Bulmer *et al.*, 1992; Marx *et al.*, 1999a).

Trophoblast cells are devoid of MHC class II antigens, but they express MHC class I molecules (Bulmer *et al.*, 1992; Hutter *et al.*, 1998). Actually, this is the expression of class I vary from species to species. Some mammals express class Ia and class Ib molecules on trophoblast cells, while other express only class Ib molecules, selectively distributed on cytotrophoblasts at the fetal-maternal interface, where it plays a role in the maternal-fetal tolerance (Kovats *et al.,* 1990).

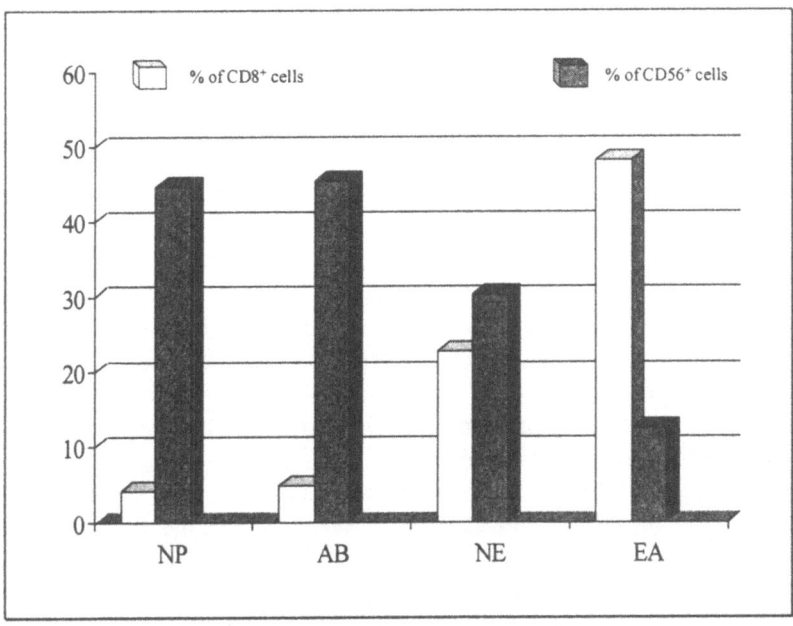

Figure 5.2. Percentages of CD8[+] and CD56[+] cells in normal pregnancy (NP) decidua basalis, decidua basalis from spontaneous abortion (AB), in normal secretory endometrium (NE), and in endometrioid adenocarcinoma (EA) (Arck *et al.,* 2000).

In the case of pregnancy, invading trophoblast cells probably receive some kind of a stop signal from immune or other decidual cells, but in the case of invasion by endometrioid carcinoma cells there is no stop signal or the signal is ineffective and can be considered as an uncontrolled invasion. Also, the activity of immune decidual cells inhibits immune reaction, resulting in the developing of specific tolerance to the invading trophoblast tissue. From these reasons, tumor-infiltrating immune cells can tolerate tumor cells, probably because the inhibitory signals emitted by tumor cells are similar or same as the signals emitted by trophoblast cells. Moreover, tumor cells are similar, to some extent, to trophoblast cells in terms of receptor phenotype, biochemism, MHC expression, cytokine and prostaglandine activity (Bubanovic, 2003a, Bubanovic *et al.,* 2003d, 2004). For example, cytokine like TGF-β and IL-10 may also be the important factors for successful pregnancy and tumor escape. By producing bioactive IL-10 and TGF-β, trophoblast, decidual and tumor cells may induce an immunodeficiency state, and escape from immune monitoring (Bubanovic, 2003a; Bubanovic *et al.,* 2003d).

5.4.1   Immunological Properties in Successful vs. Unsuccessful Pregnancy

Since immunocompetent cells are present in the decidua, these cells might be involved in the final fate of pregnancy. Arck *et al.* (2000) have detected no significant differences in total cell numbers, and the percentages of immunocompetent cells, such as CD3, CD8, CD56 and CD68 positive cells, in normal first trimester human decidua basalis versus decidua basalis from spontaneous abortion. Vassiliadou *et al.* (1999), obtained similar results after comparing T lymphocytes subsets in the normal pregnancy decidua versus the decidua from abortion. Both groups of authors have demonstrated that all T cell subpopulations were detectable in similar numbers in the decidua of normal early pregnancy, and in the decidua of spontaneous abortion (Vassiliadou *et al.,* 1999; Arck *et al.,* 2000). These findings suggest that the outcome of pregnancy does not depend on the relationships between decidual T cell subsets. Nevertheless, there is a possibility that T cells may have an altered antigenic phenotype in spontaneous aborters,. which could contribute to the pregnancy success or failure. In addition, any difference in $CD56^+$ NK cell numbers, and percentages in normal pregnancy versus abortion, also was not observed (Vassiliadou *et al.,* 1999; Arck *et al.,* 2000).

Contrary to foregoing results, several well-controlled studies have shown that certain $CD56^+$ NK cell subsets may have a key role in the pathogenesis of recurrent spontaneous (Lachapelle *et al.,* 1996; Yamamoto *et al.,* 1999). Namely, the total NK cell number is identical in the decidua of recurrent aborters and normal controls though great difference is revealed by comparing the NK cell subsets of healthy pregnant women and recurrent aborters. Flow cytometry revealed that the $CD16^-CD56^{+bright}$ NK cell subset, which is predominant in normal decidua, and endometrium, was significantly decreased in favour of an important contingent of $CD16^+CD56^{+dim}$ NK cells in all habitual aborters (Yamamoto *et al.,* 1999). Marzusch *et al.* (1997) suggested that interactions between decidual $CD16^+CD56^{+bright}$ NK cells and macrophages could play a significant role in regulating the secretion of IFN-γ and other major pro-inflammatory cytokines at the feto-maternal interface, thus contributing to the control of trophoblast invasiveness. Ho *et al.* (1996) suggest that the activation status of the cells investigated in the decidua of normal and aborted pregnancy may be different. Their findings indicate that T cells are regionally activated in first trimester normal pregnancy, and it may be the result of the stimulation by fetal antigens. The same authors have shown that the number of activated NK cells expressing CD56⁺CD16⁻CD57⁻ phenotype is considerably higher in the decidua of unembryonic pregnancy, as compared to that of normal pregnancy, although NK cell numbers of both normal and unembryonic pregnancy is similar (Ho *et al.*, 1996).

Macrophages are ubiquitous cells with a large number of functions during pregnancy. Their activity depends on many factors, thus directly reflecting the systemic microenvironmental conditions (Hunt *et al.*,. 1998). Macrophages are present in cycling and pregnant mammalian uteri and their densities and patterns of tissue distribution in this organ fluctuate in concert with the levels of circulating estrogens and progesterone. Since macrophagal production of various effector molecules also may be hormonally regulated, the hormonal influence on macrophages might be an explanation for the different percentages observed in CD68⁺ cells in normal pregnancy, abortion versus normal endometrium (Arck *et al.*, 2000).

T cells, NK cells and macrophages have been shown to communicate by producing Th1 and Th2 cytokines. In general, type 1 cytokines favours the development of a strong cellular immune response, whereas type 2 cytokines favours a strong humoral immune response (Mosmann *et al.*, 1986; Romagnani *et al.*, 1997). Th1 type cytokines stimulate abortions, while Th2 type cytokines prevent abortions. In pregnancy, a Th1 to Th2 shift is postulated, and both mouse and human data support this hypothesis (Wegmann *et al.*, 1993; Lin *et al.*, 1993). Th1 type cytokines stimulate NK cells and macrophages that are involved in abortions, while Th2 cytokines suppress them, exerting anti-abortogenic effects (Raghupathy, 1997).

Cytokines IL-10, IL-4 and TGF-β play the important role in regulating feto-maternal relationships. Lea *et al.* (1995) confirmed in their study that the patients with chronic spontaneous abortions show a weaker ability of decidual secretion of these cytokines, in comparison with healthy pregnant women. The researches on animals have shown that the TGF-β2 producing cells bear the γδTCR cells, while IL-10 and IL-4 are derived from Th2 cells and APCs (Croitoru *et al.*, 1999). The origin of γδTCR lymphocytes could be associated with the phenomenon of extrathymic lymphocyte maturation, while their activity could be critical for regulating the cytokine activity of αβTCR lymphocytes, NK cells and macrophages (Bubanovic, 2003c).

5.4.2 Immunological Properties in Pregnancy and Tumor Microenvironment

The studies comparing the involvement percentage of different subsets of immunocompetent cells in decidual tissue of pregnancies ending in spontaneous abortions, normal pregnancies and adenoid adenocarcinoma, have shown significantly fewer T and CD56$^+$ cells/mm^2 in both, benign and malignant endometrium, compared with the decidualized endometrium, and a significant increase in CD8$^+$ cells in malignant endometrium (Arck *et al.,* 2000).

Tumor infiltrating immune cells have been described in different mammalian solid tumors, whereby it is unknown whether such cells are non-specific inflammatory cells or a subset of specific host immune responses. There are studies in which TIL subpopulation were compared with the lymphocyte subpopulation in decidua. Interestingly, the observations on leukocyte percentages made in tumors are very similar to the leukocyte distribution pattern of in ectopic pregnancies. The decidua of tubal pregnancy contains the increased number of CD8$^+$ and the reduced number of CD56$^+$ cells as compared to the normal pregnancy decidua (Bulmer *et al.,* 1992; Marx *et al.,* 1999b; Arck *et al.,* 2000). The reduced number of CD56$^+$ cells accompanied by the increased number of CD8$^+$ cells, could be associated with tumor invasiveness and the failure of anti-tumor immune response. The ingress of lymphocytes into mammalian neoplasms does not appear to be a random event, and may be provoked by tumor-associated antigens and tumor secretory activity. The further course of anti-tumor response depends on the quality of the communication between lymphocytes and APCs, the degree of MHC expression on tumor cells and APCs, and cytokine network built from tumor cells and tumor infiltrating immune cells. The process in situ of the extrathymic lymphocyte maturation is not excluded, either finally; the relationship between the subsets of tumor-infiltrating immune cells is the result of selective elimination of some and favouring of other immune cells subsets. The similarities of microenvironmental conditions in the cases of controlled trophoblast invasion and unrestrained tumor invasion result in developing similar immune conditions for the infiltrated tissue. A subset composition of tumor-infiltrating immune cells does not correspond entirely to the composition of decidua in normal pregnancy, yet it is very similar to the decidua of ectopic pregnancy (*Figure* 5.2.) (Arck *et al.,* 2000). These experimental data corroborate the assumption about the similarity of immune events in trophoblast-infiltrating and tumor-infiltrating tissues, which are directed towards the immunotolerance of invasive tissues (Bubanovic, 2003a; Bubanovic *et al.,* 2003d).

5.5   EVOLUTION OF THE HYPOTHESIS

Similarly to all old and unsubstantiated hypotheses and theories, the hypothesis about malignant tumors as a phenomenon of reproduction has passed through its own evolutionary path. In the late seventies and early eighties of the 19th century, the views of Cohnheim (1882) were formed regarding the origin of malignant neoplasms. According to Cohnheim (1882), malignant tumors develop either from the embryonic tissue rests that occasionally came to be among definitive tissues of the same histogenesis. These were not included in the process of building the normal tissues or from embryonic residues transferred to another place which become heterotopic objects and therefore are not involved in intratissular relations. These embryonic residues give rise to neoplastic growth. This embryonic theory was tested by inoculating embryonic cells at all stages into an adult recipient. Results of this investigation showed that embryonic cells grew for some time and then became mature tissues. Finally, pathologists found that the mature cells did not look the same as the malignant cells (Cherezov, 1997).

The so-called "trophoblast thesis" is another idea about the connection between embryonic and tumor tissue. The "trophoblast thesis" was first put forward in 1902 by Scottish embryologist John Beard, and rediscovered about fifty years later by controversial Ernst Krebs. His often-quoted statement is that "cancer is trophoblast in spatial and temporal anomaly, hybridized with, and vascularized by, hostal or somatic cells and in irreversible and fiercely malignant antithesis to such" (Krebs Jr., 1950, 1993). Today, the Unitarian or "trophoblastic thesis" of tumors has not many supporters in the scientific world, but the idea has still survived. Maybe the most persuasive evidence against the "trophoblast thesis" is, in fact, that malignant tumors are discovered in animals like birds, reptiles and others non-mammalian classes of vertebrates.

All presented hypotheses and all yet to come are based on the similarities between tumors and embryonic or trophoblast cells like antigens phenotype of cell surface, the endocrine profiles, the production of oncofetal antigens, the insusceptibility to apoptotic signals, influence on surrounding microenvironmental immune and other cells etc. Also, the immunological properties of the two tissues are also very similar. Whether or not cancer originated from a trophoblast, the cells and the immunological events appear to act in a similar way (Bubanovic, 2003a, Bubanovic *et al.,* 2003d, 2004).

A modern concept of the connection between malignant tumors and embryonic or trophoblast tissues was promoted by Valentin Govallo and Rigdon Lentz separately, and applies to the immunological events in tumor patients and pregnancy (Govallo, 1983, 1996; Lentz, 1990). In the 1960s, Valentin Govallo began his exploration through the studies of parallelism between mother-placenta and host-tumor systems. As a result of the investigation, Govallo had the idea of using a placental extract to immunize the patient against "the fetus-like cancer". Unlike to most immune therapies that stimulate the immune system, the Govallo's therapy is designed to weaken or suppress the factors within the tumor that "turn off" normal immune responses of the host (Govallo, 1983, 1996).

Rigdon Lentz formulated an evolutionary explanation of the similarity between immunological events in pregnancy and tumor patients. He believed that pregnancy and cancer are the only two biologic conditions in which antigenic tissue is tolerated by a seemingly intact immune system. Lentz (1990, 1999) suggests that an evolved mechanism of an acquired tolerance to MHC incompatible tissue necessitated by sexual reproduction consequently provides a mechanism for the tolerance of cancer. However, many experimental works have shown that trophoblast and cancer cells are associated with an altered expression of MHC class I molecules. A poor prognosis of malignant disease has been documented in association with HLA loss and there may be a higher frequency of selective loss of HLA class I specific to metastases in comparison with the primary lesion (Geertsen *et al.,* 1996). For example, in breast cancer, the total class I loss was found in >50% of patients, with a further 35% showing selective losses, whereas only 12% tumor retained full HLA class I expression (Cabrera *et al.,* 1996). A common reason for the decreased class I expression in mammal's tumor cells is the loss of TAP expression, as well as LMP2 and 7. Also, trophoblast and tumor cells can express unusual forms and number of MHC molecules like HLA-G and HLA-C. These molecules may mediate the inhibition of antigen-specific lysis by CTL and antigen-nonspecific lysis by NK cells (Paul *et al.,* 1998).

Immunosuppression is a hallmark of advanced malignancies and successful pregnancy in humans. Over the past 40 years, many investigators have identified soluble immunosuppressive factors in blood and trophoblast or cancer tissue in humans and other mammals. The suppressive factors have also been identified that are produced by trophoblast, tumors, and decidual or tumor infiltrating immune cells. The description of immunosuppressive factors in the blood of mammals which either have cancer or which are pregnant is significant, for only in pregnancy and cancer does a seemingly normal immune system tolerate proliferative tissues. Since the similarity between immunological events in pregnancy and malignancy is so significant, the connection between these processes must be real. The facts that mammals are single vertebrates that have placenta give us the chance for comparing anti-tumor immunity in mammals and other classes of vertebrates. In addition, the detection of anti-tumor mechanisms in non-mammalian classes of vertebrates can be very usable in efforts to prove the connection between pregnancy and malignancy (Bubanovic, 2003a; Bubanovic *et al.,* 2003d, 2004).

5.5.1 Origin of Anti-tumor Immunity Failure in Vertebrates

There are many reasons to believe that the phenomenon of oncogenesis follows the same or similar postulates across all vertebrate classes. A large number of studies also indicate the possibility that the biological fundaments of oncogenesis in vertebrates are observable in invertebrates, as well. It is clearly, therefore, that one of the primary steps leading towards oncogenesis in all vertebrates is the destabilization of the genome and loss of cells' ability to repair DNA damage. Genome destabilization ending in oncogenesis is, basically, the consequence of a series of mutations or translocations, and comes largely as a result of the influence of chemical agents, viral infections, radiation and chronic inflammations including auto-immune processes. What makes an altered cell malignant is the activation of individual genes or group of genes, otherwise normally functional during embryogenesis. An incomplete and inadequate ontogenic regression in terms of both time and space results in the loss of sociability of the altered cell, destruction of its environment and the host. The described mechanism of ontogenesis has been verified in all classes of vertebrates, except for cartilaginous fish where few cases of tumors have been registered. Some of the reasons for considerable variations regarding tumor incidence across vertebrate classes, may be contained within genome stability, as well as various possibilities for repairment of the DNA damage.

There is a whole set of other factors concerning tumor genetics across vertebrate classes that could be a source of variability in the incidence of tumors in vertebrates. The purpose of this publication, namely, is bringing to light the relationship between the host's immune system and tumors, as well as indicating the great influence that the observed variability could possibly have on the incidence of manifesting tumors. It is possible to prove the failure of anti-tumor immunity in all vertebrates, though the mechanisms involved in the developing of this phenomenon and intensity of anti-tumor immunity failure may greatly differ across vertebrate classes.

Based on recent findings, several mechanisms have been identified that participate in the development of the phenomenon of anti-tumor immunity failure in vertebrates:

1. The processes of embryogenesis and mechanisms establishing central immune tolerance to "self" molecules are similar across all vertebrate classes. Therefore, the mechanisms of immunotolerance to embryonic and/or embryonic-like tissues, such as tumor tissues, may be connected with central immunotolerance.
2. There is evidence that the mechanisms of central immunotolerance are, more or less, active in adults. This phenomenon could also be involved to some extent in the development of anti-tumor immunity failure.

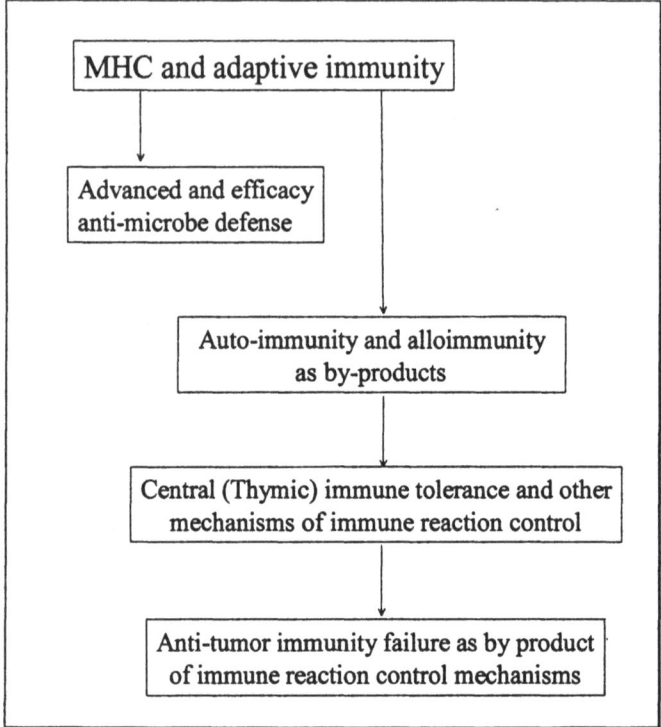

Figure 5.3. Anti-tumor immunity failure as a final by-product of MHC and adaptive
immunity.

3. The mechanisms of peripheral maturation of lymphocytes and possible
   influence of these on the quality of the immune reaction have been also
   verified in most vertebrate classes.
4. The immune system of vertebrates is closely connected with auto-
   immunity being a by-product of the adaptive immunity. "Self"-protective
   immune reaction control mechanisms that could become activated in anti-
   tumor immune reaction, co-evolved alongside auto-immunity as a "new"
   evolutionary phenomenon and a new form of selection pressure.
   Therefore, the activation of anti-tumor immunity as a unique form of
   auto-immunity process was probably followed by a parallel activation of
   protective mechanisms, i.e. the immune reaction control mechanisms
5. Unlike non-mammals, the immune system of mammals has "built-in" the
   mechanisms of tolerance to proliferative tissues like trophoblast. These
   mechanisms have developed, in the course of evolution, under a very
   strong selection pressure of alloimmunity and reproductive efficacy.
   Although very similar to the mechanisms resembling auto-immunity, the
   mechanisms of immune tolerance to trophoblast can be regarded as more

advanced and more effective. On the other hand, there is a great similarity between the mechanisms of immune tolerance to trophoblast and anti-tumor immunity failure mechanisms. These two, apparently diverse mechanisms may be regarded as a protective immune cross-reaction against the proliferative tissues of diametrically different origin (*Figure* 5.3.).

All vertebrates are, more or less, susceptible to oncogenesis, depending on the sensitivity of their DNA to the influence by various oncogenic factors. The phenomenon of anti-tumor immunity failure is relatively easily verifiable in all vertebrate classes, due to the similarities in organization and functioning of their immune systems. The phenomenon of anti-tumor immunity failure in non-mammals rests largely on the mechanisms of central immune tolerance to embryonic and/or embryonic-like cells and control mechanisms resembling auto-immunity. However, the diversification of the immune reaction control mechanisms in mammals has again produced new possibilities regarding tolerance to proliferative tissues, i.e. trophoblast and tumors. Several observations could be taken as a basis for the future research in the field of comparative tumor genetics, immunology and immunogenetics:

1) The expression of class I and class II molecules, tissue distribution of the molecules and level of polymorphism of class I and class II genes in non-mammals and mammals are substantially different. In most non-mammals, class II genes are more polymorphic in relation to class I, while class I genes are highly polymorphic in mammal genome. Tissue distribution of class II molecules in mammals is restricted on APCs, dendritic cells and B lymphocytes, while non-mammals shows the phenomenon of poor restricted or unrestricted tissue distribution of the class II molecules (Hughes *et al.*, 1993; Lawlor, 1990).

2) In all non-mammalian classes of vertebrates class I and class II genes are rambling through genome, but LMP and TAP genes are highly evolutionary conserved within class I region. In mammals, class I and class II genes are clustered on the same chromosome (except equine), but LMP and TAP genes are conserved within class II region (Kasahara *et al.*, 1996, 1997; Kaufman *et al.*, 1996).

3) The transcription of class II/LMP/TAP genes in mammals are controlled from same signals. The absence of class II genes transcription signals lockout antigens processing machinery and class I molecules peptide presentation, as well as the activation of Th1 cells and adaptive immunity effectors actions (Chaux *et al.*, 1996; Paul *et al.*, 1998). In non-mammals, antigen processing machinery is under control of class I genes transcription, because class I/LMP/TAP genes are

closely connected on the same chromosomal loci (Kasahara *et al.*, 1996, 1997; Kaufman *et al.*, 1996).

4) Anti-tumor immunity in non-mammalian (except birds) vertebrates predominantly depend on the innate immune system, while anti-tumor immunity in mammals depend on the innate and adaptive immune systems and their communication (Robert *et al.*, 1999).

5) The specificity in expression and tissue distribution of MHC genes, LMP and TAP genes transcription control, as well as the communication between native and adaptive immunity in non-mammalian vertebrates qualifies a substantially different cytokine network and immune reaction than in mammals.

6) There is a possibility that malignant cells in fishes, amphibians, reptiles and birds are more susceptible to apoptosis than mammalian malignant cells (Laurens *et al.*, 1997).

7) The high resistance on carcinogens induced genetic changes is evidenced in some experiments with lower vertebrates, leading to a conclusion that DNA from lower vertebrates shows a high level of resistance on carcinogenesis (Laurens *et al.*, 1997; Harshbarger, 1974).

8) The complex and efficient mechanisms of immune reaction control developed under the evolutionary pressure of high polymorphism of class I genes, auto-immunity and reproductive effectiveness can be included in the mechanisms of anti-tumor immunity failure in mammals.

9) Mammalian extended cytokine network can be activated/deactivated by same or similar factors under different conditions such as pregnancy and malignancy. A small number of cytokines and poor cytokine network are the characteristics of non-mammalian vertebrates. For example, cytokines like IL-10 and IL-4 are unknown in fishes and amphibians, but TGF-β is evidenced in reptiles, birds and probably in other non-mammalian classes (Paulesu *et al.*, 1997; Reboul *et al.*, 1999).

10) Th-like cells are detected in reptiles and amphibians (Wei *et al.*, 2001), as well as Th and/or Th1-like cells in birds (Vandaveer, 2002), but mammals are single vertebrates which have the advanced system of immune reaction control established on Th1 and Th2 cells, and their balanced activity. The absence or fractional awareness of Th2 model of immune reaction control probably contributes in the strong anti-tumor immunity in non-mammalian vertebrates.

11) Mammalians' immune system may be tolerant to cancer cells because they are very similar to trophoblast cells (Bubanovic, 2003a).

12) Sex hormones, steroids and other factors, which are the attributes of pregnancy and malignant processes, can impair blood-thymus barrier. It can be another mechanism of acquired thymic tolerance to foreign molecules in pregnancy and malignancy (Bubanovic, 2003b).

13) The absence of MHC and costimulatory molecules expression, prostaglandine, Th2 cytokines, sex hormones, steroids and other factors could be promoter of extrathymic lymphocytes maturation in antigen-protective manner in mammalians. It is yet one of the mechanisms that are included in trophoblast and tumor escape (Bubanovic, 2003c).

14) Unlike mammals, the mechanisms of immune reaction control in non-mammalian vertebrates probably are essentially independent from an important role of co-stimulatory molecules. Actually, co-stimulatory molecules like CD40, CD80, CD86 and OX40 were not detected in non-mammalian vertebrates, except CD80 and CD86-like molecules in birds (O'Regan *et al.*, 1999).

In conclusion, the observations about the similarities of behaviour and phenotype between tumor and embryonic/trophoblast cells are very old. If the mechanisms of anti-tumor immunity in mammals are similar or same as the mechanisms of immunoregulation in pregnancy, then the mechanisms of anti-tumor immunity in non-mammalian vertebrates may be very useful in designing new immunotherapeutic procedures. The model of cytokine network in non-mammalian anti-tumor immune response can be usable in designing of the mammalian response. Furthermore, the model of communication between the native and adoptive immune cells in non-mammalian vertebrates can also be usable in designing anti-tumor immunotherapy in mammals. Non- mammalians cytokines or some other factors might be a good adjuvant of the current anti-tumor vaccination. Finally, trophoblast or embryonic cells or their antigens can be used for anti-tumor immunization. Taking into consideration the fact that anti-tumor immunity failure in mammals is immunoreproductive phenomenon, many new possibilities for the immunotherapy of malignant diseases could be opened (Bubanovic, 2003a). Besides morphological and functional similarities, there is much evidence that immunological events in pregnancy and malignancy are closely connected or same. This possibility enables the establishing of a new definition of anti-tumor immunity failure in mammals as an immunoreproductive phenomenon, as well as new possibilities for the immunotherapy. Finally, the conclusion is that we can learn so much from our distant relatives.

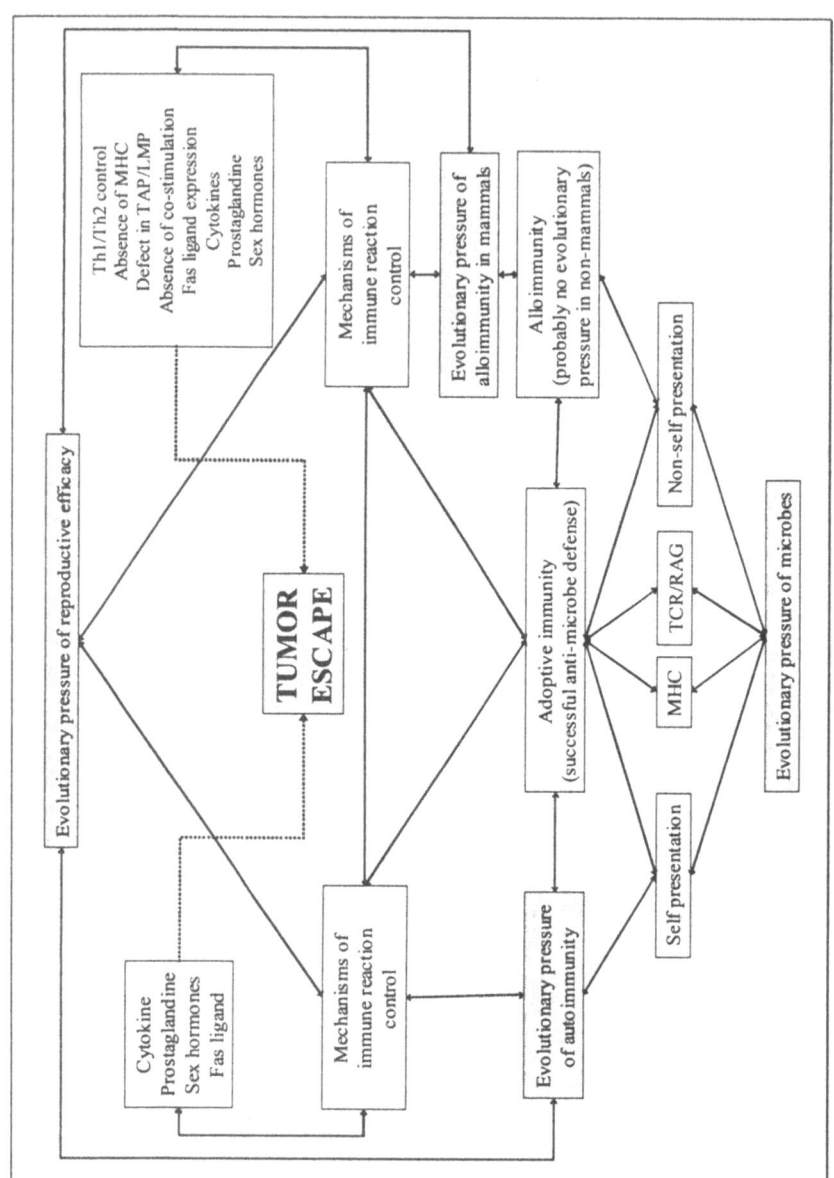

Figure 5.4. Evolution of anti-tumor immunity failure in mammals.

REFERENCES

Alford R.A., Richards J.S. (1999). "Global Amphibian Declines: A Problem in Applied Ecology". *Annu Rev Ecol Syst.* 30:133-165.

Anisimov V.N. (2003). Ageing and cancer in transgenic and mutant mice. *Front Biosci.* 1:s883-902.

Arck P.C., Hertwig E., Hagen M. *et al.* (2000). Pregnancy as a Model of Controlled Invasion Might Be Attributed to the Ratio of CD3:CD8 to CD56. *Am J Reprod Immunol.* 44:1-8.

Bailey G.S., Williams D.E., Hendricks J.D. (1996). Fish models for environmental carcinogenesis: the rainbow trout. *Envir Health Perspec.* 104:5-21.

Balls M., Ruben L.N. (1964). A review of chemical induction of neoplasms in Amphibia. *Experientia.* 20:241-247.

Baumann P.C., Harshbarger J.C., Hartman K.J. (1990). Relationship between liver tumors and age in brown bullhead populations from two Lake Erie tributaries. *Sci Total Environ.* 94:71-87.

Bell S.C., Billington W.D. (1986). Humoral immune response in murine pregnancy. V. Relationship to the differential immunogenicity of placental and fetal tissues. *J Reprod Immunol.* 9:289-295.

Bodine A.B., Luer C.A., Gangjee S. (1985). A comparative study of monooxygenase activity in elasmobranchs and mammals: activation of the model pro-carcinogen aflatoxin B1 by liver preparations of calf, nurse shark and clearnose skate. *Comp Biochem Physiol.* 82:255-257.

Bodine A.B., Luer C.A., Gangjee S.A. (1989). In vitro metabolism of the pro-carcinogen aflatoxin B1 by liver preparations of the calf, nurse shark and clearnose skate. *Comp Biochem Physiol.* 94:447-453.

Boyse E.A., Old L.J. (1969). Some aspects of normal and abnormal cell surface genetics. *Annu Rev Genet.* 3:269-290.

Bremermann H.J. (1987). The adaptive significance of sexuality. *Experientia Suppl.* 55:135-161.

Brichard V., Van Pel A., Wolfel T. *et al.* (1993). The tyrosinase gene codes for an antigen recognized by autologous cytolytic T lymphocytes on HLA-A2 melanomas. *J Exp Med.* 178:489-495.

Brown D.W., Collins W.M., Zsigray R.M. *et al.* (1984). A non-MHC genetic influence on response to Rous sarcoma virus-induced tumors in chickens. *Avian Dis.* 28:884-899.

Bubanovic I. (2003a). Origin of Anti-tumor Immunity Failure in Mammals and new Possibility for Immunotherapy. *Med Hypotheses.* 60:152-158.

Bubanovic I., (2003b). Failure of blood-thymus barrier as a mechanism of tumor and trophoblast escape. *Med Hypotheses.* 60:315-320.

Bubanovic I. (2003c). Crossroads of extrathymic lymphocytes maturation pathways. *Med Hypotheses.* 612:235-239.

Bubanovic I., Najman S. (2003d). Failure of Anti-tumor Immunity in Mammals - Evolution of the Hypothesis. The 23rd Annual Meeting of the American Society for Reproductive Immunology. New Haven, Connecticut, USA. June 18-22. *Am J Reprod Immunol.* 49:329-372, Abstract N° 49; p.:351.

Bubanovic I., Najman S. (2004). Anti-tumor Immunity Failure in Mammals - Evolution of the Hypothesis. *Acta Biotheor.* 52:57-64.

Bulmer J.N. (1992). Immune aspects of pathology of the placental bed contributing to pregnancy pathology. *Baillieres Clin Obstet Gynaecol.* 6:461-488.

Bystryn J.C., Rigel D., Friedman R.J. *et al.* (1987). Prognostic significance of hypopigmentation in malignant melanoma. *Arch Dermatol.* 123:1053-1055.

Cabrera T., Angustias Fernandez M., Sierra A. *et al.* (1996). High frequency of altered HLA class I in invasive breast carcinomas. *Human Immunol.* 50:127-134.

Chaux P., Moutet M., Faivre J. *et al.* (1996). Inflammatory cells infiltrating human colorectal carcinoma express HLA class II but not B7-1 and B7-2 costimulatory molecules of the T-cell activation. *Lab Invest.* 74:975-983.

Cherezov A.E. (1997). General theory of cancer. Tissue approach. Publishing House of Moscow University, Moscow. pp.:85-90.

Churchill A.E., Biggs P.M. (1967). Agent of Marek's disease in tissue culture. *Nature.* 215:528-530.

Clark D.A. (1991). Controversies in reproductive immunology. *Crit Rev Immunol.* 11:215-247.

Clynes R., Takechi Y., Moroi Y. (1998). Fc receptors are required in passive and active immunity to melanoma. *Proc Natl Acad Sci USA.* 95:652-656.

Cohnheim J. (1882). Vorlesungen uber allgemeine Pathologie. Berlin.

Collins W.M., Briles W.E., Zsigray R.M. *et al.* (1977). The B locus (MHC) in the chicken: association with the fate of RSV-induced tumors. *Immunogen.* 5:333-343.

Collins W.M., Brown D.W., Ward P.H. *et al.* (1985). MHC and non-MHC genetic influences on Rous sarcoma metastasis in chickens. *Immunogen.* 22:315-321.

Conroy S.E., Latchman D.S. (1996). Do heat shock proteins have a role in breast cancer? *Br J Cancer.* 74:717-721.

Cowan D.F. (1968). Diseases of captive reptiles. *J Am Vet Med Ass.* 153: 848-859.

Croitoru K., Carding S.R., Clark D.A (1999). Murine T cell determination of pregnancy outcome II. Distinct TH1 and TH2/3 populations of Vγ1+δ6+ T cells determine success or failure of pregnancy in CBA:JxDBA:2J matings. *Cell Immunol.* 196:71-79.

Cui Z.Z., Lee L.F., Liu J.L. *et al.* (1991). Structural analysis and transcriptional mapping of the Marek's disease virus gene encoding pp38, an antigen associated with transformed cells. *J Virol.* 65:6509-6515.

Curtis L.R., Zhang Q., Ezahr C. *et al.* (1995). Temperature modulated incidence of aflatoxin B1 initiated liver cancer in rainbow trout. *Fund Applied Toxicol.* 25:146-153.

Dall'Ara P. (2003). Immune system and ageing in the dog: possible consequences and control strategies. *Vet Res Commun.* 27:535-542.

Dawe C.J. (1990). Implications of aquatic animal health for human health. *Envir Health Perspec.* 86:245-255.

Dornburg R. (2003). The history and principles of retroviral vectors. *Front Biosci.* 8:818-835.

Down N.E., Peter R.E., Leatherland J.F. (1990). Seasonal changes in serum gonadotropin, testosterone, 11-ketotestosterone, and estradiol-17β levels and their relation to tumor burden in gonadal tumor-bearing carp X goldfish hybrids in the Great Lakes. *Gen Comp Endocrin.* 77:192-201.

Effron M., Griner L., Benirschke K. (1977). Nature and rate of neoplasia found in captive wild mammals, birds, and reptiles at necropsy. *J Natl Cancer Inst.* 59:185-198.

Egami N., Kyono-Hamaguchi Y., Mitani H. *et al.* (1981). Characteristics of hepatoma produced by treatment with diethylnitrosamine in the fish (Oryzias latipes). In: Harshbarger J.C., Kondo C.J., Sugimura T. *et al.* Phyletic Approaches to Cancer. Japan Scientific Press, Tokyo, pp. 217-226.

Enders A.C. (1991). Current topic: structural responses of the primate endometrium to implantation. *Placenta.* 12:309-325.

Etoh H., Hyoda-Taguchi Y., Aoki K. *et al.* (1983). Incidence of chromatoblastomas in aging goldfish (Carassius auratus). *J Nat Canc Inst.* 70:523-528.

Faisal M., Marzouk M.S., Smith C.L. *et al.* (1991). Mitogen induced proliferative responses of lymphocytes from spot (Leiostomus xanthurus) exposed to polycyclic aromatic hydrocarbon contaminated environments. *Immunopharmacol Immunotoxicol.* 13:311-327.

Fathalla M.F. (1971). Incessant ovulation-a factor in ovarian neoplasia? *Lancet.* 2:163-164.

Fitzgerald T.K. (1995). Reptiles and Cancer. *Cold Blooded News.* 22:18-21.

Forbes N.A., Cooper J.E., Higgins R. (2000). Neoplasms of Birds of Prey. In: Lumeij JS, Lierz M, Remple D, Cooper JE (eds). Raptor Biomedicine III. Lake Worth, Florida: Zoological Education Network, pp.:127-146.

Fredrickson T.N., (1987). Ovarian tumors of the hen. *Environ Health Perspect.* 73:35-51.

Frye F. (1991). Biomedical and Surgical Aspects of Captive Reptile Husbandry, 2nd ed. Malabar, FL, Krieger Publishing Co.

Frye F.L., Carney J. (1975). Parathyroid adenoma in a tortoise. *Vet Med/Small Anim Clin.* 20:582-584.

Garner M., (2001). Gross Pathology of Zoo Animals. C.L. Davis DVM Foundation Armed Forces Institute of Pathology. Washington, DC.

Geertsen R.C., Hofbauer G.F., Yue F.Y. *et al.*, (1998). Higher frequency of selective losses of HLA-A and HLA-B allospecificities in metastasis than in primary melanoma lesions. *J Invest Dermatol.* 111:497-502.

Getchell R.G., Casey J.W. Bowser P.R. (1998). Seasonal occurrence of virally induced skin tumors in wild fish. *J Aqu Anim Health.* 10:191-201.

Gilmour D.G., Collins W.M., Fredricksen T.L. *et al.* (1986). Genetic interaction between non-MHC T- and B-cell alloantigens in response to Rous sarcomas in chickens. *Immunogen.* 23:1-6.

Gonda T.J., Sheiness D.K., Bishop J.M. (1982). Transcripts from the cellular homologs of retroviral oncogenes: distribution among chicken tissues. *Mol Cell Biol.* 2:617-624.

Gould W.J., O'Connell P.H., Shivaprasad H.L. *et al.* (1993). Detection of retrovirus sequences in budgerigars with tumours. *Avian Pathol.* 22:33-45.

Govallo V.I. (1983). Paradoxes of immunology. Znaniye, Moscow. pp.:51-68.

Govallo V.I. (1996). Immunoembryotherapy. In: Transplantation of human foetal tissues and organs. Meditsina, Moscow. pp.:14-18.

Gregory R.C., Latimer K.S., Hafner S. *et al.* (2000). Undifferentiated Sarcoma in a Surinam Toad (Pipa pipa). Proceedings of International Virtual Conferences in Veterinary Medicine.

Grizzle J.M., Horowitz S.A., Strength D.R. *et al.* (1988). Caged fish as monitors of population: Effects of chlorinated effluent from a wastewater treatment plant. *Water Resource Bullten.* 24:951-959.

Harada T., Hatnanaka J., Kubota S.S. *et al.* (1990). Lymphoblastic lymphoma in medaka (Oryzias latipes). *J Fish Dis.* 13:169-173.

Harshbarger J.C. (1974). Activities Report Registry of Tumors in Lower Animals, 1965-1973. RTLA 1385. Smithsonian Institution Press, Washington, D.C.

Harshbarger J.C., (1977). Role of the registry of tumors in lower animals in the study of environmental carcinogenesis in aquatic animals. *Ann New York Acad Sci.* 298:280-289.

Harshbarger J.C., Charles A.M., Spero P.M. (1981). Collection and analysis of neoplasm in sub-homoeothermic animals from a phyletic point of view. In: Harshbarger J.C., Kondo C.J., Sugimura T. *et al.* Phyletic Approaches to Cancer. Japan Scientific Press, Tokyo, pp. 357-384.

Harshbarger J.C. (1984). Pseudoneoplasms in ectothermic animals. Natl Cancer Inst Monogr. 65:251-273.

Harshbarger J.C., Clark J.B. (1990). Epizootiology of neoplasms in bony fish of North America. *Sci Total Environ.* 94:1-32.

Harshbarger J.C. (1996). Comparative oncology. *Jpn J Cancer Res.* 87:1-6.

Hart R.W., Setlow R.B., Woodhead A.D. (1977). Evidence that pyrimidine dimers in DNA can give rise to tumors. *Proc Natl Acad Sci USA.* 74:5574-5578.

Hartwell L.H., Weinert T.A. (1989). Checkpoints: controls that ensure the order of cell cycle events. *Science.* 246:629-634.

Hayes M.A., Smith R.I. (1989). Neoplasia in Fish. In: Ferguson H.W. Systemic Pathology of Fish. Iowa State University Press. Ames, Iowa, pp.:230-247.

Hendricks J.D., Sinhuber R.O., Loveland P.M. *et al.* (1980). Hepatocarcinogenicity of glandless cottonseeds and cottonseed oil to rainbow trout (Salmo gairdnerii). *Science.* 208:309-311.

Hedrick R.P., McDowell T., Eaton W.D. *et al.* (1987). Serological relationships of five herpesviruses isolated from salmonid fishes. *J Appl Icht.* 3:87-92.

Ho H.N., Chao K.H., Chen C.K. *et al.* (1996). Activation status of T and NK cells in the endometrium throughout menstrual cycle and normal and abnormal early pregnancy. *Hum Immunol.* 49:130-136.

Hosel M., Webb D., Schroer J. *et al.* (2003). The abortive infection of Syrian hamster cells with human adenovirus type 12. *Curr Top Microbiol Immunol.* 272:415-440.

Houghton A.N., Eisinger M., Albino A.P. *et al.* (1982). Surface antigens of melanocytes and melanomas. Markers of melanocyte differentiation and melanoma subsets. *J Exp Med.* 156:1755-1766.

Houghton A.N. (1994). Cancer antigens: immune recognition of self and altered self. *J Exp Med.* 180:1-4.

Hughes A.L., Nei M. (1993). Evolutionary relationships of the classes of MHC genes. *Immunogen.* 37:337-342.

Hunt S., Miller L., Platt S. (1998). Hormonal regulation of uterine macrophages. *Dev Immunol.* 6:105-110.

Hutter H., Hammer A., Dohr G. *et al.* (1998). HLA expression at the maternal-fetal interface. *Dev Immunol.* 6:197-204.

Ilchmann G, Bergmann V. (1975). Histological and electron microscopy studies on the adenocarcinomatosis of laying hens. *Arch Exp Veterinarmed.* 29:897-907.

Jacobson E.R., Seely J.C., Novilla M.N. (1980). Lymphosarcoma associated with virus-like intranuclear inclusions in a California king snake (Colubridae:Lampropettis). *J Natl Cancer Inst.* 65:577-583.

Jacobson E.R. (1981a). Neoplastic diseases of reptiles. In: Cooper J.E., Jackson O.F. (eds): Diseases of the Reptilia. London, Academic Press. pp.: 429-468.

Jacobson E.R., Calderwood M.B., French T.W. *et al.* (1981b) Lymphosarcoma in an eastern king snake and a rhinoceros viper. *J Am Vet Med Assoc.* 179:1231-1235.

Jäger D., Jäger E., Knuth A. (2001). Immune responses to tumor antigens: implications for antigen specific immunotherapy of cancer. *J Clin Pathol.* 54:669-674.

Jhappan C., Noonan F.P., Merlino G. (2003). Ultraviolet radiation and cutaneous malignant melanoma. *Oncogene.* 22:3099-3112.

Jones D., Lee L., Liu J.L. *et al.* (1992). Marek disease virus encodes a basic-leucine zipper gene resembling the fos/jun oncogenes that is highly expressed in lymphoblastoid tumors. *Proc Natl Acad Sci USA.* 89:4042-4046.

Kaiser H.E. (1989). Comparative Aspects of Tumor Development. Kluwer Academic Publishers, Dordrecht, Netherlands. pp.:48-54.

Kasahara M., Hayashi M., Tanaka K., *et al.* (1996). Chromosomal localization of the proteasome Z subunit gene reveals an ancient chromosomal duplication involving the major histocompatibility complex. *Proc Nat Acad Sci USA.* 93:9096-9101.

Kasahara M., Nakaya J., Satta, Y. *et al.* (1997). Chromosomal duplication and the emergence of the adaptive immune system. *Trend Gen.* 13:90-92.

Kaufman J., Wallny H.J. (1996). Chicken MHC molecules, disease resistance and the evolutionary origin of birds. *Cur Top Microbiol Immunol.* 212:129-141.

Kent M.L., Dawe S.C. (1990). *et al.* Experimental transmission of a plasmacytoid leukemia of chinook salmon, Oncorhynchus tshawytscha. *Cancer Res.* 50:5679-5681.

Kieser D., Kent M.L., Groff J.M. *et al.* (1991). An epizootic of an epitheliotropic lymphoblastic lymphoma in coho salmon Oncorhynchus kisutch. *Dis Aqu Org.* 11:1-8.

King A., Hiby S.E., Verma S. *et al.* (1997). Uterine NK cells and trophoblast HLA class I molecules. *Am J Reprod Immunol.* 37:459-462.

King A., Loke Y.W. (1991). On the nature and function of human uterine granular lymphocytes. *Immunol Today.* 12:432-435.

Kovats S., Main E.K., Librach C. *et al.* (1990). A class I antigen, HLA-G, expressed in human trophoblasts. *Science.* 248:220-223.

Krebs E.T. Jr., Krebs E.T., Beard H.H. (1950). He trophoblastic thesis of malignancy. *Medical Record.* 163:148-170.

Krebs E.T. Jr., (1993). The Letter. Townsend Letter for Doctors. pp. 175.

Lachapelle M.H., Miron P., Hemmings R. *et al.* (1996). Endometrial T, B, and NK cells in patients with recurrent spontaneous abortion. Altered profile and pregnancy outcome. *J Immunol.* 15:4027-4034.

Laiho M., Latonen L. (2003). Cell cycle control, DNA damage checkpoints and cancer. *Ann Med.* 35:391-397.

Laurens N.R. (1997). Cancer resistance in amphibia. *Dev Com Immunol.* 21:102-106.

Lawlor D.A., Zemmour J., Ennis P.D., *et al.*, (1990). Evolution of class-I MHC genes and proteins: from natural selection to thymic selection. *An Rev Immunol.* 8:23-29.

Lea R.G., Underwood J., Flanders K.C. *et al.* (1995). A subset of patients with recurrent spontaneous abortion is deficient in TGFβ-2-producing 'suppressor cells' in uterine tissue near the placental attachment site. *Am J Reprod Immunol.* 34:52-64.

Lentz M.R. (1990). The phylogeny of oncology. *Mol Biother.* 2:137-144.

Lentz M.R. (1999). The Role of Therapeutic Apheresis in the Treatment of Cancer. *Therapeutic Apheresis.* 3:40-49.

Letcher J. (1992). Intracoelomic use of tricaine methanesulfonate for anesthesia of bullfrogs (Rana catesbeiana) and leopard frogs (Rana pipiens). *Zoo Biology.* 11:243-251.

Lin H., Mosmann T.R., Guilbert L. *et al.* (1993). Synthesis of T helper 2 type cytokines at the maternal–fetal interface. *J Immunol.* 151:4562-4573.

Lucke' B. (1942). Tumors of the nerve sheaths in fish of the snapper family (Lutianidae). *Arch Pathol.* 34:133-150.

Lunger P.D., Hardy W.D., Clark H.F. (1974). C-type particles in a reptilian tumor. *J Natl Cancer Inst.* 52:1231-1235.

Maccubbin, A.E., N. Ersing. (1991). Tumors in fish from the Detroit River. *Hydrobiol.* 219:301-306.

Malitschek B., Fornzler D., Shartl M. (1995). Melanoma formation in Xiphophorus: a model system for the role of receptor tyrosine kinases in tumorogrnesis. *Bio Essays.* 17:1017-1023.

Malumbres M., Carnero A. (2003). Cell cycle deregulation: a common motif in cancer. *Prog Cell Cycle Res.* 5:5-18.

Marx L., Arck P., Kapp M. *et al.* (1999a). Leukocyte populationshormone receptors and apoptosis in eutopic and ectopic first trimester human pregnancies. *Hum Reprod.* 14:1111-1117.

Marx L., Arck P., Kieslich C. *et al.* (1999b). Decidual mast cells might be involved in the onset of human first trimester abortion. *Am J Reprod Immunol.* 41:34-40.

Marzusch K., Buchholz F., Ruck P. *et al.* (1977). IL-12 and IL-2-stimulated release of IFN-γ by uterine CD56^{++} large granular lymphocytes is amplified by decidual macrophages. *Hum Reprod.* 12:921-924.

Masahito P., Ishikawa T., Okamoto N. *et al.* (1985). Nephoroblastoma in the Japanese eel, Anguilla japonica Temminck and Schlegel. *Canc Res.* 52:2575-2579.

Medarova Z., Elwood B.W., Taylor L.T. Jr. (2002). Alloantigen System L Affects the Outcome of Rous Sarcomas. *Exp Biol Med.* 227:158-163.

Mikkers H., Berns A. (2003). Retroviral insertional mutagenesis: tagging cancer pathways. *Adv Cancer Res.* 88:53-99.

Mitsumori K. (2002). Evaluation on carcinogenicity of chemicals using transgenic mice. *Toxicology.* 27:241-244.

Mosmann T.R., Cherwinski H., Bond M.W. *et al.* (1986). Two types of murine helper T cell clone. I. Definition according to profiles of lymphokine activities and secreted proteins. *J Immunol.* 136:2348-2357.

Naftzger C., Takechi K., Kohda H. *et al.* (1996). Immune response to a differentiation antigen induced by altered antigen: a study of tumor rejection and autoimmunity. *Proc Natl Acad Sci USA.* 93:14809-14814.

Nazerian K., Solomon J.J., Witter R.L. *et al.* (1968). Studies on the etiology of Marek's disease. II. Finding of a herpesvirus in cell culture. *Proc Soc Exp Biol Med.* 127:177-182.

Nunez O., Hendricks J.D., Arbogast D.N. *et al.* (1989). Promotion of aflatoxin B1 hepatocarcionogenesis in rainbow trout (Oncorhynchus mykiss). *Toxicol Pathol.* 19:11-23.

O'Regan M.N., Parsons K.R., Tregaskes C.A. *et al.* (1999). A chicken homologue of the co-stimulating molecule CD80 which binds to mammalian CTLA-4. *Immunogen.* 49:68-71.

Orós J., Torrent A., Espinosa de los Monteros A. (2001). Multicentric Lymphoblastic Lymphoma in a Loggerhead Sea Turtle (Caretta caretta). *Vet Pathol.* 38:464-467.

Overwijk W.W., Lee D.S., Surman D.R. *et al.* (1999). Vaccination with a recombinant vaccinia virus encoding a "self" antigen induces autoimmune vitiligo and tumor cell destruction in mice: requirement for CD4$^+$ T lymphocytes. *Proc Natl Acad Sci USA.* 96:2982-2987.

Palo J., Duchesne J., Wikstrom J. (1977). Malignant diseases among patients with multiple sclerosis. *J Neurol.* 216: 217-222.

Paul P., Rouas-Freiss N., Khalil-Daher I. *et al.* (1998). HLA-G expression in melanoma: A way for tumor cells to escape from immunosurveillance. *Proc Nat Acad Sci USA.* 95:4510-4515.

Paul R., Remes K., Lakkala T. *et al.* (1994). Spontaneous remission in acute myeloid leukaemia. *Br J Haematol.* 86:210-212.

Paulesu L. (1997). Cytokines in mammalian reproduction and speculation about their possible involvement in nonmammalian viviparity. *Micros Res Tech.* 38:188-194.

Payne L.N., Gillespie A.M., Howes K. (1992). Myeloid leukaemogenicity and transmission of the HPRS-103 strain of avian leukosis virus. *Leukemia.* 6:1167-1176.

Perlmutter A., Potter H. (1988). Hyperthermic suppression of a genetically programmed melanoma in hybrids of fishes: genus Xiphophorus. *J Can Res Clin Oncol.* 114:359-362.

Pfeiffer C.J., Nagai T., Fujimura M. *et al.* (1979). Spontaneous regressive epitheliomas in the Japanese newt, Cynops pyrrhogaster. *Canc Res.* 39:1904-1910.

Raghupathy R. (1997). Th1 type immunity is incompatible with successful pregnancy. *Immunol Today.* 18:478-482.

Rast J.P., Anderson M.K., Strong S.J. (1997). α, β, γ, and δ T cell antigen receptor genes arose early in vertebrate phylogeny. *Immunity.* 6:1-11.

Reboul J., Gardiner K., Monneron D. *et al.* (1999). Comparative genomic analysis of the IFN/IL-10 receptor gene cluster. *Gen Res.* 9:242-250.

Reece R.L. (1992). Observations on naturally occurring neoplasms in birds in the state of Victoria, Australia. *Avian Pathol.* 21:3-32.

Robert J., Guiet C., Du Pasquier L. (1995). Ontogeny of the alloimmune response against a transplanted tumor in Xenopus laevis. *Differentiation.* 59:135-139.

Robert J., Cohen N. (1999). Evolution of immune surveillance and tumor immunity: studies in Xenopus. *Immunol Rev.* 166: 231-243.

Romagnani S., Parronchi P., D'Elios M.M. *et al.* (1997). An update on human Th1 and Th2 cells. *Int Arch Allergy Immunol.* 113:153-156.

Romagnano A., Jacobson E.R., Boon G.D. et al (1996). Lymphosarcoma in a green iguana (Iguana iguana). *J Zoo Wildl Med.* 27:83-89.

Rosen P., Woodhead A.D. (1980). High ionic strength: its significance in immunosurveillance against tumor cells in sharks and rays (elasmobranchs). *Med Hypotheses.* 6:441-446.

Rosenberg S.A., White D.E. (1996). Vitiligo in patients with melanoma: normal tissue antigens can be targets for cancer immunotherapy. *J Immunother Emphasis Tumor Immunol.* 19:81-84.

Ruben L.N., Edwards B.F., Risting J. (1977). Temperature and variation of the function of complement and antibody of Amphibia. *Experientia.* 33:1522-1523.

Ruben N.L., Clothier H.R., Balls M. *et al.* (1997). Cancer resisitance in amphibia. *Dev Comp Immunol.* 21:102-102.

Rubin H. (2001). Selected Cell and Selective Microenvironment in Neoplastic Development. *Canc Res.* 61:799-807.

Ruddle F.H. (1973). Parasexual approaches to mammalian gene regulation. *Harvey Lect.* 69:103-124.

Ruddon W.R. (1995). Cancer Biology. Third edition. Oxford University Press. New York - Oxford.

Sakai C., Kawakami Y., Law L.W. *et al.* (1997). Melanosomal proteins as melanoma-specific immune targets. *Melanoma Res.* 7:83-95.

Salih H.R., Nussler V. (2001). Immune escape versus tumour tolerance: how do tumours evade immune surveillance? *Eur J Med Res.* 27: 323-332.

Schierman L.W., Watanabe D.H., McBride R.A. (1977).Genetic control of Rous sarcoma regression in chickens: linkage with the major histocompatibility complex. *Immunogen.* 5:325-332.

Schultze A.E., Mason G.L., Clyde V.L. (1999). Lymphosarcoma with leukemic blood profile in a savannah monitor lizard (Varanus exanthematicus). *J Zoo Wild Med.* 30:158-164.

Schumberger H.G., Lucke B. (1948). Tumor of fishes, amphibians, and reptiles. *Canc Res.* 8:657-753.

Schwemmler J.W. (1991). Carcinogenesis as reversal of eukaryotic symbiogenesis: The aposybiosis thory of cancer endocytobiosis. *Cell Res.* 7:163-199.

Schwemmler J.W. (1998). Basic Cancer Programs - Genes, Signals, Metabolites - Unified Holistic Theory of Evolution. Karger.

Scott H.H., Beattie J. (1927). Neoplasm in a porose crocodile. *J Pathol Bacteriol.* 30:61-66.

Shartl A., Hornung U., Nanda I. *et al.* (1997). Susceptibility to the development of pigment cell tumors in a clone of the Amazon molly (Poecilia formosa), introduced through a micro-chromosome. *Canc Res.* 57:2993-3000.

Shlumberger H.G. (1957). Tumors characteristics for certain animal species. *Canc Res.* 17:823-832.

Siciliano M.J., Perlmutter A., Clark E. (1971). Effect of sex on the development of melanoma in hybrid fish of the genus Xiphophorus. *Canc Res.* 31:725-729.

Setlow R.B., Woodhead A.D., Grist E. (1989). Animal model for ultraviolet radiation-induced melanoma: platyfish-swordtail hybrid. *Proc Natl Acad Sci USA.* 86:8922-8926.

Sleeman J., Campbell T., Turner O. (1999). Soft Tissue Sarcoma and Possible Eosinophilic Leukemia in a Tiger Salamander, Ambystoma tigrinum. *Ass Rept Amph Vet.* 9:26-29.

Smith A.C., Little H.F. (1969). Liver lesions produced by hydatid-like cysts in an elasmobranch, the electric ray, Torpedo californica. *Natl Cancer Inst Monogr.* 31:251-254.

Smith L.E., Nagar S., Kim G.J. *et al.* (2003). Radiation-induced genomic instability: radiation quality and dose response. *Health Phys.* 85:23-29.

Stanley B. (1999). Granulocytic sarcoma in a King Cobra. Twenty-seventh Annual Southeastern Veterinary Pathology Conference, Tifton, GA, USA.

Stavely-O'Carrol K., Sotomayor E., Montgomery J., *et al.* (1998). Induction of antigen-specific T cell anergy: an early event in the course of tumour progression. *Proc Natl Acad Sci USA.* 95:1178-1183.

Stewart H.L. (1972). Cancer and comparative pathology. *Prog Exp Tumor Res.* 16:142-150.

Stolk A. (1964). Succinic dehydrogenase activity in the nucleolus of the normal and tumorous liver cells of the common iguana (Iguana iguana). *Acta Morphol.* 5:302-315.

Takashima T. (1976). Hepatoma and cutaneous fibrosarcoma in hatchery-reared trout and salmon related to gonadal maturation. *Prog Exp Tumor Res.* 20:351-366.

Taylor M., Smith D.A. (2001). Long term effects of internal papillomatosis in Amazona spp. Proc Eur Assoc of Avian Vets. Munich, 122-123.

Tomlinson I.P.M., Novelli M.R., Bodmer W.F. (1996). The mutation rate and cancer. *Proc Natl Acad Sci USA.* 93:14800-14803.

Tomlinson I.P.M., Bodmer W.F. (1999). Selection, the mutation rate and cancer: ensuring that the tail does not wag the dog. *Nat Med.* 5:11-12.

Vandaveer S.S., Erf G.F., Durdik J.M. (2001). Avian T helper one/two immune response balance can be shifted toward inflammation by antigen delivery to scavenger receptors. *Brit Poul Sci.* 80:172-181.

Vassiliadou N., Searle R.F., Bulmer J.N. (1999). Elevated expression of activation molecules by decidual lymphocytes in women suffering spontaneous early pregnancy loss. *Hum Reprod.* 14:1194-1200.

Vijayasaradhi S., Bouchard B., Houghton A.N. (1990). The melanoma antigen gp75 is the human homologue of the mouse b (brown) locus gene product. *J Exp Med.* 171:1375-1380.

von Mensdorff-Pouilly S., Gourevitch M.M., Kenemans P., *et al.* (1996). Humoral immune response to polymorphic epithelial mucin 1 in patients with benign and malignant breast tumours. *Eur J Cancer.* 32:1325-1331.

Wadsworth J.R. Hill W.C.O. (1956). Selected tumors from the London Zoo menagerie. *Univ Penn Vet Ext Quart.* 141:70-73.

Wales J.H., Sinnhuber R.O., Hendricks J.D. *et al.* (1978). Aflatoxin B1 induction of hepatocellular carcinoma in the embryos of rainbow trout (Salmo gairdneri). *J Nat Canc Inst.* 60:1133-1139.

Wang R.F., Appella E., Kawakami Y. (1996). Identification of TRP-2 as a human tumor antigen recognized by cytotoxic T lymphocytes. *J Exp Med.* 184:2207-2216.

Weber L.W., Bowne W.B., Wolchok J.D. *et al.* (1998). Tumor immunity and autoimmunity induced by immunization with homologous DNA. *J Clin Invest.* 102:1258-1264.

Wegmann T.G., Lin H., Guilbert L. *et al.* (1993). Bidirectional cytokine interactions in the maternal-fetal relationship: is successful pregnancy a TH2 phenomenon? *Immunol Today.* 14:353-356.

Wei Y.Q., Huang M.J., Yang L., *et al.* (2001). Immunogene therapy of tumors with vaccine based on Xenopus homologous vascular endothelial growth factor as a model antigen. *Proc Nat Acad Sci USA.* 98:11545-1150.

Well M., Rodiger K. (1992). Cholangioma in a green iguana. *Kleintierpraxis.* 93:415-417.

Wellings S.R. (1969). Neoplasia and primitive vertebrate phylogeny: echinoderms, prevertebrates, and fishes-A review. *Natl Cancer Inst Monogr.* 31:59-128.

Wilson H., Graham J., Roberts R. *et al.* (2000). Integumentary neoplasms in Psittacine birds: treatment strategies. Proc Assoc of Avian Vets Annual Conference. AAV. Lake Worth, Florida, 211-214.

Wohlgemuth W. (1957). Eine allgemeine theorestische cancerologische konzeption in anlenhnung an Warburg. Das Deutche Gesundheitswesen: *Zoo Med.* 12:793-798.

Xie Q., Anderson A.S., Morgan R.W. (1996). Marek's disease virus (MDV) ICP4, pp38, and meq genes are involved in the maintenance of transformation of MDCC-MSB1 MDV-transformed lymphoblastoid cells. *J Virol.* 70:1125-1131.

Yamamoto T., Takahashi Y., Kase N. *et al.* (1999). Role of decidual natural killer (NK) cells in patients with missed abortion: differences between cases with normal and abnormal chromosome. *Clin Exp Immunol.* 116:449-452.

Young J., Leggett B., Gustafson C. (1993). Genomic instability occurs in colorectal carcinomas but not in adenomas. *Hum Mutat.* 2:351-354.

Zeigel R.F., Clark H.F. (1969). Electron microscopic observations on a "C"-type virus in cell cultures derived from a tumor-bearing viper. *J Natl Cancer Inst.* 43:1097-1102.

Zhou Y., Fisher S.J., Janatpour M., Genbacev O. *et al.* (1997). Human cytotrophoblasts adopt a vascular phenotype as they differentiate. A strategy for successful endovascular invasion? *J Clin Invest.* 99:2139-2151.

Zwart P., Harshbarger J.C. (1972). Hematopoietic neoplasms in lizards: report of a typical case in Hydrosaurus amboinensis and of a probable case in (Varanus salvator). *Int J Cancer.* 9:548-553.

INDEX

Th1 cells, 6,7,39,44,45,47,68,80,95-100,112,113,119-121,123,138,139,146-153,157,158,160,162,165,166,170,171,173-175,179,180,183,184,226,231,237,238
Th2 cells, 6,7,39,44,45,47,68,80,95-100,112,113,119-121,123,138,139,146-153,157,158,160,162,165,166,170,171,173-175,179,180,183,184,226,231,237,238
Th3 cells, 7,80,143,152,158,180
Thp cells, 109,112-114
Thymic involution, 117,118,122,123,180,181
Thymus, 4-8,11,13,22,66,117,118-123,139,146,154,176,179-181,213,217,218,238
TJ6, 175
TLX molecules, 136
TNF-α, 12,15,39-41,44,91,94-98,109-113,125,137,147-149,152,155,157-159,162,165,166,175,178
TNF-β, 15,39,123,165,166,174,175
Triakis scyllia, 14
Trophoblast, 21,24,37,67,68,73-77,108,118,123,136-141,143-145,147-149,151-160,163-170,175,178-184

Tumor associated antigens, 86,248
Tumor Infiltrating Lymphocytes (TIL), 100,107,110,115,117,118,232
Tumor suppresor genes, 202
Turtle, 13,197,208,209
Tyrosinase, 227
Umbilical cord, 69
Urochordata, 4
UVR (UV-light/radiation), 210,218
V domain, 9
Vaccinia virus, 103
Viral Antigens (VA), 104
Vitamin D3, 170
Vitiligo, 226
Viviparity, 15,28,66-75,78,79,181,182
Xenopus laevis, 7,8,11,13,18,24-26,34,217,218
Xenotransplant(ation), 5,62
Xiphophorus helleri, 208,209,215,216
X-ray, 210,215
YAC-1 Lymphoma Cell Line, 169
Yeast, 196,200
Yolk Sac, 69,70,75
Zebrafish, 11,16,18,25-27
Zona pellucida, 153